Cardiff Libraries
www.cardiff.gov.uk/libraries
Llyfrgelloedd Caer
www.caerdydd.gov.uk/llyfrc

D0244965

AS & A2 Biology

Exam Board: AQA

Complete Revision and Practice

ACC. No: 03149051

AS-Level Contents

A2-Level Contents

Published by CGP

From original material by Richard Parsons.

Editors:
Claire Boulter, Ellen Bowness, Katie Braid, Joe Brazier, Charlotte Burrows, Katherine Craig, Rosie Gillham, Andy Park, Jane Towle, Karen Wells, Dawn Wright.

Contributors:
Gloria Barnett, Jessica Egan, James Foster, Barbara Green, Derek Harvey, Brigitte Hurwitt, Liz Masters, Stephen Phillips, Claire Ruthven, Adrian Schmit, Sophie Watkins, Anna-fe Williamson.

Proofreaders:
Glenn Rogers, Sue Hocking.

ISBN: 978 1 84762 423 9

With thanks to Laura Stoney for the copyright research.

Groovy website: www.cgpbooks.co.uk
Jolly bits of clipart from CorelDRAW®
Printed by Elanders Ltd, Newcastle upon Tyne.

Photocopying – it's dull, grey and sometimes a bit naughty. Luckily, it's dead cheap, easy and quick to order more copies of this book from CGP – just call us on 0870 750 1242. Phew!

Text, design, layout and original illustrations © Coordination Group Publications Ltd. (CGP) 2009
All rights reserved.

The Scientific Process

'How Science Works' is all about the scientific process — how we develop and test scientific ideas. It's what scientists do all day, every day (well, except at coffee time — never come between a scientist and their coffee).

Scientists Come Up with Theories — Then Test Them...

Science tries to explain **how** and **why** things happen — it **answers questions**. It's all about seeking and gaining **knowledge** about the world around us. Scientists do this by **asking** questions and **suggesting** answers and then **testing** them, to see if they're correct — this is the **scientific process**.

1) **Ask** a question — make an **observation** and ask **why or how** it happens. E.g. why is trypsin (an enzyme) found in the small intestine but not in the stomach?
2) **Suggest** an answer, or part of an answer, by forming a **theory** (a possible **explanation** of the observations) e.g. pH affects the activity of enzymes. (Scientists also sometimes form a **model** too — a **simplified picture** of what's physically going on.)
3) Make a **prediction** or **hypothesis** — a **specific testable statement**, based on the theory, about what will happen in a test situation. E.g. trypsin will be active at pH 8 (the pH of the small intestine) but inactive at pH 2 (the pH of the stomach).
4) Carry out a **test** — to provide **evidence** that will support the prediction (or help to disprove it). E.g. measure the rate of reaction of trypsin at various pH levels.

The evidence supported Quentin's Theory of Flammable Burps.

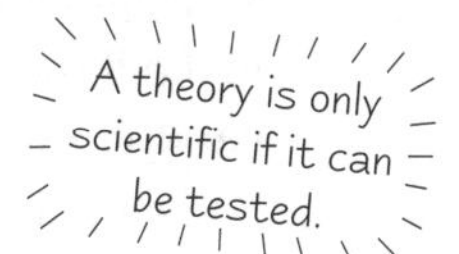
A theory is only scientific if it can be tested.

...Then They Tell Everyone About Their Results...

The results are **published** — scientists need to let others know about their work. Scientists publish their results in **scientific journals**. These are just like normal magazines, only they contain **scientific reports** (called papers) instead of the latest celebrity gossip.

1) Scientific reports are similar to the **lab write-ups** you do in school. And just as a lab write-up is **reviewed** (marked) by your teacher, reports in scientific journals undergo **peer review** before they're published.
2) The report is sent out to **peers** — other scientists who are experts in the **same area**. They examine the data and results, and if they think that the conclusion is reasonable it's **published**. This makes sure that work published in scientific journals is of a **good standard**.
3) But peer review **can't guarantee** the science is **correct** — other scientists still need to **reproduce** it.
4) Sometimes **mistakes** are made and flawed work is published. Peer review **isn't perfect** but it's probably the best way for scientists to self-regulate their work and to publish **quality reports**.

...Then Other Scientists Will Test the Theory Too

Other scientists read the published theories and results, and try to **test the theory** themselves. This involves:

- Repeating the **exact same experiments**.
- Using the theory to make **new predictions** and then testing them with **new experiments**.

If the Evidence Supports a Theory, It's Accepted — for Now

1) If all the experiments in all the world provide good evidence to back it up, the theory is thought of as **scientific 'fact'** (for now).
2) But it will never become **totally indisputable** fact. Scientific **breakthroughs or advances** could provide new ways to question and test the theory, which could lead to **new evidence** that **conflicts** with the current evidence. Then the testing starts all over again...

And this, my friend, is the **tentative nature of scientific knowledge** — it's always **changing** and **evolving**.

The Scientific Process

So scientists need evidence to back up their theories. They get it by carrying out experiments, and when that's not possible they carry out studies. But why bother with science at all? We want to know as much as possible so we can use it to try and improve our lives (and because we're nosy).

Evidence Comes from Lab Experiments...

1) Results from **controlled experiments** in **laboratories** are **great**.
2) A lab is the easiest place to **control variables** so that they're all **kept constant** (except for the one you're investigating).
3) This means you can draw meaningful **conclusions**.

For example, if you're investigating how temperature affects the rate of an enzyme-controlled reaction you need to keep everything but the temperature constant, e.g. the pH of the solution, the concentration of the solution etc.

...and Well-Designed Studies

1) There are things you **can't** investigate in a lab, e.g. whether stress causes heart attacks. You have to do a study instead.
2) You still need to try and make the study as controlled as possible to make it **more reliable**. But in reality it's **very hard** to control **all the variables** that **might** be having an effect.
3) You can do things to help, e.g. have **matched groups** — **choose two groups** of people (those who have quite stressful jobs and those who don't) who are **as similar as possible** (same mix of ages, same mix of diets etc.). But you can't easily rule out every possibility.

Samantha thought her study was very well designed — especially the fitted bookshelf.

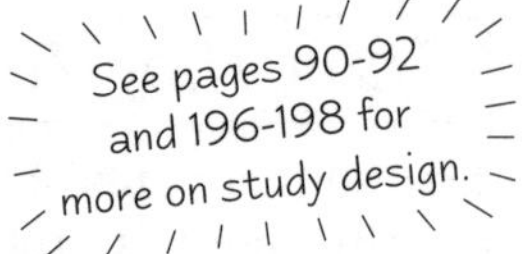

See pages 90-92 and 196-198 for more on study design.

Society Makes Decisions Based on Scientific Evidence

1) Lots of scientific work eventually leads to **important discoveries** or breakthroughs that could **benefit humankind**.
2) These results are **used by society** (that's you, me and everyone else) to **make decisions** — about the way we live, what we eat, what we drive, etc.
3) All sections of society use scientific evidence to make decisions, e.g. politicians use it to devise policies and individuals use science to make decisions about their own lives.

Other factors can **influence** decisions about science or the way science is used:

Economic factors

- Society has to consider the **cost** of implementing changes based on scientific conclusions — e.g. the **NHS** can't afford the most expensive drugs without **sacrificing** something else.
- Scientific research is **expensive** so companies won't always develop new ideas — e.g. developing new drugs is costly, so pharmaceutical companies often only invest in drugs that are likely to make them **money**.

Social factors

- **Decisions** affect **people's lives** — E.g. scientists may suggest **banning smoking** and **alcohol** to prevent health problems, but shouldn't **we** be able to **choose** whether **we** want to smoke and drink or not?

Environmental factors

- Scientists believe **unexplored regions** like remote parts of rainforests might contain **untapped drug** resources. But some people think we shouldn't **exploit** these regions because any interesting finds may lead to **deforestation** and **reduced biodiversity** in these areas.

So there you have it — how science works...

Hopefully these pages have given you a nice intro to how science works, e.g. what scientists do to provide you with 'facts'. You need to understand this, as you're expected to know how science works — for the exam and for life.

AS-Level Biology

Exam Board: AQA

Disease

What a lovely way to start a book... two pages on disease. Nice.

Pathogens Cause Infectious Diseases

1) A pathogen is any **organism** that **causes disease**.
2) Pathogens include **microorganisms** and some larger organisms, such as **tapeworms**.
3) Pathogenic microorganisms include some **bacteria**, some **fungi** and all **viruses**.

Pathogens can Penetrate an Organism's Interface with the Environment

Pathogens need to **enter** the body to cause disease — they **get in** through an organism's **surface of contact** (**interface**) with the **environment**, e.g. nose, eyes, a cut. An organism has **three** main interfaces with the environment — the **gas-exchange system**, the **skin** and the **digestive system**.

Gas-Exchange System — If you breathe in **air** that contains **pathogens**, most of them will be trapped in **mucus** lining the lung epithelium (the outer layer of cells in the passages to the lungs). These cells also have **cilia** (hair-like structures) that **beat** and **move** the mucus up the trachea to the mouth, where it's removed. Unfortunately, some pathogens are still able to reach the **alveoli** where they can **invade** cells and cause **damage**.

Skin — If you **damage** your skin, **pathogens** on the surface can enter your **bloodstream**. The blood **clots** at the area of damage to **prevent** pathogens from entering, but some may get in **before** the clot forms.

Digestive System — If you **eat** or **drink food** that contains **pathogens**, most of them will be **killed** by the **acidic** conditions of the **stomach**. However, some may **survive** and pass into the intestines where they can invade **cells** of the **gut wall** and cause disease.

Pathogens Cause Disease by Producing Toxins and Damaging Cells

Despite our **protective mechanisms**, pathogens can still **successfully enter** our bodies. Once inside, they **cause disease** in **two** main ways:

1) Production of toxins — Many bacteria **release toxins** (harmful molecules) into the body. For example, the bacterium that causes **tetanus** produces a toxin that **blocks** the function of certain **nerve cells**, causing **muscle spasms**.
2) Cell damage — Pathogens can physically damage host cells by:
   - **Rupturing** them to **release nutrients** (proteins etc.) inside them.
   - **Breaking down** nutrients inside the cell for their own use. This starves and eventually **kills** the **cell**.
   - **Replicating** inside the cells and **bursting** them when they're released, e.g. some **viruses** do this.

The cell the pathogen has invaded and is reproducing inside is called the host cell.

Lifestyle can Affect Your Risk of Developing Some Diseases

1 Coronary heart disease (**CHD**) is a disease that affects your **heart** (see p. 44 for more). There are plenty of lifestyle factors that **increase** your **risk** of developing CHD:

1) **Poor diet** — a diet high in **saturated fat** or **salt** increases the risk.
2) **Smoking**, **lack of exercise** and **excessive alcohol intake** — these can all lead to **high blood pressure**, which can **damage** the heart and the blood vessels, **increasing** the risk of CHD

See p. 44 for more on risk factors and CHD.

2 Cancer is the result of **uncontrolled cell division** (see p. 65). Factors that **increase** the **risk** of developing cancer include:

1) **Smoking** — the main cause of **mouth**, **throat** and **lung cancer** is smoking.
2) **Excessive exposure to sunlight** — excessive exposure can cause **skin cancer**. Using **sunbeds** and sunbathing **without sunscreen** increases the risk.
3) **Excessive alcohol intake** — this can increase the risk of many types of cancer, especially **liver cancer**.

Disease

*It's never Too Late to **Change** Your **Lifestyle***

Changing your **lifestyle** for the **better** (e.g. reducing your alcohol intake, doing more exercise, eating healthily etc.) doesn't mean you'll **never** develop these diseases, but it can **reduce the risk**. So it's never too late to change. For example, studies have shown that the risk to a **smoker** of developing **lung cancer** is reduced if they stop smoking.

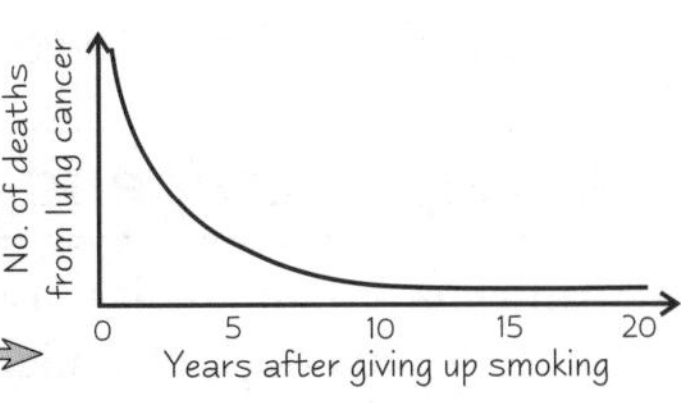

*You Need to be Able to **Interpret Data** About **Risk Factors** and **Disease***

In the exam you could be asked to **interpret data** about **lifestyle** and the **risk** of disease. Here's an example for you:

A study looked at the **link** between **body mass index** (BMI) and the **relative risk of developing cancer**. BMI is a measure of **obesity**. At the start of the study the BMI of **1.2 million women** aged 50-64 was taken, and after **five years** the **number** of these women **with cancer** was recorded. The relative risk of developing cancer was then worked out by taking into account **other factors** about the women, including **daily alcohol intake**, **smoking status** and **physical activity**. The graph on the right shows the results.

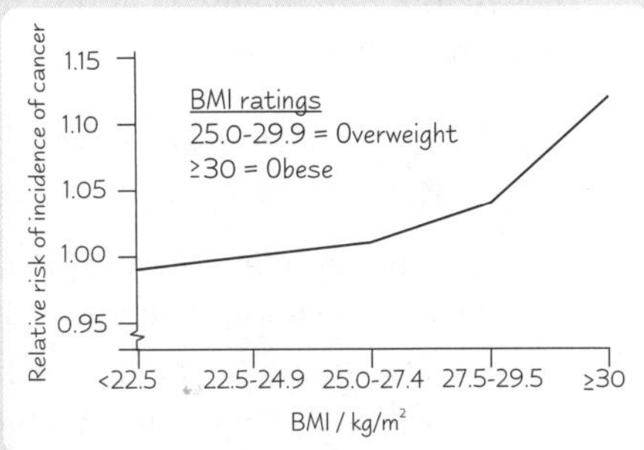

1) <u>You might be asked to **describe the data...**</u>
The graph shows a **positive correlation** between **BMI** and the relative **risk** of cancer in women. The **higher** the BMI, the **higher** the risk of developing cancer. It's **not** a linear (straight-line) relationship though — the **risk increases** much more **quickly** for women with a **BMI over 27.5**.

2) <u>...or **draw conclusions**</u>
   - The relative risk of developing cancer is **greatly increased** for women who are **overweight** or **obese**.
   - Be careful — you **can't conclude** that obesity **causes** cancer, only that they're **linked**. Another factor could be involved. For example, obese people are more likely to have **diets high in saturated fat** — it may be the saturated fat that causes the higher risk of cancer, not the actual increase in body mass.
   - You **can't conclude** that the risk of developing a **specific form** of cancer (e.g. throat cancer) increases with increasing BMI. The data doesn't deal with specific forms of cancer separately. So you can only conclude that the risk of developing cancer in **general** increases with increasing BMI.

3) <u>... or **evaluate the methodology**</u>
   - The **sample size** is **large** — **1.2 million women**. This makes the results **more reliable**.
   - The study took into account **other lifestyle factors** (e.g. alcohol intake, smoking) that can **affect** the risk of developing cancer too. This also makes the results **more reliable**.

See pages 90-92 and 196-198 for more about interpreting data.

Practice Questions

Q1 State four lifestyle factors that can affect your chances of getting CHD.

Q2 If you give up smoking, what happens to your risk of developing lung cancer?

Exam Questions

Q1 Describe three ways in which a pathogen may damage host cells. [3 marks]

Q2 Sketch a graph to show the likely correlation between the amount of time spent sunbathing and the incidence of skin cancer. [2 marks]

<u>Pizza, beer and telly a bad lifestyle? — depends who you ask...</u>

Phew, they were a tough two pages to start with, but they're done now and it's all fun, fun, fun from here on in. There's loads more interpreting data to come, so 'describe, draw conclusions and evaluate the methodology' might get a bit repetitive — but those pesky AQA examiners love asking you to interpret, so it's all good practice.

The Immune System

So, you can reduce your risk of getting some non-infectious diseases by changing your lifestyle. But infectious diseases are a whole different kettle of fish. Luckily we have an army of cells that help to protect us — the immune system.

Foreign Antigens Trigger an Immune Response

Antigens are **molecules** (usually proteins or polysaccharides) found on the **surface** of **cells**. When a pathogen invades the body, the antigens on its cell surface are **identified as foreign**, which activates cells in the immune system. There are **four** main stages involved in the immune response:

1) Phagocytes Engulf Pathogens

A **phagocyte** (e.g. a macrophage) is a type of **white blood cell** that carries out **phagocytosis** (engulfment of pathogens). They're found in the **blood** and in **tissues** and are the first cells to respond to a pathogen inside the body. Here's how they work:

1) A phagocyte **recognises** the **antigens** on a pathogen.
2) The cytoplasm of the phagocyte moves round the pathogen, **engulfing** it.
3) The pathogen is now contained in a **phagocytic vacuole** (a bubble) in the cytoplasm of the phagocyte.
4) A **lysosome** (an organelle that contains **lysosomal enzymes**) **fuses** with the phagocytic vacuole. The lysosomal enzymes **break down** the pathogen.
5) The phagocyte then **presents** the pathogen's antigens — it sticks the antigens on its **surface** to **activate** other immune system cells.

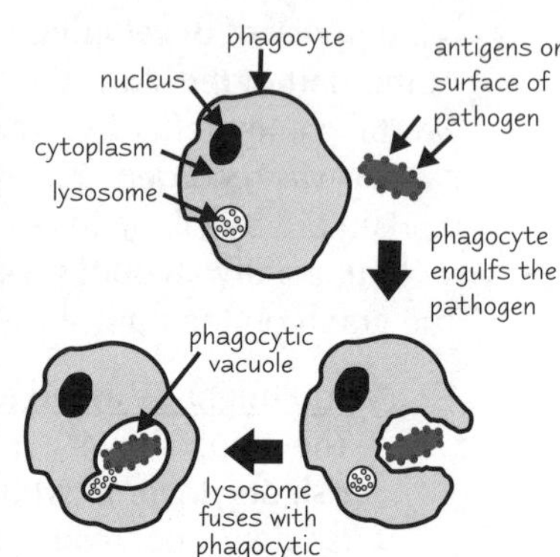

2) Phagocytes Activate T-cells

A **T-cell** is another type of **white blood cell**. It has **proteins** on its surface that **bind** to the **antigens** presented to it by **phagocytes**. This **activates** the T-cell. Different types of T-cells respond in different ways:

1) Some **release substances** to **activate B-cells**.
2) Some **attach** to antigens on a pathogen and **kill** the cell.

3) T-cells Activate B-cells, Which Divide into Plasma Cells

B-cells are also a type of **white blood cell**. They're covered with **antibodies** — proteins that **bind antigens** to form an **antigen-antibody complex**. Each B-cell has a **different shaped antibody** on its membrane, so different ones bind to **different shaped antigens**.

1) When the antibody on the surface of a B-cell meets a **complementary shaped** antigen, it binds to it.
2) This, together with substances released from T-cells, **activates** the B-cell.
3) The activated B-cell **divides** into **plasma cells**.

4) Plasma Cells Make More Antibodies to a Specific Antigen

Plasma cells are **identical** to the B-cell (they're **clones**). They secrete loads of the **antibody** specific to the antigen. Antibody **functions** include:

1) Coating the pathogen to make it easier for a **phagocyte** to engulf it.
2) Coating the pathogen to **prevent** it from **entering** host cells.
3) **Binding to** and **neutralising** (inactivating) toxins produced by the pathogen.

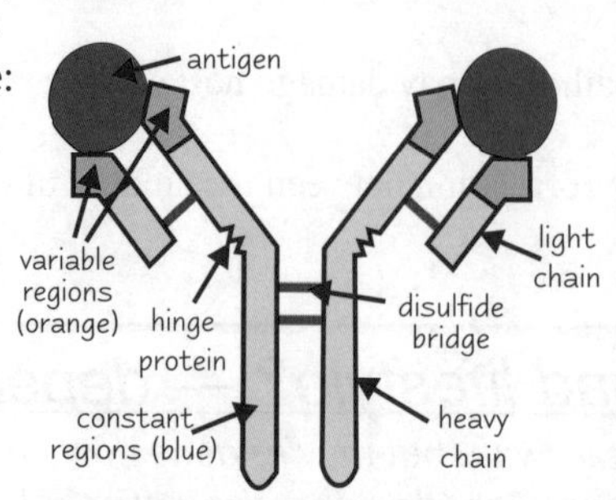

Antibodies are **proteins** — they're made up of chains of **amino acid** monomers linked by **peptide bonds** (see p. 14 for more on proteins). The **specificity** of an antibody depends on its **variable regions**. Each antibody has a **different shaped** variable region (due to different **amino acid sequences**) that's **complementary** to one **specific antigen**. The **constant regions** are the **same** in all antibodies.

The Immune System

The *Immune Response* Can be Split into *Cellular* and *Humoral*

Just to add to your fun, the **immune response** is often split into **two** — the **cellular response** and the **humoral response**.

1) Cellular — The **T-cells** and **other** immune system **cells** that they **interact** with, e.g. phagocytes, form the cellular response.
2) Humoral — **B-cells** and the production of **antibodies** form the **humoral response**.

Both types of response are **needed** to remove a pathogen from the body and the responses **interact** with each other, e.g. T-cells help to **activate** B-cells, and antibodies **coat** pathogens making it **easier** for **phagocytes** to **engulf** them.

The *Immune Response* for Antigens can be *Memorised*

Neil's primary response — to his parents.

The Primary Response

1) When an antigen enters the body for the **first time** it activates the immune system. This is called the **primary response**.
2) The primary response is **slow** because there **aren't many B-cells** that can make the antibody needed to bind to it.
3) Eventually the body will produce **enough** of the right antibody to overcome the infection. Meanwhile the infected person will show **symptoms** of the disease.
4) After being exposed to an antigen, both T- and B-cells produce **memory cells.** These memory cells **remain in the body** for a **long** time. Memory T-cells remember the **specific antigen** and will recognise it a second time round. Memory B-cells record the specific **antibodies** needed to bind the antigen.

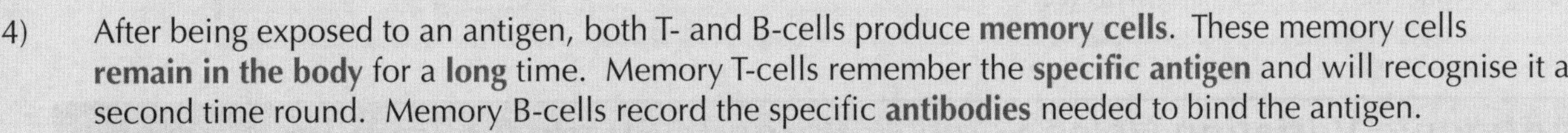

5) The person is now **immune** — their immune system has the **ability** to respond **quickly** to a 2nd infection.

The Secondary Response

1) If the **same pathogen** enters the body again, the immune system will produce a **quicker**, **stronger** immune response — the **secondary response**.
2) **Memory B-cells** divide into **plasma cells** that produce the right antibody to the antigen. **Memory T-cells** divide into the **correct type of T-cells** to kill the cell carrying the antigen.
3) The secondary response often gets rid of the pathogen **before** you begin to show any **symptoms**.

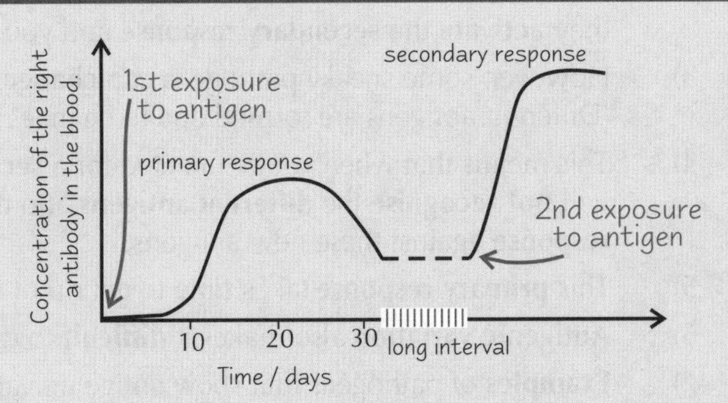

Practice Questions

Q1 What are antigens?

Q2 What is phagocytosis?

Q3 What are the functions of T-cells and B-cells?

Exam Questions

Q1 Describe the function of antibodies. [3 marks]

Q2 Describe and explain how a secondary immune response differs to a primary immune response. [6 marks]

Memory cells — I need a lot more to cope with these pages...

If memory cells are mentioned in the exam remember that they are still types of T-cells and B-cells. They just hang around a lot longer than most T-cells and B-cells. When the antigen enters the body for a second time they can immediately divide into more of the specific T-cells and B-cells that can kill the pathogen or release antibodies against it.

Vaccines and Antibodies in Medicine

The primary response gives us immunity against a disease, but only after you've become infected. If only there was a way to stimulate memory cell production without getting the disease... well, there is — vaccination.

Vaccines can Protect Individuals and Populations Against Disease

1) While your B-cells are busy **dividing** to build up their numbers to deal with a pathogen (i.e. the **primary response** — see previous page), you **suffer** from the disease. **Vaccination** can help avoid this.
2) Vaccines **contain antigens** that cause your body to **produce memory cells** against a particular pathogen, **without** the pathogen **causing disease**. This means you become **immune** without getting any **symptoms**.
3) Vaccines protect individuals that have them and, because they reduce the **occurrence** of the disease, those **not** vaccinated are also less likely to catch the disease (because there are fewer people to catch it from). This is called **herd immunity**.
4) Vaccines always contain antigens — these may be **free** or attached to a **dead** or **attenuated** (weakened) **pathogen**.
5) Vaccines may be **injected** or taken **orally**. The **disadvantages** of taking a vaccine orally are that it could be **broken down** by **enzymes** in the gut or the **molecules** of the vaccine may be **too large** to be **absorbed** into the blood.
6) Sometimes **booster** vaccines are given later on (e.g. after several years) to **make sure** that memory cells are produced.

The oral vaccine was proving hard to swallow.

Antigenic Variation Helps Some Pathogens Evade the Immune System

1) **Antigens** on the surface of pathogens **activate** the **primary response**.
2) When you're **infected** a **second time** with the **same pathogen** (which has the **same antigens** on its surface) they **activate** the **secondary response** and you don't get ill.
3) However, some sneaky pathogens can **change** their surface antigens. This is called **antigenic variation**. (Different antigens are formed due to changes in the **genes** of a pathogen.)
4) This means that when you're infected for a **second time**, the **memory cells** produced from the **first infection** will **not recognise** the **different antigens**. So the immune system has to start from scratch and carry out a **primary response** against these new antigens.
5) This **primary response** takes **time** to get rid of the infection, which is why you get **ill again**.
6) **Antigenic variation** also makes it **difficult** to develop **vaccines** against some pathogens for the same reason.
7) **Examples** of pathogens that show antigenic variation include **HIV**, ***S. pneumoniae*** bacteria and the **influenza virus**. You need to **learn** how it works in influenza:

Antigenic variation in the influenza virus

1) The **influenza virus** causes **influenza** (flu).
2) **Proteins** (**neuraminidase** and **haemagglutinin**) on the **surface** of the influenza virus act as **antigens**, **triggering** the immune system.
3) These antigens can **change regularly**, forming **new strains** of the virus.
4) **Memory cells** produced from **infection** with **one strain** of flu will **not recognise** other strains with **different antigens**.
5) This means your immune system produces a **primary response** every time you're infected with a **new strain** (carrying different antigens).
6) So this means you can **suffer from flu** more than once — each time you're infected with a **new strain**.

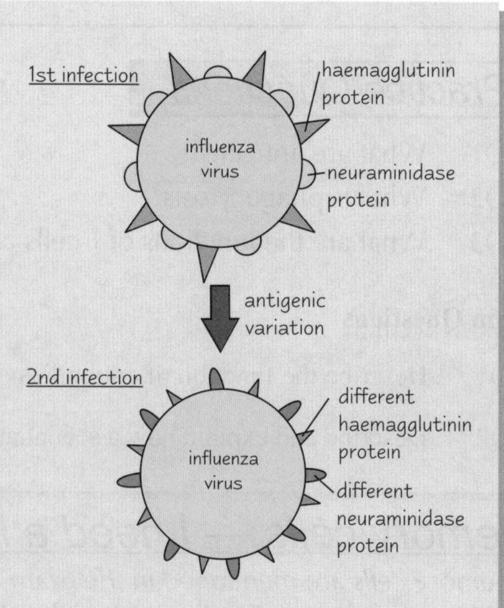

Vaccines and Antibodies in Medicine

Monoclonal Antibodies can be used to *Target Specific Substances* or *Cells*

1) **Monoclonal antibodies** are antibodies **produced** from a **single group of genetically identical B-cells** (plasma cells). This means that they're all **identical** in **structure**.
2) As you know, antibodies are **very specific** because their binding sites have a **unique structure** that only one particular antigen will fit into (one with a **complementary shape**).
3) You can make monoclonal antibodies **that bind to anything** you want, e.g. a cell antigen or other substance, and they will only bind to (target) this molecule.

EXAMPLE: TARGETING CELLS — CANCER

1) **Different cells** in the body have **different** surface **antigens**.
2) Cancer cells have antigens called **tumour markers** that are **not** found on normal body cells.
3) **Monoclonal antibodies** can be made that will bind to the tumour markers.
4) You can also attach **anti-cancer drugs** to the antibodies.
5) When the antibodies come into **contact** with the cancer cells they will **bind** to the tumour markers.
6) This means the drug will **only accumulate** in the body where there are **cancer cells**.
7) So, the **side effects** of an antibody-based drug are lower than other drugs because they accumulate near **specific cells**.

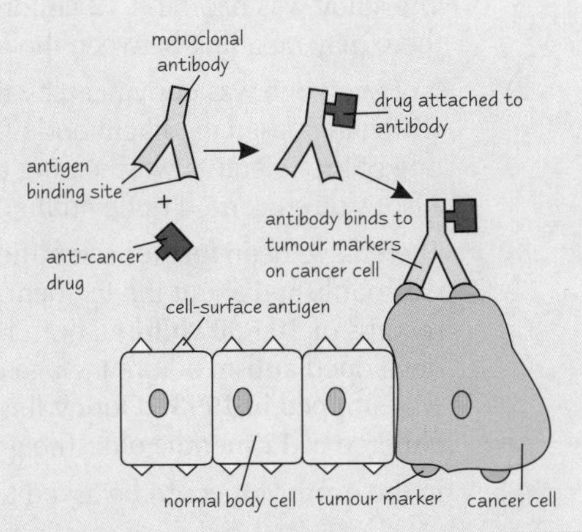

EXAMPLE: TARGETING SUBSTANCES — PREGNANCY TESTS

Pregnancy tests detect the hormone **human chorionic gonadotropin (hCG)** that's found in the **urine** of pregnant women:

1) The application area contains **antibodies for hCG** bound to a **coloured bead** (**blue**).
2) When urine is applied to the application area any hCG will **bind** to the antibody on the beads, forming an **antigen-antibody complex**.
3) The urine **moves** up the stick to the **test strip**, **carrying** any **beads** with it.
4) The test strip contains **antibodies to hCG** that are stuck in place (**immobilised**).
5) If there **is hCG present** the test strip turns **blue** because the **immobilised** antibody binds to any **hCG** — concentrating the hCG-antibody complex with the **blue beads** attached. If **no hCG** is present, the beads will **pass through** the test area **without** binding to anything, and so it **won't** go blue.

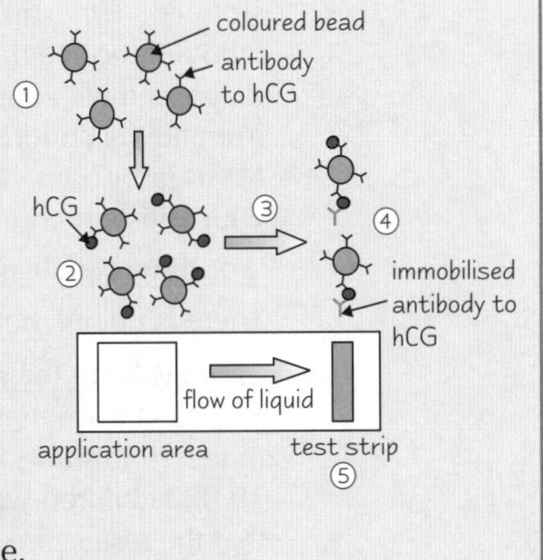

Practice Questions

Q1 How do vaccines cause immunity?

Q2 What are monoclonal antibodies?

Exam Questions

Q1 Explain why it is possible to suffer from the flu more than once. [4 marks]

Q2 Describe how monoclonal antibodies can be used to target a drug to cancer cells. [4 marks]

An injection of dead bugs — roll on my next vaccine...

Monoclonal antibodies are really useful — they can even be made against other antibodies. For example, people with asthma produce too many of a type of antibody that causes inflammation in the lungs. Monoclonal antibodies can be made to bind this type of antibody, so it can no longer cause inflammation, which can reduce the symptoms of asthma sufferers.

Interpreting Vaccine and Antibody Data

If someone claims anything about a vaccine or an antibody the claim has to be validated (confirmed) before it's accepted.

New Knowledge *About* ***Vaccines*** *and* ***Antibodies*** *is* ***Validated*** *by* ***Scientists***

When a **study** presents evidence for a **new theory** (e.g. that a vaccine has a dangerous side effect) it's important that other scientists come up with **more evidence** in order to **validate** (confirm) the theory. To validate the theory other scientists may **repeat** the study and try to **reproduce** the results, or **conduct other studies** to try to prove the same theory (see p. 1).

EXAMPLE 1: The MMR Vaccine

1) In 1998, a study was published about the **safety** of the **measles, mumps** and **rubella (MMR) vaccine**. The study was based on **12 children** with **autism** (a life-long developmental disability) and concluded that there may be a **link** between the MMR vaccine and autism.
2) Not everyone was convinced by this study because it had a **very small sample size** of 12 children, which increased the likelihood of the results being due to **chance**. The study may have been **biased** because one of the scientists was helping to gain evidence for a **lawsuit** against the MMR vaccine manufacturer. Also, studies carried out by different scientists found no link between autism and the MMR vaccine.
3) There have been **further scientific studies** to sort out the **conflicting** evidence. In **2005**, a **Japanese** study was published about the incidence of autism in Yokohama (an area of Japan). They looked at the medical records of **30 000 children** born between **1988 and 1996** and counted the number of children that developed **autism** before the age of seven. The **MMR jab** was first **introduced in Japan in 1989** and was **stopped in 1993**. During this time the MMR vaccine was administered to children at **12 months old**. The graph shows the results of the study.
4) In the exam you could be asked to **evaluate evidence** like this.

See pages 90-92 and 196-198 for more about evaluating data.

- You might be asked to **explain the data...**
  The graph shows that the number of children diagnosed with autism continued to **rise** after the MMR vaccine was **stopped**. For example, from all the children born in 1992, who did receive the MMR jab, about 60 out of 10 000 were diagnosed with autism before the age of seven. However, from all the children born in 1994, who did not receive the MMR jab, about 160 out of 10 000 of them were diagnosed with autism before the age of seven.
- ...or **draw conclusions**
  There is **no link** between the MMR vaccine and autism.
- ... or **evaluate the methodology**
  You can be much more confident in this study, compared to the 1998 study, because the **sample size** was so **large** — 30 000 children were studied. A larger sample size means that the results are less likely to be due to **chance**.

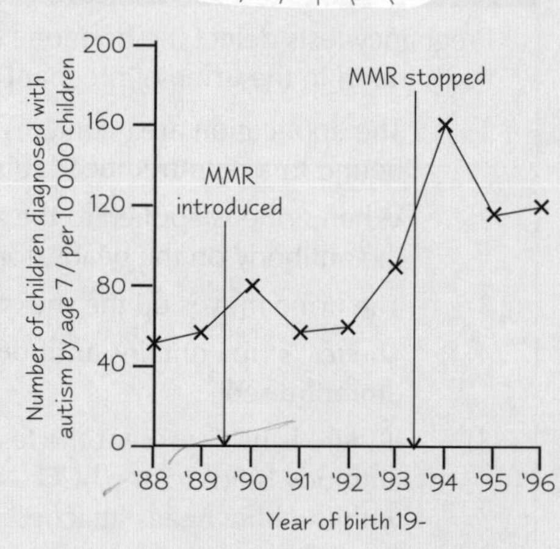

EXAMPLE 2: Herceptin® — Monoclonal Antibodies

About **20%** of **women with breast cancer** have tumours that produce more than the usual amount of a **receptor** called **HER2**. **Herceptin®** is a **drug** used to treat this type of breast cancer — it contains **monoclonal antibodies** that **bind the HER2 receptor** on a **tumour cell** and **prevent** the cells from growing and dividing.

In **2005**, a study **tested** Herceptin® on women who had already undergone **chemotherapy** for HER2-type **breast cancer**. **1694** women took the **drug** for a **year** after chemotherapy and another **1694** women were **observed** for the **same time** (the control group). The results are shown in the graph on the right.

Describe the data: Almost **twice as many** women in the **control group** developed breast cancer again or died **compared** to the group taking Herceptin®.

Draw conclusions: A **one-year treatment** with Herceptin®, after chemotherapy, **increases** the disease-free survival rate for women with HER2-type breast cancer.

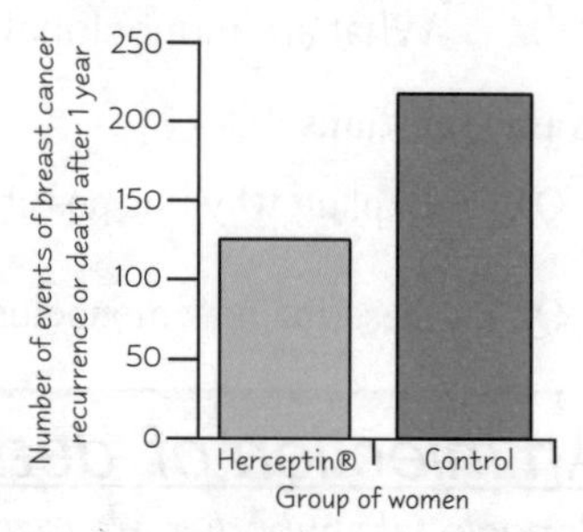

Interpreting Vaccine and Antibody Data

We use **Scientific Knowledge** to Make **Decisions**

When **new scientific information** about **vaccines** and **monoclonal antibodies** has been **validated** by scientists, **society** (organisations and the public) can **use** this information to make **informed decisions**. **Two examples** are given below:

EXAMPLE 1: The MMR Vaccine

Scientific knowledge:

The **validity** of the 1998 study that linked MMR and autism is in doubt. **New studies** have shown **no link** between the vaccine and autism.

Decision:

Scientists and doctors still recommended that parents **immunise** their **children** with the **MMR vaccine**.

EXAMPLE 2: Herceptin® — Monoclonal Antibodies

Scientific knowledge:

Early studies about Herceptin® showed **severe heart problems** could be a **side effect** of the drug.

Decision:

All patients receiving Herceptin® must be **monitored** for heart problems, e.g. by having **heart tests** done.

Use of **Vaccines** and **Antibodies** Raises **Ethical Issues**

Ethical issues surrounding vaccines include:

1) All vaccines are **tested on animals** before being tested on humans — some people **disagree** with animal testing. Also, **animal based substances** may be used to **produce** a vaccine, which some people disagree with.
2) **Testing** vaccines on **humans** can be **tricky**, e.g. volunteers may put themselves at **unnecessary risk** of contracting the disease because they think they're fully protected (e.g. they might have unprotected sex because they have had a new HIV vaccine and think they're protected — and the vaccine might not work).
3) Some people **don't** want to take the vaccine due to the **risk** of **side effects**, but they are **still protected** because of **herd immunity** (see p. 8) — other people think this is **unfair**.
4) If there was an **epidemic** of a **new disease** (e.g. a new influenza virus) there would be a rush to **receive** a vaccine and **difficult decisions** would have to be made about **who** would be the **first** to receive it.

Ethical issues surrounding monoclonal antibody therapy often involve animal rights issues. **Animals** are used to **produce the cells** from which the monoclonal antibodies are produced. Some people **disagree** with the use of animals in this way.

Practice Questions

Q1 Suggest one ethical issue surrounding vaccines.

Q2 Suggest one ethical issue surrounding monoclonal antibodies.

Exam Question

Q1 The graph on the right shows the number of laboratory reports of *Haemophilus influenzae* type b (Hib), in England and Wales, from 1990 to 2004. Hib affects children and can lead to meningitis and pneumonia.

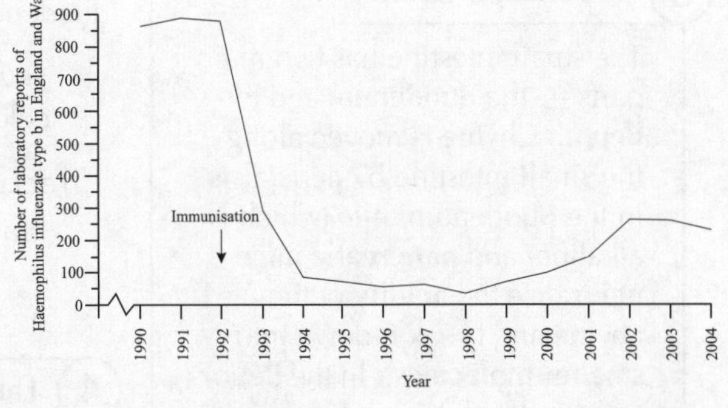

a) Why did the number of cases of Hib decrease after 1992? [2 marks]

b) Due to a shortage of the normal vaccine in 2000-2001, a different type of Hib vaccine was used. What effect did this have on the number of cases of Hib? [1 mark]

Some scientists must have to validate the taste of chocolate — nice job...

After the 1998 study, some parents were worried about giving their kids the MMR vaccine, so the number of children given the vaccine fell. With fewer children in each community protected by the vaccine, herd immunity decreased. This meant that more people were vulnerable to the diseases, so the number of cases of measles, mumps and rubella went up.

The Digestive System

With the digestive system in contact with the environment and us left open to pathogens, eating can be a pretty risky game. It's dead important though — without it we'd lack all the energy and important nutrients we need.

Digestion Breaks Down Large Molecules into Smaller Molecules

1) Many of the molecules in our **food** are **polymers**.
2) These are **large**, complex molecules composed of long chains of **monomers** — small **basic molecular units**.
3) **Proteins** and some **carbohydrates** are **polymers**. In carbohydrates, the monomers are called **monosaccharides**. They contain the elements **carbon**, **hydrogen** and **oxygen**. In proteins the monomers are called **amino acids**. They contain **carbon**, **hydrogen**, **oxygen**, **nitrogen**.

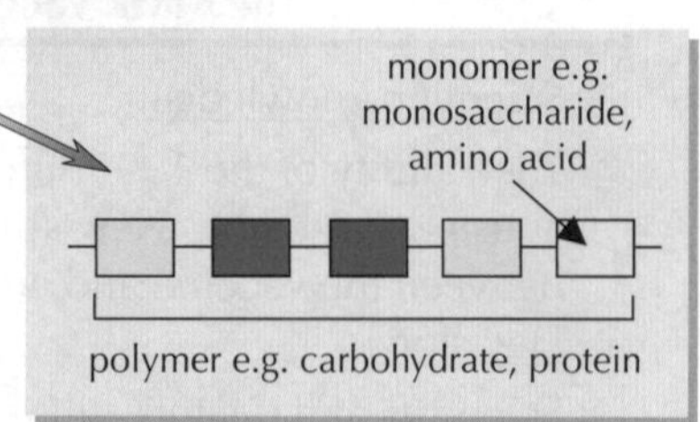

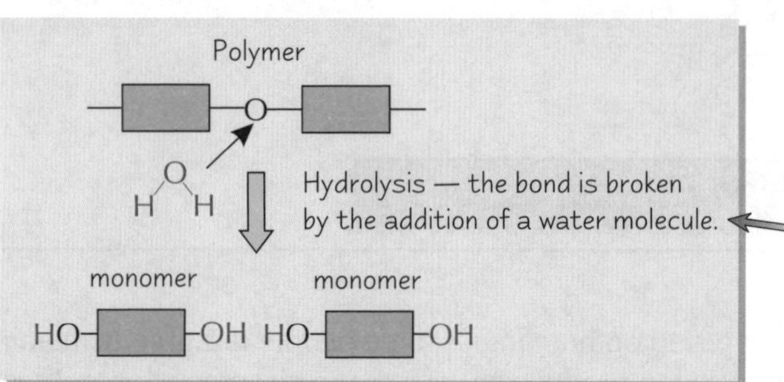

4) The polymers in our food are **insoluble** — they can't be directly **absorbed** into our bloodstream and **assimilated** (made) into new products.
5) The polymers have to be **hydrolysed** (broken down) into **smaller**, more **soluble** molecules by **adding water**.
6) This process happens during **digestion**.
7) **Hydrolysis** is catalysed by **digestive enzymes**.

Each Part of the Digestive System Has a Specific Function

All the organs in the **digestive system** have a **role** in **breaking down** food and **absorbing nutrients**:

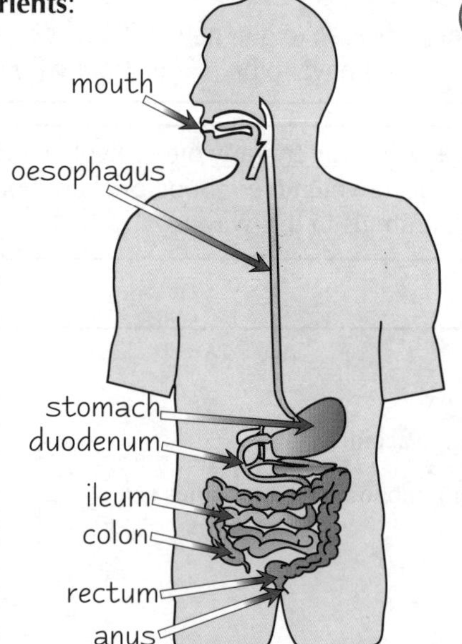

1 Oesophagus

The **tube** that takes **food** from the **mouth** to the **stomach** using waves of **muscle contractions** called **peristalsis**. **Mucus** is secreted from tissues in the walls, to **lubricate** the food's passage downwards.

2 The Stomach

The stomach is a **small sac**. It has lots of **folds**, allowing the stomach to **expand** — it can hold up to 4 litres of food and liquid. The entrance and exit of the stomach are controlled by **sphincter muscles**. The stomach walls produce **gastric juice**, which helps break down food. Gastric juice consists of **hydrochloric acid** (HCl), **pepsin** (an enzyme) and **mucus**. **Pepsin** hydrolyses **proteins**, into smaller polypeptide chains. It only works in **acidic conditions** (provided by the HCl). **Peristalsis** of the stomach turns food into an acidic fluid called **chyme**.

3 The Small Intestine

The small intestine has two main parts — the **duodenum** and the **ileum**. **Chyme** is moved along the small intestine by **peristalsis**. In the duodenum, **bile** (which is alkaline) and **pancreatic juice neutralise** the **acidity** of the chyme and break it down into **smaller molecules**. In the ileum, the small, soluble molecules (e.g. glucose and amino acids) are **absorbed** through structures called **villi** that line the gut wall. Molecules are absorbed by **diffusion, facilitated diffusion** and **active transport** (see pages 28-30 for more).

4 Large intestine

The large intestine (colon) absorbs **water**, **salts** and **minerals**. Like other parts of the digestive system, it has a **folded wall** — this provides a **large surface area** for absorption. **Bacteria** that **decompose** some of the undigested nutrients are found in the large intestine.

5 Rectum

Faeces are stored in the rectum and then pass through **sphincter** muscles at the **anus** during **defecation**. Nice.

The Digestive System

The Pancreas and Salivary Glands Play Important Roles in Digestion

Glands along the **digestive system** release **enzymes** to help **break down** food.
You need to know about two of these glands — the **salivary glands** and the **pancreas**.

The Salivary Glands

There are three main pairs of salivary glands in the mouth. They secrete **saliva** that consists of **mucus**, **mineral salts** and **salivary amylase** (an enzyme). Salivary amylase breaks down **starch** into **maltose**, a disaccharide (see p. 16). Saliva has other roles in digestion — e.g. it helps to **lubricate** food, making it easier to **swallow**.

The Pancreas

The pancreas releases **pancreatic juice** into the **duodenum** (see previous page) through the **pancreatic duct**. Pancreatic juice contains **amylase**, **trypsin**, **chymotrypsin** and **lipase** (the functions of these enzymes are listed below). It also contains **sodium hydrogencarbonate**, which **neutralises** the acidity of **hydrochloric acid** from the **stomach**.

Enzymes Help us to Digest Food Molecules

Digestive enzymes can be divided into three classes.

1) **Carbohydrases** catalyse the hydrolysis of **carbohydrates**.
2) **Proteases** catalyse the hydrolysis of **proteins**.
3) **Lipases** catalyse the hydrolysis of **lipids**.

The table shows some more specific enzyme reactions.

LOCATION	ENZYME	CLASS	HYDROLYSES	INTO
salivary glands	**amylase**	carbohydrase	starch	maltose
stomach	**pepsin**	protease	protein	peptides
pancreas	**amylase**	carbohydrase	starch	maltose
	trypsin	protease	protein	peptides
	chymotrypsin	protease	protein	peptides
	carboxypeptidase	protease	peptides	amino acids
	lipase	lipase	lipids	fatty acids + glycerol
ileum	**maltase**	carbohydrase	maltose	glucose
	sucrase	carbohydrase	sucrose	glucose + fructose
	lactase	carbohydrase	lactose	glucose + galactose
	peptidase	protease	peptides	amino acids

Practice Questions

Q1 What is hydrolysis?

Q2 Describe the structure and function of the stomach.

Q3 Describe the structure and function of the small intestine.

Q4 What do lipases break down?

Exam Question

Q1 There are several glands associated with the digestive system.

a) Describe how the pancreas aids digestion. [6 marks]

b) Name one other gland associated with the digestive system. [1 mark]

Gastric juice — not for me thanks, think I'll stick to orange...

So when I eat a biscuit, first it gets chomped up by my teeth. Then it's broken down by carbohydrases and lipases, before all those smaller, more soluble molecules are absorbed and, most likely, deposited on my hips. Then there's the messy business of defecation. Learn all this stuff on the digestive system inside out — it might just save your life.

Proteins

Protein is found in loads of different foods like chicken, eggs and nuts. But what are proteins? How are they made? What do they look like? Well, for your enjoyment, here are the answers to all those questions and many, many more...

Proteins** are Made from **Long Chains** of **Amino Acids

1) The **monomers** of proteins are **amino acids**.
2) A **dipeptide** is formed when **two** amino acids join together.
3) A **polypeptide** is formed when **more than two** amino acids join together.
4) **Proteins** are made up of **one or more polypeptides**.

Grant's cries of "die peptide, die" could be heard for miles around. He'd never forgiven it for sleeping with his wife.

Different Amino Acids** Have **Different Variable Groups

All amino acids have the same general structure — a **carboxyl group** (-COOH) and an **amino group** ($-NH_2$) attached to a **carbon** atom. The **difference** between different amino acids is the **variable** group (**R** on diagram) they contain.

Structure of an Amino Acid

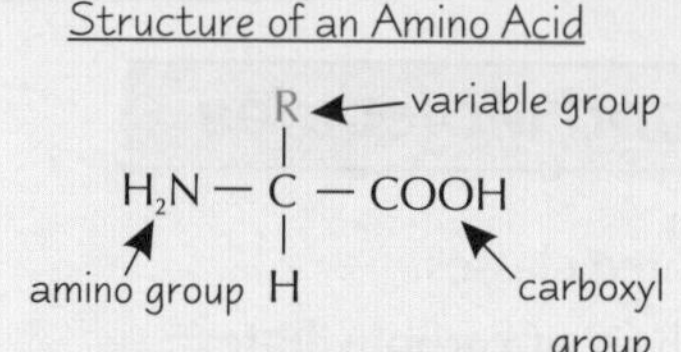

E.g. Structure of Glycine

$H_2N - CH_2 - COOH$

Glycine is the smallest amino acid — the R group is a hydrogen atom.

Polypeptides** are Formed by **Condensation Reactions

Amino acids are linked together by **condensation** reactions to form polypeptides. A molecule of **water** is **released** during the reaction. The bonds formed between amino acids are called **peptide bonds**. The reverse reaction happens during digestion.

amino acid 1 + amino acid 2 ⇌ dipeptide (condensation / hydrolysis)

H_2O — a molecule of water is formed during condensation.

peptide bond

Proteins** Have **Four Structural Levels

Proteins are **big**, **complicated** molecules. They're much easier to explain if you describe their structure in four 'levels'. These levels are a protein's **primary**, **secondary**, **tertiary** and **quaternary** structures.

<u>**Primary Structure**</u> — this is the **sequence** of **amino acids** in the **polypeptide chain**.

<u>**Secondary Structure**</u> — the polypeptide chain doesn't remain flat and straight. **Hydrogen bonds** form between the amino acids in the chain. This makes it automatically **coil** into an **alpha** (α) **helix** or **fold** into a **beta** (β) **pleated sheet** — this is the secondary structure.

<u>**Tertiary Structure**</u> — the coiled or folded chain of amino acids is often **coiled** and **folded further**. **More bonds** form between different parts of the polypeptide chain. For proteins made from a **single** polypeptide chain, the tertiary structure forms their **final 3D structure**.

<u>**Quaternary Structure**</u> — some proteins are made of **several different polypeptide chains** held together by **bonds**. The **quaternary structure** is the way these polypeptide chains are assembled together. For proteins made from more than one polypeptide chain (e.g. haemoglobin, insulin, collagen), the quaternary structure is the protein's **final 3D structure**.

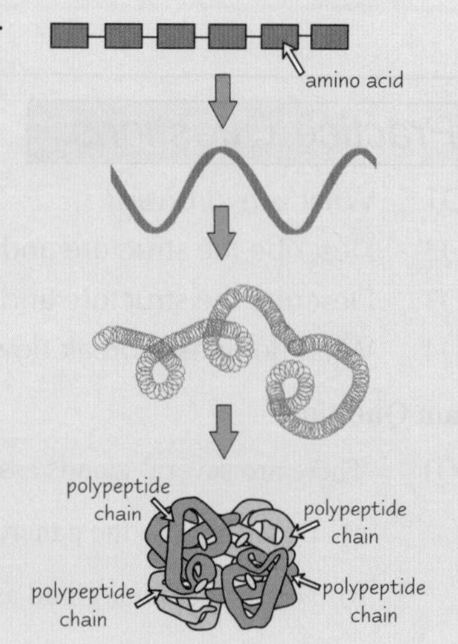

A protein's **shape** determines its **function**. E.g. **haemoglobin** is a **compact**, **soluble protein**, which makes it easy to **transport**. This makes it great for **carrying oxygen** around the body (see p. 58). **Collagen** has three polypeptide chains **tightly coiled** together, which makes it **strong**. This makes it a great **supportive tissue** in animals.

Proteins

Proteins have a *Variety* of *Functions*

There are **loads** of different **proteins** found in **living organisms**. They've all got **different structures** and **shapes**, which makes them **specialised** to carry out particular **jobs**. For example:

1) **Enzymes** — they're usually roughly **spherical** in shape due to the **tight folding** of the polypeptide chains. They're **soluble** and often have roles in **metabolism**, e.g. some enzymes break down large food molecules (**digestive enzymes**, see p. 13) and other enzymes help to **synthesise** (make) large molecules.

2) **Antibodies** — are involved in the **immune response**. They're made up of **two light** (short) polypeptide chains and **two heavy** (long) polypeptide chains bonded together. Antibodies have **variable regions** (see p. 6) — the **amino acid sequences** in these regions **vary** greatly.

3) **Transport proteins** — are present in **cell membranes** (p. 29). They contain **hydrophobic** (water hating) and **hydrophilic** (water loving) amino acids, which cause the protein to **fold up** and form a **channel**. These proteins **transport molecules** and **ions across** membranes.

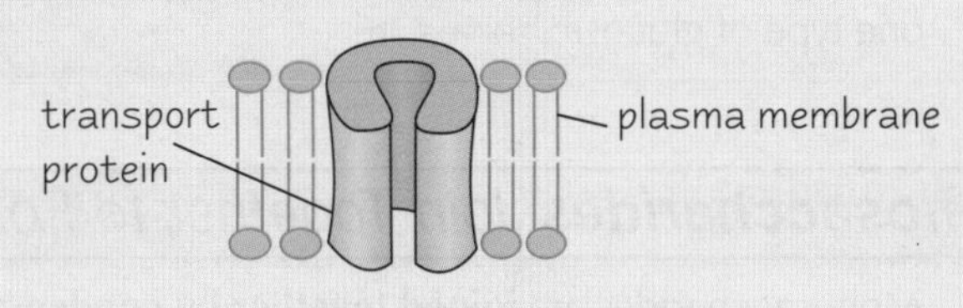

4) **Structural proteins** — are physically **strong**. They consist of **long polypeptide chains** lying **parallel** to each other with **cross-links** between them. Structural proteins include **keratin** (found in hair and nails) and **collagen** (found in connective tissue).

Use the *Biuret Test* for *Proteins*

If you needed to find out if a substance, e.g. a **food sample**, contained **protein** you'd use the **biuret test**.

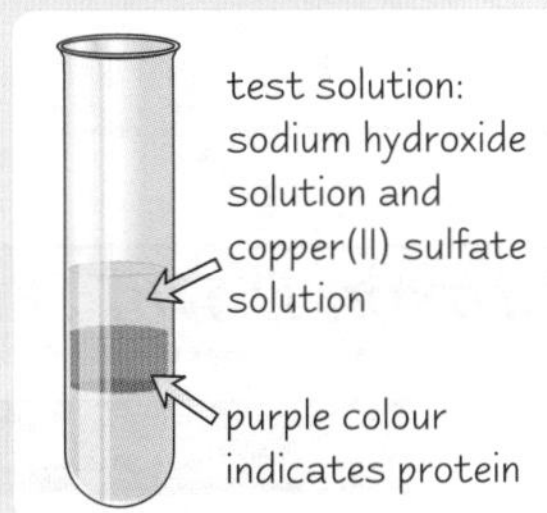

There are **two stages** to this test.

1) The test solution needs to be **alkaline**, so first you add a few drops of **sodium hydroxide solution**.
2) Then you add some **copper(II) sulfate solution**.

- If protein **is** present a **purple layer** forms.
- If there's **no protein**, the solution will **stay blue**. The colours are pale, so you need to look carefully.

Practice Questions

Q1 What groups do all amino acid molecules have in common?

Q2 Give three functions of proteins.

Q3 Describe how you would test for the presence of protein in a sample.

Exam Questions

Q1 Describe how a dipeptide is formed. [5 marks]

Q2 Describe the structure of a protein, explaining the terms primary, secondary, tertiary and quaternary structure. No details of the chemical nature of the bonds are required. [9 marks]

Condensation — I can see the reaction happening on my car windows...

Protein structure is hard to imagine. I think of a Slinky — the wire's the primary structure, it coils up to form the secondary structure and if you coil the Slinky round your arm that's the tertiary structure. When a few Slinkies get tangled up that's like the quaternary structure. Oh, I need to get out more. I wish I had more friends and not just this stupid Slinky for company.

Carbohydrates

Carbohydrates are present in foods like pasta, potatoes, bread and cakes — basically all of the yummy stuff. You need to know how they're made in the first place. So go grab a potato... and read on...

Carbohydrates are Made from **Monosaccharides**

1) As you know, most carbohydrates are **polymers**.
2) All carbohydrates contain the elements **C**, **H** and **O**.
3) The **monomers** that they're made from are **monosaccharides**, e.g. **glucose**, **fructose** and **galactose**. You need to learn the structure of one type of glucose:

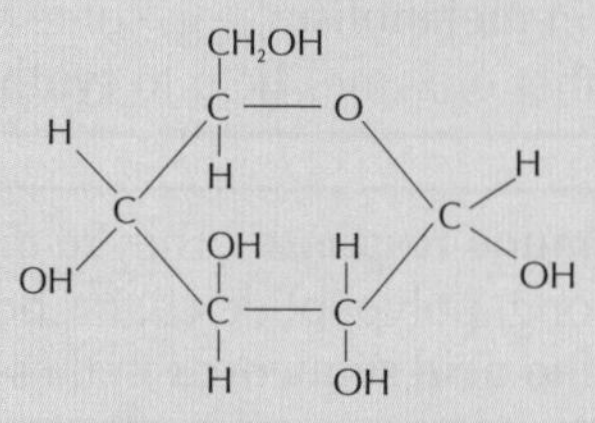

an α-glucose molecule

1) Glucose is a **hexose sugar** — a monosaccharide with **six carbon** atoms in each molecule.
2) There are two forms of glucose — **alpha** (α) and **beta** (β) glucose (see p. 60 for more on β-glucose). For this Unit you need to learn the **structure** of α-glucose.

Monosaccharides Join Together to Form **Disaccharides** and **Polysaccharides**

1) Monosaccharides are **joined together** by **condensation reactions**.
2) During a condensation reaction a molecule of **water** is **released** and a **glycosidic bond** forms between the two monosaccharides.
3) A **disaccharide** is formed when **two monosaccharides** join together. A **polysaccharide** is formed when **more than two monosaccharides** join together.

<u>Example</u>

Two α-**glucose** molecules are joined together by a **glycosidic bond** to form **maltose**.

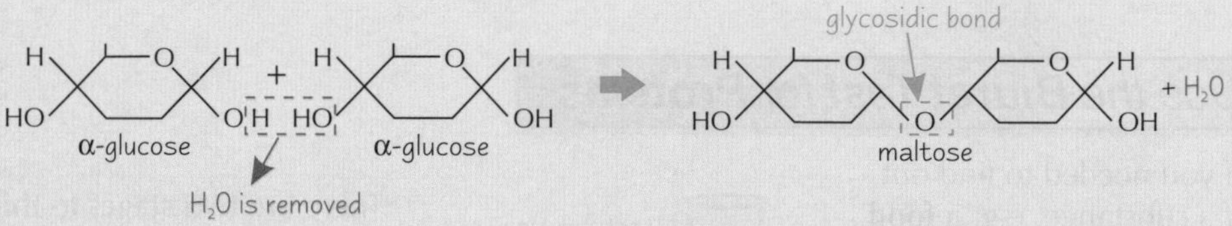

α-glucose + α-glucose → maltose + H_2O

H_2O is removed

glycosidic bond

Disaccharides and **Polysaccharides** are **Broken Down** During **Digestion**

Disaccharides and polysaccharides are often present in the **food** we eat, so we need to be able to break them down. Luckily we have **enzymes** released by the **intestinal epithelium** (see p. 13), that **hydrolyse** (break down) disaccharides and polysaccharides.

You need to learn the **monosaccharides** that make up **maltose**, **sucrose** and **lactose** and the **enzymes** that **hydrolyse** them. Luckily for you we've put it all in this pretty purple table.

Disaccharide	Hydrolysed by...	Into...
maltose	maltase	glucose + glucose
sucrose	sucrase	glucose + fructose
lactose	lactase	glucose + galactose

Lactose-Intolerance is Caused by a **Lack** of the Digestive Enzyme **Lactase**

1) **Lactose** is a **sugar** found in milk.
2) It's digested by an **enzyme** called **lactase**, found in the intestines.
3) If you **don't** have enough of the enzyme lactase, you won't be able to break down the lactose in milk properly — a condition called **lactose-intolerance**.
4) Undigested lactose is fermented by bacteria and can cause a whole host of **intestinal complaints** such as **stomach cramps**, excessive **flatulence** (wind) and **diarrhoea**.
5) Milk can be artificially treated with purified lactase to make it suitable for lactose-intolerant people.
6) It's fairly **uncommon** to be lactose **tolerant** though — around 15% of Northern Europeans, 50% of Mediterraneans, 95% of Asians and 90% of people of African descent are lactose intolerant.

Carbohydrates

Use the **Benedict's Test** for **Sugars**

Sugar is a general term for **monosaccharides** and **disaccharides**. All sugars can be classified as **reducing sugars** or **non-reducing sugars**. If you carry out an **experiment** on the **digestion** of **carbohydrates** you'll need to **test** for sugars — to do this you use the **Benedict's test**. The test **differs** depending on the **type** of sugar you're testing for.

REDUCING SUGARS

1) Reducing sugars include **all monosaccharides** and **some disaccharides**, e.g. maltose.
2) You add **Benedict's reagent** (which is **blue**) to a sample and **heat it**. If the sample contains reducing sugars it gradually turns **brick red** due to the formation of a **red precipitate**.

NON-REDUCING SUGARS

1) To test for **non-reducing sugars**, like sucrose, first you have to break them down into monosaccharides.
2) You do this by **boiling** the test solution with **dilute hydrochloric acid** and then **neutralising** it with **sodium hydrogencarbonate**. Then just carry out the **Benedict's test** as you would for a reducing sugar.
3) Annoyingly, if the result of this test is **positive** the sugar could be reducing **or** non-reducing. To **check** it's non-reducing you need to do the **reducing sugar test** too (to rule out it being a reducing sugar).

Starch is Made from **Two Polysaccharides**

1) Starch is made up of a mixture of two polysaccharides — **amylose** and **amylopectin** (see p. 60).
2) Both are composed of **long chains** of **α-glucose** linked together by glycosidic bonds, formed in condensation reactions.
3) When starch is digested, it's first broken down into **maltose** by **amylase** — an enzyme released by the **salivary glands** and the **pancreas** (see p. 13).
4) Maltose is then broken down into **α-glucose molecules** by **maltase**, which is released by the **intestinal epithelium**.

Jamelia would never even dream of leaving the house without her collar starched.

Use the **Iodine Test** for **Starch**

If you do any **experiment** on the **digestion** of **starch** and want to find out if any is **left**, you'll need the **iodine test**.

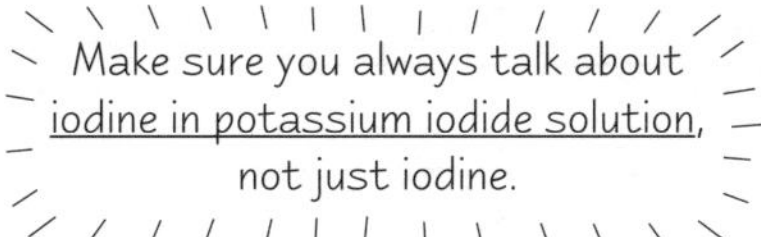

Make sure you always talk about iodine in potassium iodide solution, not just iodine.

Just add **iodine dissolved in potassium iodide solution** to the test sample. If there **is starch present**, the sample changes from **browny-orange** to a dark, **blue-black** colour.

Practice Questions

Q1 Draw the structure of α-glucose.

Q2 What type of bond holds monosaccharide molecules together in a disaccharide?

Q3 Name the enzymes that break down: i) sucrose, ii) maltose, iii) lactose.

Q4 Name the two polysaccharides present in starch.

Exam Questions

Q1 Describe the cause and symptoms of lactose intolerance. [4 marks]

Q2 Describe the test used to identify a non-reducing sugar.
Include the result you would expect to see if the test was positive. [6 marks]

Reducing sugars — who on earth would want to do that?

Just to confuse matters, in addition to α-glucose, there's also β-glucose. But you don't have to worry about that for now. It ain't fun learning the structure of glucose... you basically have to copy it down, then cover the page and test yourself until you know it off by heart. But at least then there'll be no nasty surprises in the exam...

Enzyme Action

In our digestive system, enzymes help to break down all the stuff we eat. Without them we wouldn't get the nutrients and energy from our food very quickly at all... which wouldn't be good. So here's how they work....

Enzymes are Biological Catalysts

Enzymes **speed up chemical reactions** by acting as **biological catalysts**.

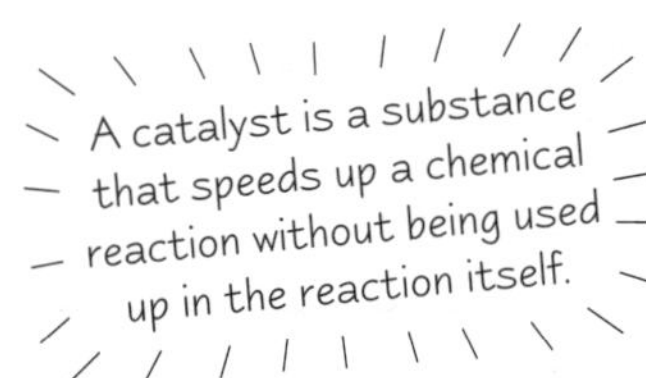

A catalyst is a substance
that speeds up a chemical
reaction without being used
up in the reaction itself.

1) They catalyse **metabolic reactions** in your body, e.g. digestion and respiration. Even your **phenotype** (physical appearance) is due to enzymes that catalyse the reactions that cause growth and development (see p. 52).
2) Enzymes are **proteins** (see p. 15).
3) Enzymes have an **active site**, which has a **specific shape**. The active site is the part of the enzyme where the **substrate** molecules (the substance that the enzyme interacts with) **bind to**.
4) Enzymes are **highly specific** due to their tertiary structure (see next page).

Enzymes Lower the Activation Energy of a Reaction

In a chemical reaction, a certain amount of **energy** needs to be supplied to the chemicals before the reaction will **start**. This is called the **activation energy** — it's often provided as **heat**. Enzymes **lower** the amount of activation energy that's needed, often making reactions happen at a **lower temperature** than they could without an enzyme. This **speeds up** the **rate of reaction**.

When a substrate fits into the enzyme's active site it forms an **enzyme-substrate complex** — it's this that lowers the activation energy. Here are two reasons why:

1) If two substrate molecules need to be **joined**, being attached to the enzyme holds them **close together**, **reducing** any **repulsion** between the molecules so they can bond more easily.
2) If the enzyme is catalysing a **breakdown reaction**, fitting into the active site puts a **strain** on bonds in the substrate, so the substrate molecule **breaks up** more easily.

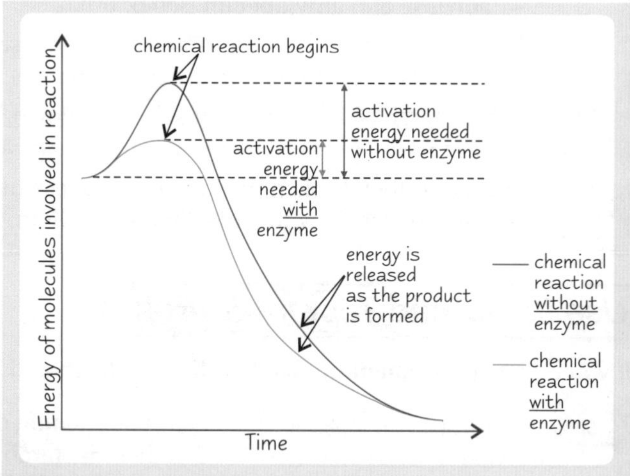

The 'Lock and Key' Model is a Good Start...

Enzymes are a bit picky — they only work with substrates that fit their active site. Early scientists studying the action of enzymes came up with the '**lock and key**' model. This is where the **substrate fits** into the **enzyme** in the same way that a **key fits** into a **lock**.

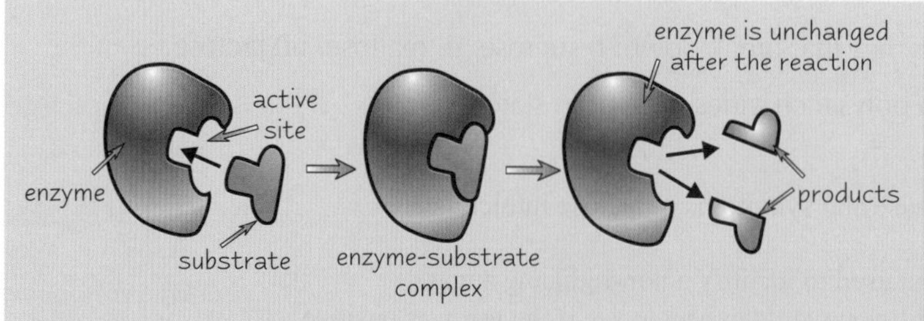

Scientists soon realised that the lock and key model didn't give the full story. The enzyme and substrate do have to fit together in the first place, but new evidence showed that the **enzyme-substrate complex changed shape** slightly to complete the fit. This **locks** the substrate even more tightly to the enzyme. Scientists modified the old lock and key model and came up with the '**induced fit**' model.

Enzyme Action

...but the 'Induced Fit' Model is a Better Theory

The '**induced fit**' model helps to explain why enzymes are so **specific** and only bond to one particular substrate. The substrate doesn't only have to be the right shape to fit the active site, it has to make the active site **change shape** in the right way as well. This is a prime example of how a widely accepted theory can **change** when **new evidence** comes along. The 'induced fit' model is still widely accepted — for now, anyway.

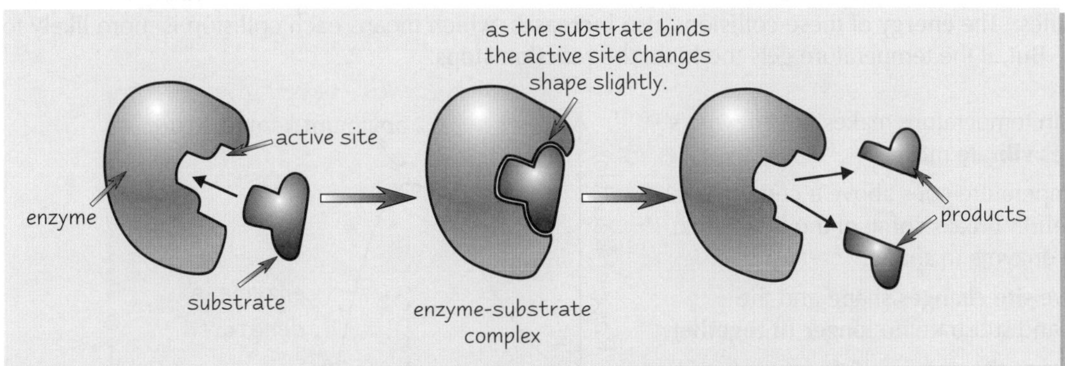

The 'Luminous Tights' model was popular in the 1980s but has since been found to be grossly inappropriate.

Enzyme Properties relate to their Tertiary Structure

1) Enzymes are **very specific** — they usually only catalyse **one** reaction, e.g. maltase only breaks down maltose, sucrase only breaks down sucrose.
2) This is because **only one substrate will fit** into the active site.
3) The active site's **shape** is determined by the enzyme's **tertiary structure** (which is determined by the enzyme's **primary structure**).
4) Each **different enzyme** has a **different tertiary structure** and so a **different shaped active site**. If the substrate shape doesn't match the active site, the reaction won't be catalysed.
5) If the tertiary structure of a protein is **altered** in any way, the **shape** of the active site will **change**. This means the **substrate won't fit** into the active site and the enzyme will no longer be able to carry out its function.
6) The tertiary structure of an enzyme may be **altered** by changes in **pH** or **temperature** (see next page).
7) The **primary structure** (amino acid sequence) of a protein is determined by a **gene**. If a mutation occurs in that gene (see p. 53), it could change the tertiary structure of the enzyme **produced**.

Practice Questions

Q1 What is an enzyme?

Q2 What is the name given to the amount of energy needed to start a reaction?

Q3 What is an enzyme-substrate complex?

Q4 Why can an enzyme only bind one substance?

Exam Questions

Q1 Describe the 'lock and key' model of enzyme action and explain how the 'induced fit' model is different. [7 marks]

Q2 Explain how a change in the amino acid sequence of an enzyme may prevent it from functioning properly. [2 marks]

But why is the enzyme-substrate complex?

So enzymes lower the activation energy of a reaction. I like to think of it as an assault course (bear with me). Suppose the assault course starts with a massive wall — enzymes are like the person who gives you a leg up over the wall (see?). Without it you'd need lots of energy to get over the wall yourself and complete the rest of the course. Unlikely.

Factors Affecting Enzyme Activity

Well, here we are again... more about enzymes. You can't just bung an enzyme into a reaction and expect it to work. They're temperamental things, bless 'em, and require special conditions...

***Temperature** has a **Big Influence** on Enzyme Activity*

Like any chemical reaction, the **rate** of an enzyme-controlled reaction **increases** when the **temperature's increased**. More heat means **more kinetic energy**, so molecules **move faster**. This makes the enzymes **more likely** to **collide** with the substrate molecules. The **energy** of these collisions also **increases**, which means each collision is more likely to **result** in a **reaction**. But, if the temperature gets too high, the **reaction stops**.

1) The rise in temperature makes the enzyme's molecules **vibrate more**.
2) If the temperature goes above a certain level, this vibration **breaks** some of the **bonds** that hold the enzyme in shape.
3) The **active site changes shape** and the enzyme and substrate **no longer fit together**.
4) At this point, the enzyme is **denatured** — it no longer functions as a catalyst.

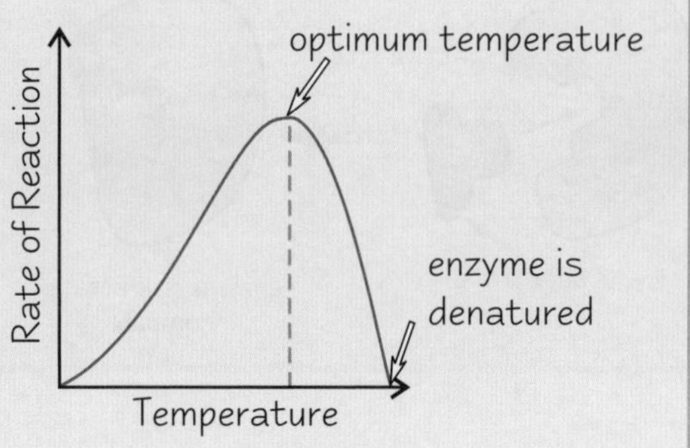

Every enzyme has an optimum temperature. For most human enzymes it's around 37 °C, but some enzymes, like those used in biological washing powders, can work well at 60 °C.

***pH** Also Affects Enzyme Activity*

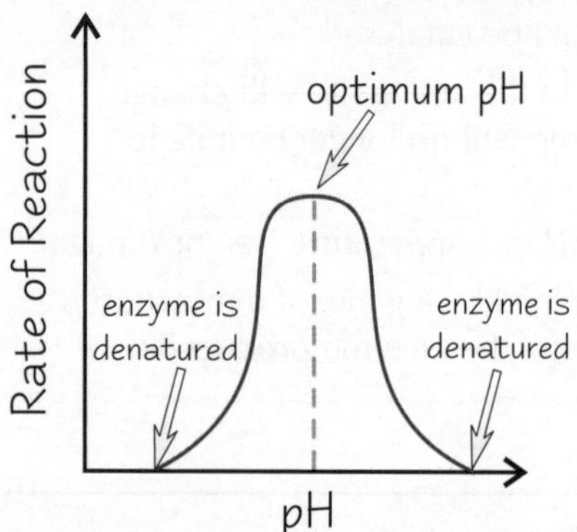

All enzymes have an **optimum pH value**. Most human enzymes work best at pH 7 (neutral), but there are exceptions. **Pepsin**, for example, works best at pH 2 (acidic), which is useful because it's found in the stomach. Above and below the optimum pH, the H^+ and OH^- ions found in acids and alkalis can mess up the **ionic bonds** and **hydrogen bonds** that hold the enzyme's tertiary structure in place. This makes the **active site change shape**, so the enzyme is **denatured**.

Substrate Concentration** Affects the Rate of Reaction **Up to a Point

The **higher** the substrate concentration, the **faster** the reaction — more substrate molecules means a **collision** between substrate and enzyme is **more likely** and so more active sites will be used.
This is only true up until a '**saturation**' **point** though. After that, there are so many substrate molecules that the enzymes have about as much as they can cope with (all the **active sites are full**), and adding more **makes no difference**.

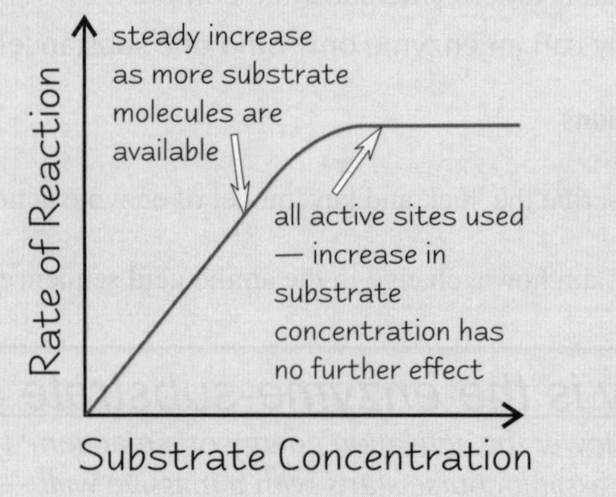

Factors Affecting Enzyme Activity

Enzyme Activity can be **Inhibited**

Enzyme activity can be prevented by **enzyme inhibitors** — molecules that **bind to the enzyme** that they inhibit. Inhibition can be **competitive** or **non-competitive**.

COMPETITIVE INHIBITION

1) **Competitive inhibitor** molecules have a **similar shape** to that of **substrate** molecules.
2) They **compete** with the substrate molecules to **bind** to the **active site**, but **no reaction** takes place.
3) Instead they **block** the active site, so **no substrate** molecules can **fit** in it.
4) How much the enzyme is inhibited depends on the **relative concentrations** of the inhibitor and substrate.
5) If there's a **high concentration** of the **inhibitor**, it'll take up nearly **all** the **active sites** and hardly any of the substrate will get to the enzyme.

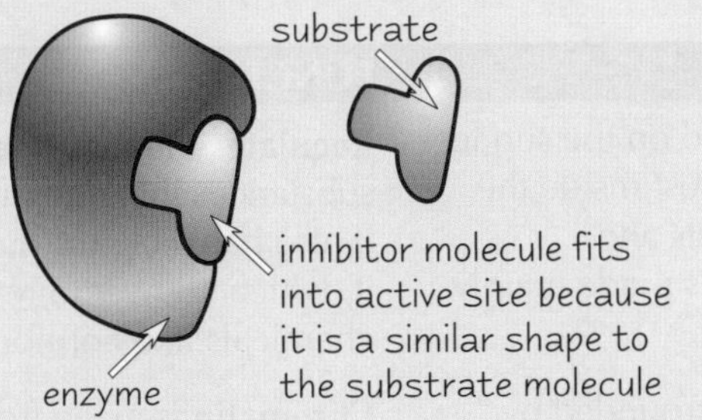

NON-COMPETITIVE INHIBITION

1) **Non-competitive inhibitor** molecules bind to the enzyme **away from its active site**.
2) This causes the active site to **change shape** so the substrate molecules can no longer bind to it.
3) They **don't** 'compete' with the substrate molecules to bind to the active site because they are a **different shape**.
4) **Increasing** the concentration of **substrate won't** make any difference — enzyme activity will still be inhibited.

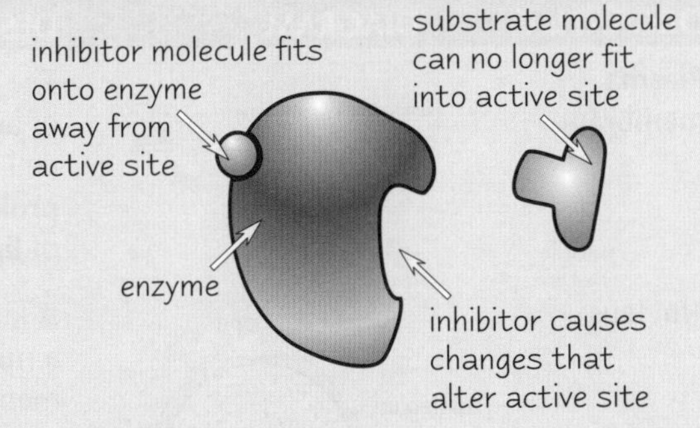

Practice Questions

Q1 Draw a graph to show the effect of temperature on enzyme activity.

Q2 Draw a graph to show the effect of pH on enzyme activity.

Q3 Explain the effect of increasing substrate concentration on the rate of an enzyme-catalysed reaction.

Exam Questions

Q1 When doing an experiment on enzymes, explain why it is necessary to control the temperature and pH of the solutions involved. [8 marks]

Q2 Inhibitors prevent enzymes from working properly. They can be competitive or non-competitive.

a) Explain how a competitive inhibitor works. [3 marks]

b) Explain how a non-competitive inhibitor works. [2 marks]

Activity — mine is usually inhibited by pizza and a movie...

Human enzymes work well under normal body conditions — a neutral pH and body temp of 37 °C. Many poisons are enzyme inhibitors, e.g. cyanide. Even though there are thousands of enzymes in our bodies, inhibiting just one of them can cause severe problems. Some drugs are enzyme inhibitors though, e.g. Viagra, aspirin and penicillin, so they're not all bad.

Animal Cell Structure

You've probably taken one look at the big table on these pages and got the fear. Don't worry though — there's nothing too taxing here... And besides, who doesn't love cells?

All Cells Contain **Organelles**

Organelles are **parts of** cells. Each one has a **specific function**.

If you examine a cell through an **electron microscope** (see p. 24) you can see its **organelles** and the **internal structure** of most of them. Most of what's known about cell structure has been discovered by electron microscope studies. The diagram on the right shows the major parts of an **animal cell**.

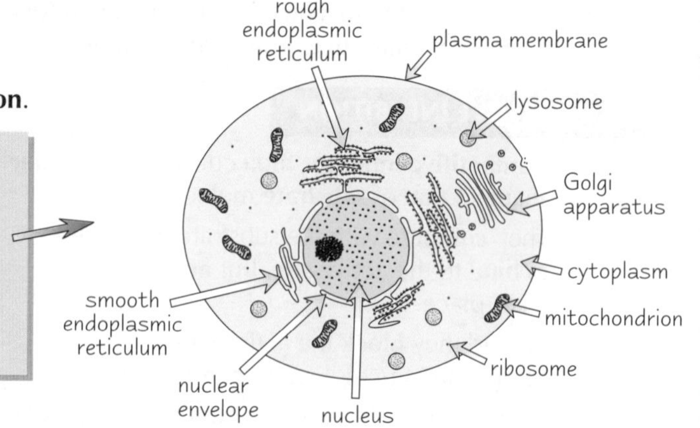

Different Organelles have **Different Functions**

This giant table contains a big list of organelles — you need to know the **structure** and **function** of them all. Sorry. Most organelles are surrounded by **membranes**, which sometimes causes confusion — don't make the mistake of thinking that a diagram of an organelle is a diagram of a whole cell. They're not cells — they're **parts of** cells.

ORGANELLE	DIAGRAM	DESCRIPTION	FUNCTION
Plasma membrane	plasma membrane; cytoplasm	The membrane found on the surface of **animal cells** and just inside the cell wall of **plant cells** and **prokaryotic cells**. It's made mainly of **lipids** and **protein**.	**Regulates the movement** of substances into and out of the cell. It also has **receptor molecules** on it, which allow it to respond to chemicals like hormones.
Nucleus	nuclear envelope; nucleolus; nuclear pore; chromatin	A large organelle surrounded by a **nuclear envelope** (double membrane), which contains many **pores**. The nucleus contains **chromatin** and often a structure called the **nucleolus**.	**Chromatin** is made from proteins and DNA (DNA **controls the cell's activities**). The pores allow substances (e.g. RNA) to move between the nucleus and the cytoplasm. The **nucleolus** makes **ribosomes** (see below).
Lysosome		A **round organelle** surrounded by a **membrane,** with no clear internal structure.	Contains **digestive enzymes**. These are kept separate from the cytoplasm by the surrounding membrane, and can be used to **digest invading cells** or to **break down** worn out components of the cell.
Ribosome	small subunit; large subunit	A **very small organelle** that floats free in the cytoplasm or is attached to the rough endoplasmic reticulum.	The **site** where **proteins** are **made**.
Endoplasmic Reticulum (ER)	a) b) ribosome; fluid	There are two types of endoplasmic reticulum: the **smooth endoplasmic reticulum** (diagram **a**) is a system of membranes enclosing a fluid-filled space. The **rough endoplasmic reticulum** (diagram **b**) is similar, but is **covered in ribosomes**.	The **smooth endoplasmic reticulum synthesises** and **processes lipids**. The **rough endoplasmic reticulum folds** and **processes proteins** that have been made at the ribosomes.

Animal Cell Structure

ORGANELLE	DIAGRAM	DESCRIPTION	FUNCTION
Golgi Apparatus		A group of fluid-filled **flattened sacs.**	It **processes** and **packages** new lipids and proteins. It also **makes lysosomes.**
Microvilli	microvilli; plasma membrane	These are **folds** in the plasma membrane.	They're found on cells involved in processes like absorption, such as epithelial cells in the small intestine (see p. 12). They **increase** the **surface area** of the plasma membrane.
Mitochondrion	outer membrane; inner membrane; crista; matrix	They're usually oval-shaped. They have a **double membrane** — the inner one is folded to form structures called **cristae**. Inside is the **matrix**, which contains enzymes involved in respiration.	The **site of aerobic respiration.** They're found in large numbers in cells that are very **active** and require a lot of **energy**.

*Cells Have **Different Organelles** Depending on Their **Function***

In the exam, you might get a question where you need to apply your knowledge of the **organelles** in a cell to explain why it's particularly **suited** to its **function**. Here are some tips:

- Think about how the **structure** of the cell might affect its **job** — e.g. if it's part of an **exchange surface** it might have organelles that **increase** the **surface area** (e.g. microvilli). If it **carries things** it might have **lost** some of its organelles to make **more room**.
- Think about **what** the cell **needs** to do its **job** — e.g. if the cell uses a lot of **energy**, it'll need lots of **mitochondria**. If it makes a lot of **proteins** it'll need a lot of **ribosomes**.

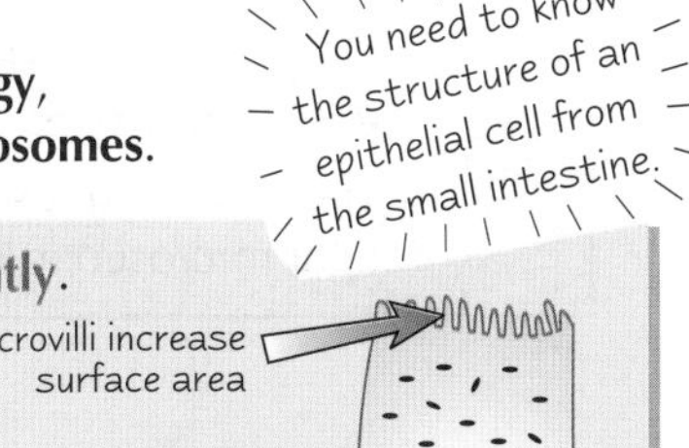

Example **Epithelial cells** in the **small intestine** are adapted to **absorb food efficiently**.

1) The walls of the small intestine have lots of finger-like projections called **villi** to **increase surface area**.
2) The **cells** on the surface of the villi have **microvilli** to increase surface area even more.
3) They also have **lots of mitochondria** — to provide **energy** for the transport of digested food molecules into the cell.

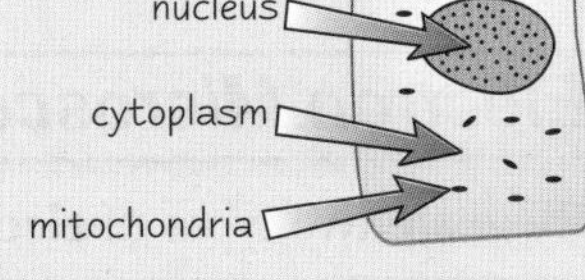

Practice Questions

Q1 What is the function of a ribosome?

Q2 What is the function of a mitochondrion?

Exam Questions

Q1 Pancreatic cells make and secrete hormones (made of protein) into the blood. From production to secretion, list the organelles involved in making hormones. [4 marks]

Q2 Cilia are hair-like structures found on lung epithelial cells. Their function is to beat and move mucus out of the lungs. Beating requires energy. Suggest how ciliated cells are adapted to their function in terms of the organelles they contain. Explain your answer. [2 marks]

The function of an organelle — to play music...

Not the most exciting pages in the world but you need to know what all the organelles listed do. I'm afraid they'll keep popping up throughout the rest of the book — microvilli are important in digestion, the plasma membrane is essential for controlling the movement of things in and out of the cell and all the DNA stuff happens in the nucleus.

Analysis of Cell Components

If you were born over a century ago then you wouldn't have had to learn all this stuff about organelles because people wouldn't have known anything about them. But then better microscopes were invented and here we are.

Magnification is Size, Resolution is Detail

We all know that microscopes produce a **magnified image** of a sample, but **resolution** is just as important...

1) MAGNIFICATION is how much **bigger** the image is than the specimen (the sample you're looking at). It's calculated using this formula:

$$\text{magnification} = \frac{\text{length of image}}{\text{length of specimen}}$$

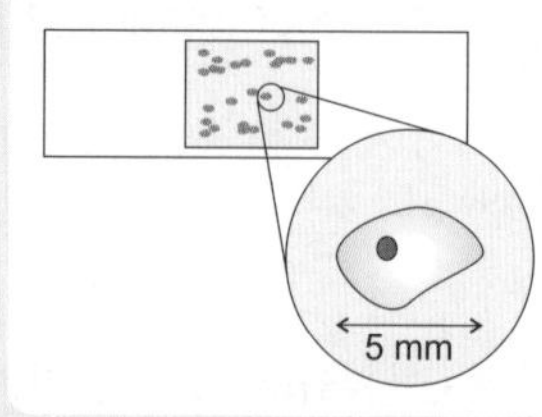

For example:
If you have a magnified image that's 5 mm wide and your specimen is 0.05 mm wide the magnification is:
5 ÷ 0.05 = **× 100**.

2) RESOLUTION is how **detailed** the image is. More specifically, it's how well a microscope **distinguishes** between **two points** that are **close together**. If a microscope lens can't separate two objects, then increasing the magnification won't help.

There are Two Main Types of Microscope — Light and Electron

Light microscopes

1) They use **light** (no surprises there).
2) They have a **lower resolution** than electron microscopes.
3) They have a maximum resolution of about **0.2 micrometres** (μm).
4) The maximum useful **magnification** of a light microscope is about **× 1500**.

Electron microscopes

1) They use **electrons** instead of light to form an image.
2) They have a **higher resolution** than light microscopes so give a **more detailed image**.
3) They have a maximum resolution of about **0.0001 micrometres** (μm). (About 2000 times higher than light microscopes.)
4) The maximum useful **magnification** of an electron microscope is about **× 1 500 000**.

A micrometre (μm) is one millionth of a metre, or 0.001 mm.

Electron Microscopes are either 'Scanning' or 'Transmission'

There are **two** types of **electron microscope**:

Transmission electron microscopes (TEMs)

1) TEMs use **electromagnets** to focus a **beam of electrons**, which is then transmitted **through** the specimen.
2) **Denser** parts of the specimen absorb **more electrons**, which makes them look **darker** on the image you end up with.
3) TEMs are good because they give **high resolution images**.
4) But they can only be used on **thin specimens**.

Nancy's New Year resolution was to get a hair cut.

Scanning electron microscopes (SEMs)

1) SEMs **scan** a beam of electrons across the specimen.
2) This **knocks off** electrons from the **specimen**, which are gathered in a **cathode ray tube** to form an **image**.
3) The images you end up with show the **surface** of the specimen and they can be **3-D**.
4) SEMs are good because they can be used on **thick specimens**.
5) But they give **lower resolution images** than TEMs.

Analysis of Cell Components

Cell Fractionation Separates Organelles

Suppose you wanted to look at some **organelles** under an **electron microscope**. First you'd need to **separate** them from the **rest of the cell** — you can do this by **cell fractionation**. There are **three** steps to this technique:

1 Homogenisation — Breaking Up the Cells

Homogenisation can be done in several **different ways**, e.g. by vibrating the cells or by grinding the cells up in a blender. This **breaks up** the **plasma membrane** and **releases** the **organelles** into solution.

2 Filtration — Getting Rid of the Big Bits

Next, the homogenised cell solution is **filtered** through a **gauze** to separate any **large cell debris** or **tissue debris**, like connective tissue, from the organelles. The organelles are much **smaller** than the debris, so they pass through the gauze.

3 Ultracentrifugation — Separating the Organelles

After filtration, you're left with a solution containing a **mixture** of organelles. To separate a particular organelle from all the others you use **ultracentrifugation**.

1) The cell fragments are poured into a **tube**. The tube is put into a **centrifuge** (a machine that separates material by spinning) and is spun at a **low speed**. The **heaviest organelles**, like nuclei, get flung to the **bottom** of the tube by the centrifuge. They form a **thick sediment** at the bottom — the **pellet**. The rest of the organelles stay suspended in the fluid above the sediment — the **supernatant**.
2) The supernatant is **drained off**, poured into **another tube**, and spun in the centrifuge at a **higher speed**. Again, the **heaviest organelles**, this time the mitochondria, form a pellet at the bottom of the tube. The supernatant containing the rest of the organelles is drained off and spun in the centrifuge at an **even higher speed**.
3) This process is **repeated** at higher and higher speeds, until all the organelles are **separated out**. Each time, the pellet at the bottom of the tube is made up of lighter and lighter organelles.

As the ride got faster, everyone felt their nuclei sink to their toes...

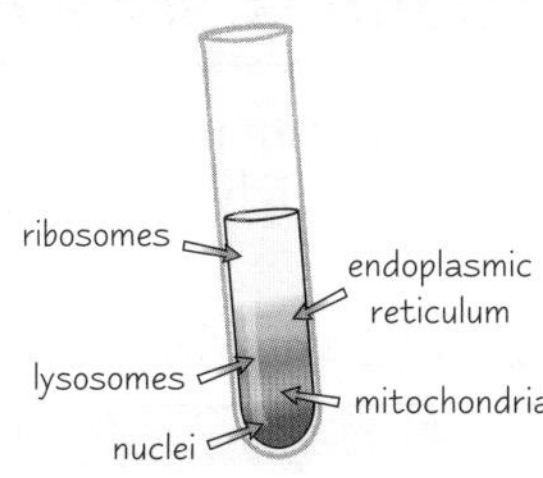

The organelles are separated in order of mass (from heaviest to lightest) — this order is usually: nuclei, then mitochondria, then lysosomes, then endoplasmic reticulum, and finally ribosomes.

Practice Questions

Q1 What is meant by a microscope's magnification?

Q2 What is meant by a microscope's resolution?

Exam Questions

Q1 Describe the difference between SEMs and TEMs and give one limitation of each. [6 marks]

Q2 Describe how you would separate organelles from a cell sample using cell fractionation.
Explain why each step is done. [8 marks]

Cell fractionation — sounds more like maths to me...

So, if you fancy getting up close and personal with mitochondria remember to homogenise, filter and ultracentrifuge first. Easy peasy. Then you have to decide if you want to use an SEM or TEM to view them, taking into account each of their limitations. Finally, you get to look at pretty pictures of what's inside our cells and make sounds like 'oooh' or 'ahhh'...

Plasma Membranes

The plasma membrane is basically the cell boundary. To understand how substances get across this boundary (so they can enter or leave the cell) you have to know what it's made of. All I can say is... this section does get better.

Substances are Exchanged Across Plasma Membranes

In order to survive and carry out their functions, cells need to **take in** substances like glucose and oxygen, and **get rid of** substances like urea and carbon dioxide. The **plasma membrane** is a complex structure that controls what substances **enter** or **leave** the cell.

Plasma Membranes are Mostly Made of Lipids

The **structure** of all **membranes** is basically the same. They're composed of **lipids** (mainly **phospholipids**), **proteins** and **carbohydrates** (usually attached to proteins or lipids).

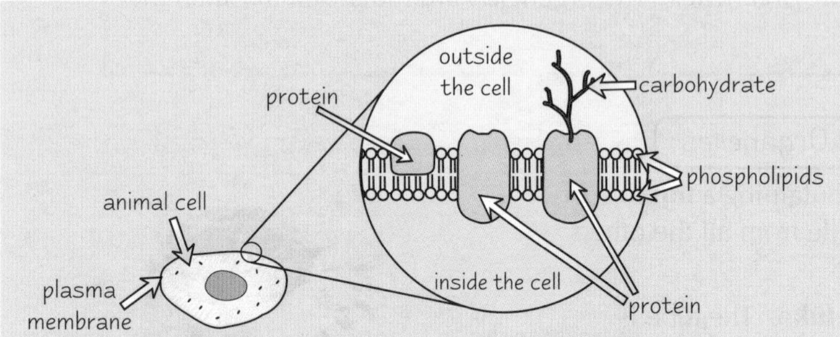

In 1972, the **fluid mosaic model** was suggested to describe the arrangement of molecules in the membrane. In the model, **phospholipid molecules** form a continuous, double layer (**bilayer**). This layer is 'fluid' because the phospholipids are constantly moving. **Protein molecules** are scattered through the layer, like tiles in a **mosaic**.

Triglycerides are a Kind of Lipid

The lipids in membranes **aren't** triglycerides, but you need to know about triglycerides before you can understand phospholipids. Triglycerides have **one** molecule of **glycerol** with **three fatty acids** attached to it.

Structure of a Triglyceride

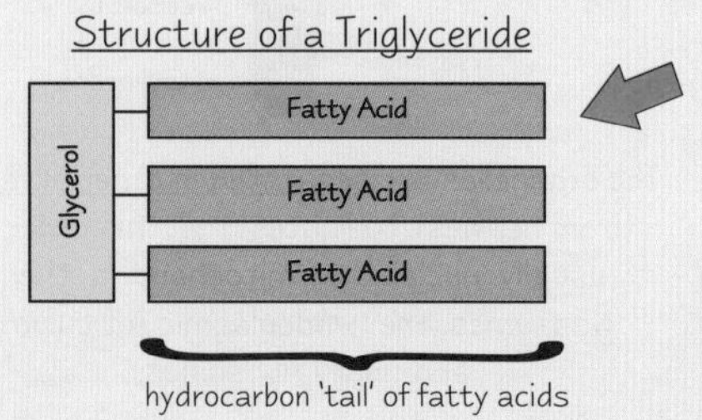

Fatty acid molecules have long 'tails' made of **hydrocarbons**. The tails are '**hydrophobic**' (they repel water molecules). These tails make lipids insoluble in water. All **fatty acids** have the same basic structure, but the **hydrocarbon tail varies**.

Basic Structure of a Fatty Acid

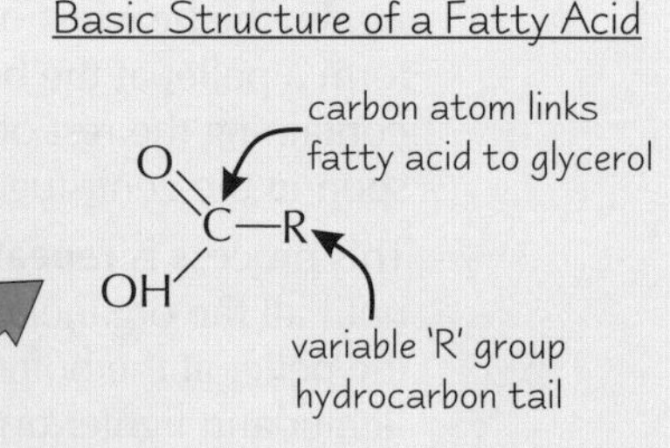

Triglycerides are Formed by Condensation Reactions

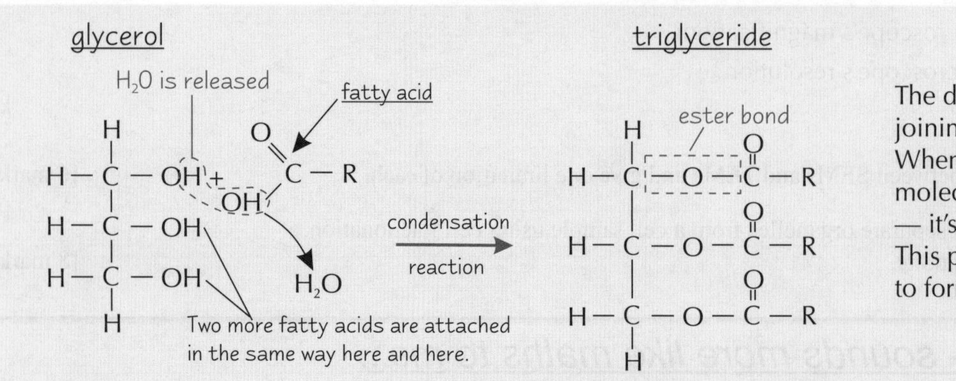

The diagram shows a **fatty acid** joining to a **glycerol molecule.** When the **ester bond** is formed a molecule of **water** is **released.** — it's a **condensation reaction.** This process happens twice more to form a **triglyceride.**

Plasma Membranes

Fatty Acids** can be **Saturated** or **Unsaturated

There are **two** kinds of fatty acids — **saturated** and **unsaturated**. The difference is in their **hydrocarbon tails**.

Saturated fatty acids **don't** have any **double bonds** between their **carbon atoms**. The fatty acid is 'saturated' with hydrogen.

saturated hydrocarbon tail

Unsaturated fatty acids **do** have double bonds between **carbon atoms**, which cause the chain to kink.

unsaturated hydrocarbon tail

Phospholipids** are **Similar** to **Triglycerides

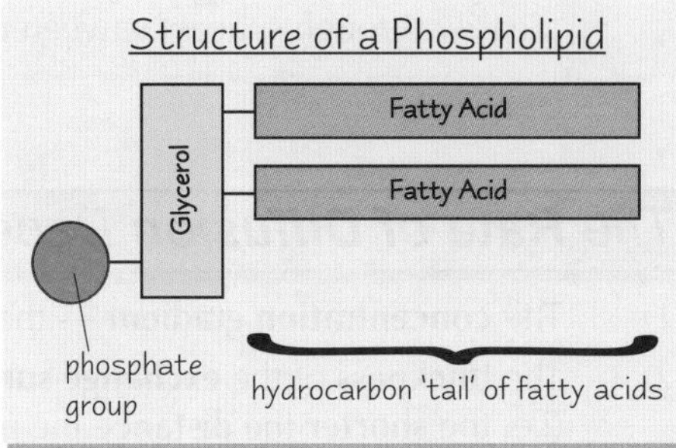

As you know, the lipids in **plasma membranes** are mainly **phospholipids**.

1) Phospholipids are pretty similar to triglycerides except one of the fatty acid molecules is replaced by a **phosphate group**.
2) The phosphate group is **hydrophilic** (attracts water). The fatty acid tails are **hydrophobic** (repel water). This is important in the plasma membrane (see p. 31 to find out why).

*Use the **Emulsion Test** for **Lipids***

If you wanted to find out if there was any **fat** in a particular **food** you could do the **emulsion test**:

1) **Shake** the test substance with **ethanol** for about a minute, then **pour** the solution into **water**.
2) Any lipid will show up as a **milky emulsion**.
3) The more lipid there is, the more noticeable the milky colour will be.

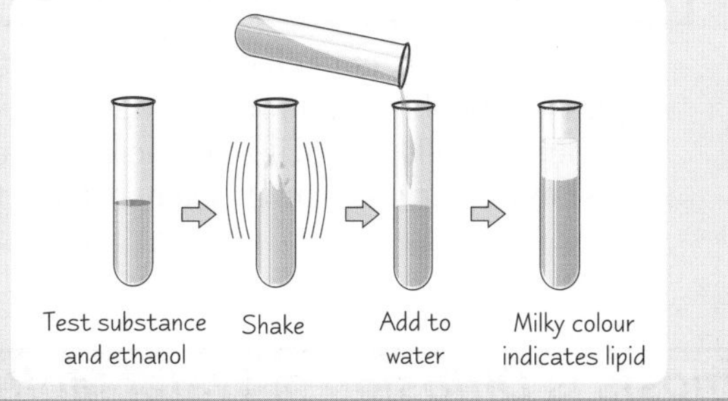

Practice Questions

Q1 State three components of the plasma membrane.

Q2 Describe how you would test for lipids in a solution.

Exam Question

Q1 Explain why the plasma membrane can be described as having a fluid-mosaic structure. [2 marks]

Q2 Plasma membranes contain phospholipids.

a) Describe the structure of a phospholipid. [3 marks]

b) Explain the difference between a saturated fatty acid and an unsaturated fatty acid. [2 marks]

The test for lipids — stick them in a can of paint...

Not really. Otherwise you might upset your Biology teacher a bit. Instead, why not sit and contemplate all those phospholipids jumping around in your plasma membranes... their water-loving, phosphate heads poking out of the cell and into the cytoplasm, and their water-hating, hydrocarbon tails forming an impenetrable layer in between...

Exchange Across Plasma Membranes

Ooooh it's starting to get a bit more exciting... here's how some substances can get across the plasma membrane without using energy. Just what you've always wanted to know, I bet.

Diffusion is the Passive Movement of Particles

1) Diffusion is the net movement of particles (molecules or ions) from an area of **higher concentration** to an area of **lower concentration**.
2) Molecules will diffuse **both ways**, but the **net movement** will be to the area of **lower concentration**. This continues until particles are **evenly distributed** throughout the liquid or gas.
3) The **concentration gradient** is the path from an area of higher concentration to an area of lower concentration. Particles diffuse **down** a concentration gradient.
4) Diffusion is a **passive process** — **no energy** is needed for it to happen.
5) Particles can diffuse **across plasma membranes**, as long as they can **move freely** through the membrane. E.g. oxygen and carbon dioxide molecules are **small enough** to pass easily through spaces between phospholipids.

The Rate of Diffusion Depends on Several Factors

1) The **concentration gradient** — the **higher** it is, the **faster** the rate of diffusion.
2) The **thickness** of the **exchange surface** — the **thinner** the exchange surface (i.e. the **shorter** the **distance** the particles have to travel), the **faster** the rate of diffusion.
3) The **surface area** — the **larger** the surface area (e.g. of the plasma membrane), the **faster** the rate of diffusion.

Microvilli Increase Surface Area for Faster Diffusion

Some cells (e.g. epithelial cells in the small intestine) have **microvilli** — projections formed by the plasma membrane folding up on itself (see p. 23). Microvilli give the cell a **larger surface area** — in human cells microvilli can increase the surface area by about **600 times**. A larger surface area means that **more particles** can be **exchanged** in the same amount of time — **increasing** the **rate of diffusion**.

Osmosis is Diffusion of Water Molecules

1) Osmosis is the **diffusion of water molecules** across a **partially permeable membrane**, from an area of **higher water potential** (i.e. higher concentration of water molecules) to an area of **lower water potential** (i.e. lower concentration of water molecules).
2) **Water potential** is the potential (likelihood) of water molecules to diffuse out of or into a solution.
3) A **partially permeable membrane** allows some molecules through it, but not all.
4) The **plasma membrane** is **partially permeable**. Water molecules are small and can diffuse easily through the **plasma membrane**, but large solute molecules can't.
5) **Pure water** has the **highest water potential**. All solutions have a **lower** water potential than pure water.
6) If two solutions have the **same water potential**, they're said to be **isotonic**.

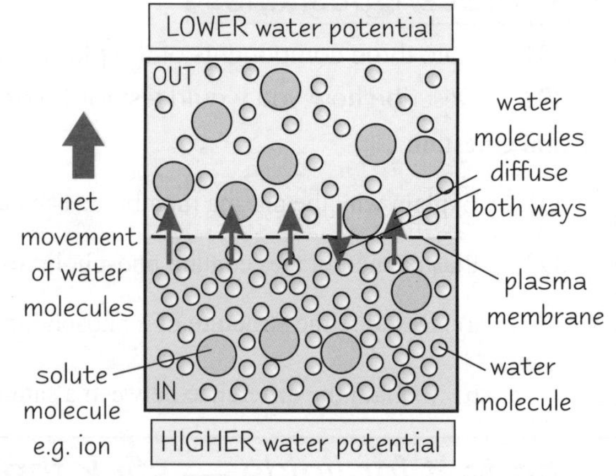

Exchange Across Plasma Membranes

Facilitated Diffusion uses Carrier Proteins and Protein Channels

1) Some **larger molecules** (e.g. amino acids, glucose) and **charged atoms** (e.g. chloride ions) **can't diffuse directly through** the phospholipid bilayer of the cell membrane.
2) Instead they diffuse through **carrier proteins** or **protein channels** in the cell membrane — this is called **facilitated diffusion**.
3) Like diffusion, facilitated diffusion moves particles **down** a **concentration gradient**, from a higher to a lower concentration.
4) It's also a passive process — it **doesn't** use **energy**.

Andy needed all his concentration for this particular gradient...

Carrier proteins move **large molecules** into or out of the cell, down their concentration gradient. **Different carrier proteins** facilitate the diffusion of **different molecules.**

1) First, a large molecule **attaches** to a carrier protein in the membrane.
2) Then, the protein **changes shape**.
3) This **releases** the molecule on the **opposite side** of the membrane.

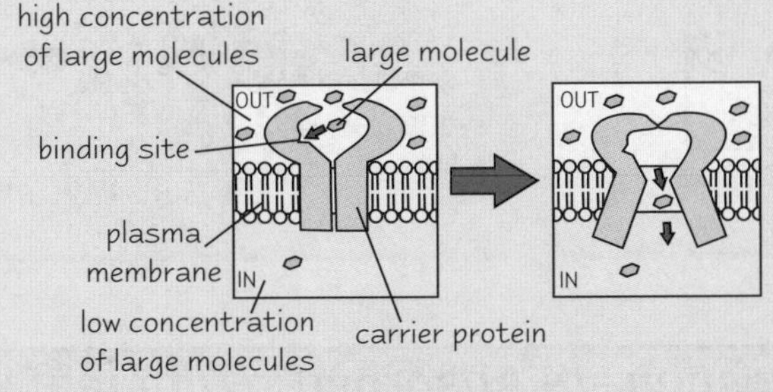

Protein channels form **pores** in the membrane for **charged particles** to diffuse through (down their concentration gradient). **Different protein channels** facilitate the diffusion of **different charged particles.**

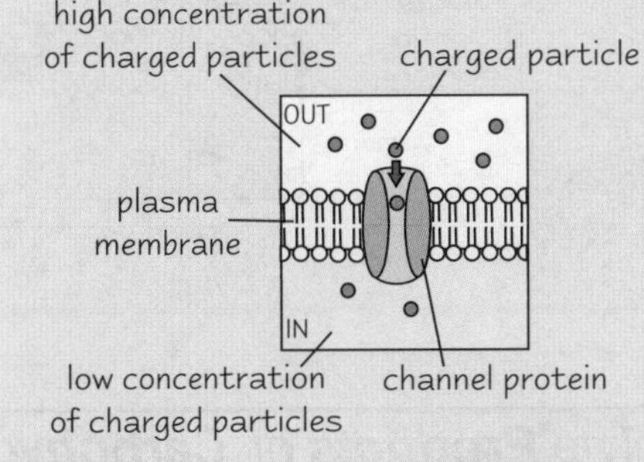

Practice Questions

Q1 Diffusion is a passive process. What does this mean?

Q2 Give two factors that affect the rate of diffusion.

Q3 How do microvilli increase the rate of diffusion?

Q4 What is facilitated diffusion?

Exam Question

Q1 Pieces of potato of equal mass were put into different concentrations of sucrose solution for three days. The difference in mass for each is recorded in the table on the right.

Concentration of sucrose / %	1	2	3	4
Mass difference / g	0.4	0.2	0	– 0.2

a) Explain why the pieces of potato in 1% and 2% sucrose solutions gained mass. [3 marks]

b) Suggest a reason why the mass of the piece of potato in 3% sucrose solution stayed the same. [1 mark]

c) What would you expect the mass difference for a potato in a 5% solution to be? Explain your answer. [4 marks]

All these molecules moving about — you'd think they'd get tired...

Right, I think I get it. If you're a small molecule, like oxygen, you can just cross the membrane by simple diffusion. If you're a water molecule you can also cross the membrane by diffusion, but you call it a fancy name – osmosis. And if you're a large or charged molecule you have a little help from a channel or carrier protein. There's a transport process to suit everyone...

Exchange Across Plasma Membranes

Diffusion and osmosis are passive processes, so for those of you feeling a bit more active here's a page all about... you guessed it... active transport.

Active Transport Moves Substances ***Against*** a Concentration Gradient

Active transport uses **energy** to move **molecules** and **ions** across plasma membranes, **against** a **concentration gradient**.

Carrier proteins are also involved in active transport:

1) The process is pretty similar to facilitated diffusion — a molecule **attaches** to the carrier protein, the protein **changes shape** and this moves the molecule **across** the membrane, **releasing it** on the other side.
2) The only difference is that **energy** is used (from **ATP** — a common source of energy used in the cell), to move the solute against its concentration gradient.

The diagram shows the active transport of **calcium**.

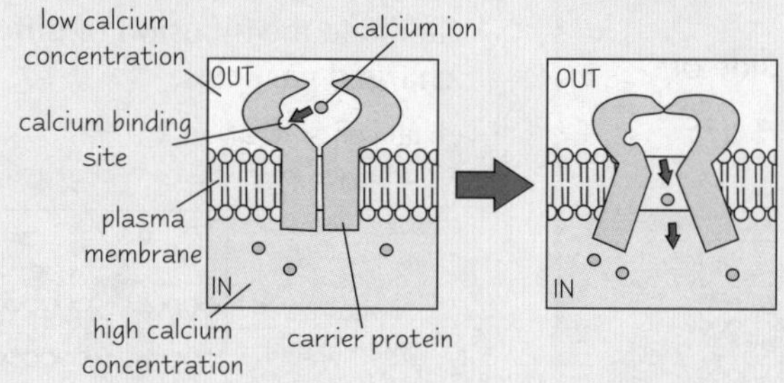

Co-transporters are a type of **carrier protein.**

1) They bind **two** molecules at a time.
2) The concentration gradient of one of the molecules is used to move the other molecule **against** its own concentration gradient.

The diagram shows the co-transport of **sodium ions** and **glucose**. Sodium ions move into the cell **down** their concentration gradient. This moves glucose into the cell too, **against** its concentration gradient.

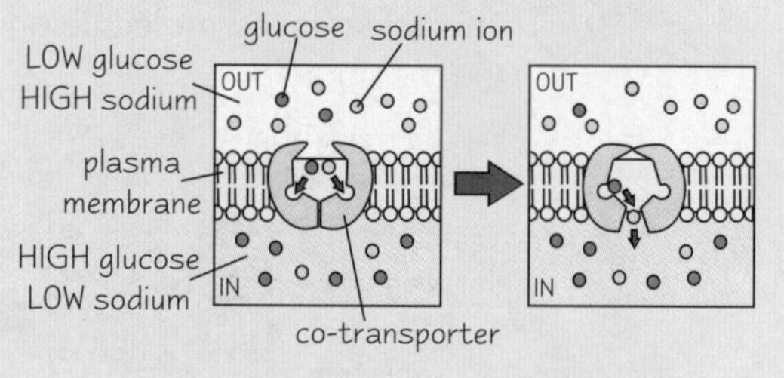

The ***Products*** of ***Carbohydrate Digestion*** are ***Absorbed*** in ***Different Ways***

All these processes — **diffusion**, **facilitated diffusion** and **active transport** — are **essential** in the body.
For example, to **absorb** the **products** of **carbohydrate digestion** (e.g. **glucose**) across the **intestinal epithelium cells**:

Some **glucose diffuses** across the **intestinal epithelium** into the **blood**

When carbohydrates are first broken down, there's a **higher concentration** of **glucose** in the **small intestine** than in the **blood** — there's a **concentration gradient**. Glucose moves across the **epithelial cells** of the small intestine into the blood by **diffusion**. When the concentration in the lumen becomes **lower** than in the blood diffusion **stops**.

Some **glucose** enters the **intestinal epithelium** by **active transport** with **sodium ions**

The remaining glucose is absorbed by **active transport**. Here's how it all works:

1) **Sodium ions** are **actively transported out** of the small intestine epithelial **cells**, into the **blood**, by the **sodium-potassium pump**. This creates a **concentration gradient** — there's now a higher concentration of sodium ions in the small intestine lumen than inside the cell.
2) This causes sodium ions to diffuse from the small intestine lumen into the cell, down their concentration gradient. They do this via the **sodium-glucose co-transporter proteins**.
3) The co-transporter carries **glucose** into the cell with the sodium. As a result the concentration of **glucose** inside the cell **increases**.
4) Glucose diffuses out of the cell, into the blood, down its concentration gradient through a protein channel, by **facilitated diffusion**.

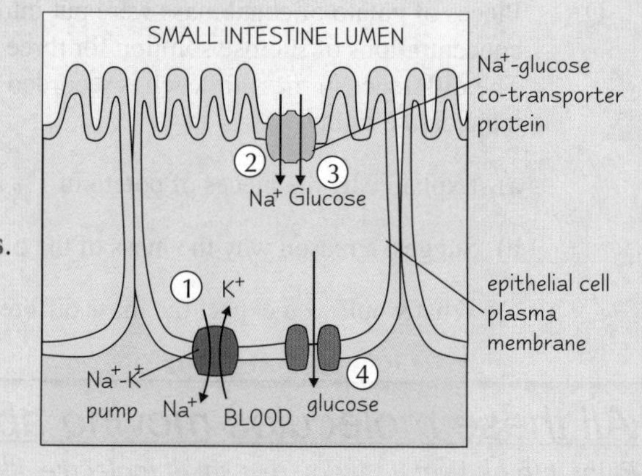

Exchange Across Plasma Membranes

*You can use the **Fluid Mosaic Model** to Explain **Membrane Properties***

You might get a question in the exam where you need to use your **knowledge** of the **fluid mosaic model** to explain why the **plasma membrane** has various **properties**. You can't go far wrong if you learn these **five** points.

1) The membrane is a good barrier against most water-soluble molecules

Phospholipids are the major component of the membrane bilayer. The **hydrophobic tails** of the phospholipids make it difficult for **water-soluble** substances, such as sodium ions and glucose, to get through.

2) The membrane controls what enters and leaves

Protein channels and **carrier proteins** in the membrane allow the passage of large or charged **water-soluble** substances that would otherwise find it difficult to cross the membrane. **Different cells** have **different** protein channels and carrier proteins — e.g. the membrane of a **nerve cell** has many **sodium-potassium carrier proteins** (which help to conduct nerve impulses) and **muscle cells** have **calcium protein channels** (which are needed for muscle contraction).

3) The membrane allows cell communication

Membranes contain **receptor proteins**. These allow the cell to **detect chemicals** released from other cells. The chemicals **signal** to the cell to **respond** in some way, e.g. the hormone insulin binds to receptors in the membranes of liver cells — this tells the liver cells to absorb glucose. This cell communication is vital for the body to **function properly**. **Different cells** have **different receptors** present in their membranes.

4) The membrane allows cell recognition

Some **proteins** and **lipids** in the plasma membrane have short **carbohydrate chains** attached to them — they're called **glycoproteins** and **glycolipids**. These molecules tell **white blood cells** that the cell is **your own**. White blood cells only attack cells that they don't recognise as **self** (e.g. those of **microorganisms** like bacteria).

5) The membrane is fluid

The **phospholipids** in the plasma membrane are **constantly moving** around. The more **unsaturated** fatty acids there are in the phospholipid bilayer, the **more fluid** it becomes. **Cholesterol** molecules fit in between the phospholipids of the bilayer — the **more** cholesterol molecules there are, the **less fluid** the membrane becomes. Cholesterol is important as it makes the cell membrane more **rigid** and prevents it from **breaking up**.

Practice Questions

Q1 What is active transport?

Q2 Describe how carrier proteins actively transport substances across the cell membrane.

Q3 Which molecule provides the energy for active transport?

Q4 Give three properties of membranes.

Exam Questions

Q1 Describe and explain how the glucose produced from starch digestion is absorbed into the blood by diffusion and active transport. [10 marks]

Q2 Use the fluid mosaic model of membrane structure to explain how the membrane controls what enters and leaves the cell. [5 marks]

Revision — like working against a concentration gradient...

Don't worry if it takes you a while to learn these pages — there's quite a lot to cover. It's a good idea to learn it bit by bit. Don't move on to co-transport until you fully understand active transport in normal carrier proteins. Then make sure you can explain each of the properties of the plasma membrane using the fluid mosaic model. They don't ask for much, do they...

Cholera

It's not all cherries and pie though — exchange of substances in and out of cells can be disrupted by diseases like cholera...

The **Cholera Bacterium** is a **Prokaryotic** Organism

There are **two** types of organisms — **prokaryotic** and **eukaryotic**.
Prokaryotic organisms are made up of **prokaryotic cells** (i.e. they're single-celled organisms), and eukaryotic organisms are made up of **eukaryotic cells**.

1) Eukaryotic cells are **complex** and include all **animal** and **plant cells**.
2) Prokaryotic cells are **smaller** and **simpler**, e.g. bacteria.

You need to know the **structure** of a prokaryotic cell and what all the different organelles inside are for.

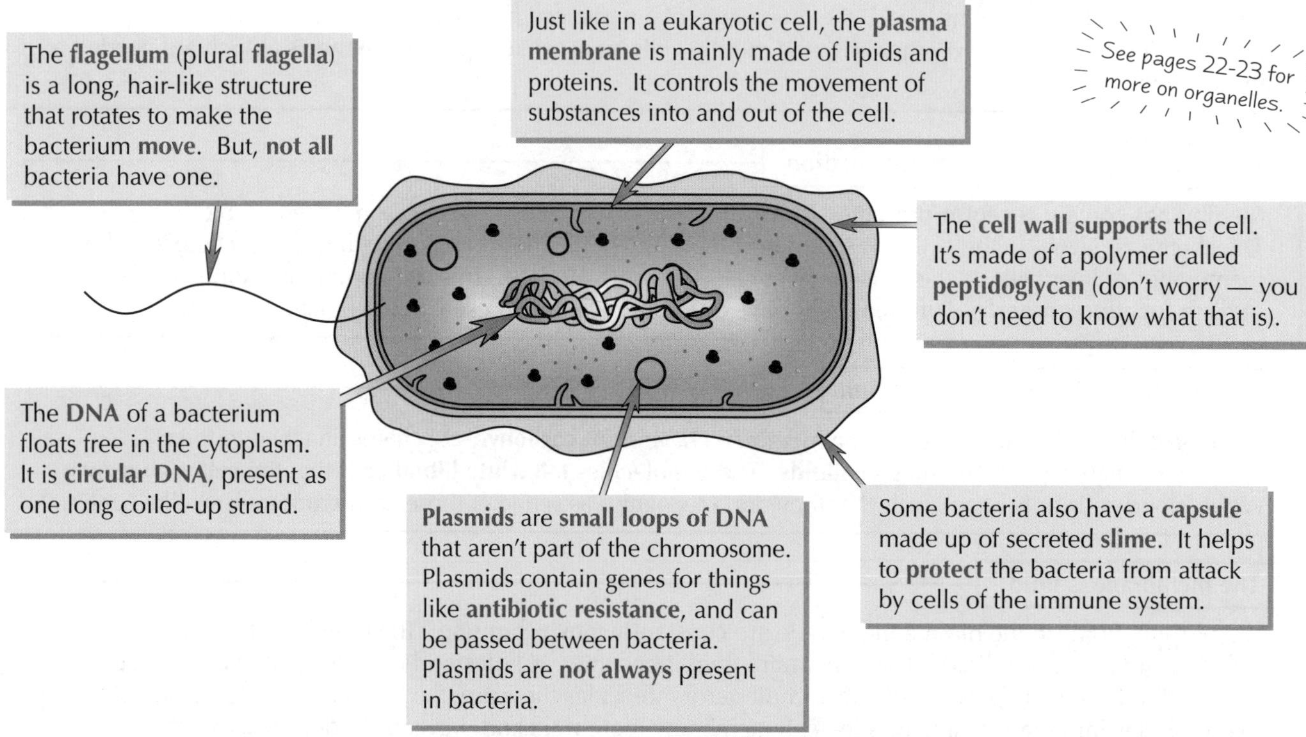

The **flagellum** (plural **flagella**) is a long, hair-like structure that rotates to make the bacterium **move**. But, **not all** bacteria have one.
Just like in a eukaryotic cell, the **plasma membrane** is mainly made of lipids and proteins. It controls the movement of substances into and out of the cell.
The **cell wall supports** the cell. It's made of a polymer called **peptidoglycan** (don't worry — you don't need to know what that is).
The **DNA** of a bacterium floats free in the cytoplasm. It is **circular DNA**, present as one long coiled-up strand.
Plasmids are **small loops of DNA** that aren't part of the chromosome. Plasmids contain genes for things like **antibiotic resistance**, and can be passed between bacteria. Plasmids are **not always** present in bacteria.
Some bacteria also have a **capsule** made up of secreted **slime**. It helps to **protect** the bacteria from attack by cells of the immune system.

See pages 22-23 for more on organelles.

Cholera Bacteria Produce a **Toxin** That Affects **Chloride Ion Exchange**

Cholera bacteria produce a **toxin** when they infect the body. This toxin causes a fair old bit of havoc...

1) The toxin causes **chloride ion protein channels** in the plasma membranes of the small intestine epithelial cells to **open**.
2) Chloride ions move **into the small intestine lumen**. The build up of chloride ions **lowers** the **water potential** of the lumen.
3) **Water** moves **out** of the **blood**, across the epithelial cells, and **into the small intestine lumen** by **osmosis** (to even up the water concentration).
4) The massive increase in water secretion into the intestine lumen leads to **really, really, really bad diarrhoea** — causing the body to become extremely **dehydrated**.

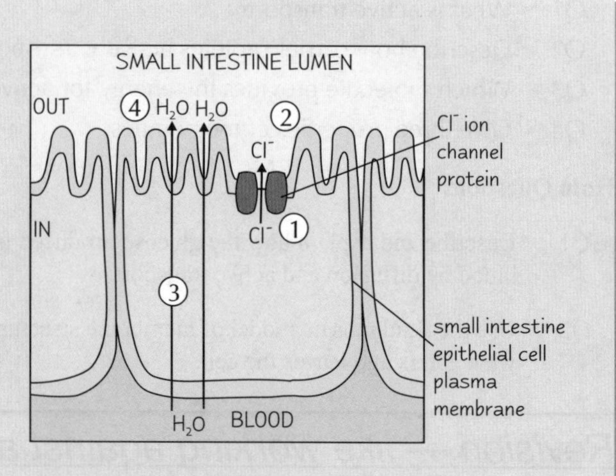

Cholera

Oral Rehydration Solutions are used to Treat Diarrhoeal Diseases

People suffering from **diarrhoeal diseases** like cholera need to **replace** all the **fluid** that they've **lost** in the diarrhoea. The quickest way to do this is by inserting a **drip** into a person's **vein**. However, not everywhere in the world has access to drips, so **oral rehydration solutions** are used instead.

Oral Rehydration Solutions (ORSs)

1) An oral rehydration solution is a **drink** that contains large amounts of **salts** (such as sodium ions and chloride ions) and **sugars** (such as glucose and sucrose) dissolved in water.
2) **Sodium** ions are included to increase **glucose** absorption (sodium and glucose are **co-transported** into the epithelium cells in the intestine — see p. 30).
3) Getting the **concentration** of the ORS right is essential for effective treatment.
4) An ORS is a very **cheap** treatment and the people administering it **don't** require much **training**. This makes it great for treating diarrhoeal diseases in **developing countries** (where they're a huge problem).

New Oral Rehydration Solutions can be Tested on Humans

ORS are so important in treating diarrhoeal disease that research into the development of **new**, **improved** ORS is always being carried out. But before a new ORS can be put into use, scientists have to show that it's **more effective** than the old ORS and that it's **safe**. This is done by **clinical testing** on humans.

There are some **ethical issues** associated with **trialling ORS**

1) **Diarrhoeal diseases** mostly affect **children**, so many **trials** involve **children**. **Parents** decide whether the child will **participate** in the trial. The child doesn't make their **own decision** — some people think this is unethical.
2) But scientists believe the treatment must be trialled on children if it's to be shown to be **effective** against a **disease** that mainly affects children.
3) Clinical trials usually involve a **blind trial**. This is where some patients who are admitted into hospital with diarrhoeal diseases are given the **standard ORS** and others are given the **new ORS**. This means that the two can be **compared**. It's called a blind trial because the patients **don't know** which treatment they've been given. Some people don't agree with this — they think that people have the **right** to **know** and **decide** on the **treatment** that they're going to have.
4) Scientists argue that a blind trial is important to eliminate any **bias** that may **skew** the **data** as a result of **patients knowing** which treatment they've received.
5) When a new ORS is first trialled, there's no way of knowing whether it'll be **better** than the current ORS — there is a **risk** of the patient **dying** when the original, better treatment was available.

Practice Questions

Q1 Give an example of a prokaryotic organism.

Q2 What is the function of a flagellum?

Q3 What are plasmids?

Q4 Suggest a reason why oral rehydration solutions contain sodium ions.

Exam Question

Q1 Explain how infection with the cholera bacterium leads to diarrhoea. [5 marks]

Q2 Give one argument for and one argument against trialling new oral rehydration solutions on children. [2 marks]

Diarrhoea is hereditary — it runs in your jeans...

Well, I don't know about you but all I can think about now is poo. With cholera, it's actually the poo that can kill you, so to speak, because you lose so much water with it. Antibiotics can help to clear the cholera bacterium, but if you aren't rehydrated straightaway you could die because your cells need plenty of water to carry out all their chemical reactions.

Lung Function

To the examiners, lung function is more than just breathing in and out — they love it (and don't even get them started on gas exchange in the alveoli). So unsurprisingly it's a good idea if you know a bit about the lungs... take a deep breath...

Lungs are Specialised Organs for Gas Exchange

Humans need to get **oxygen** into the blood (for respiration) and they need to **get rid** of **carbon dioxide** (made by respiring cells). This is where **breathing** (or **ventilation** as it's sometimes called) and the **lungs** come in.

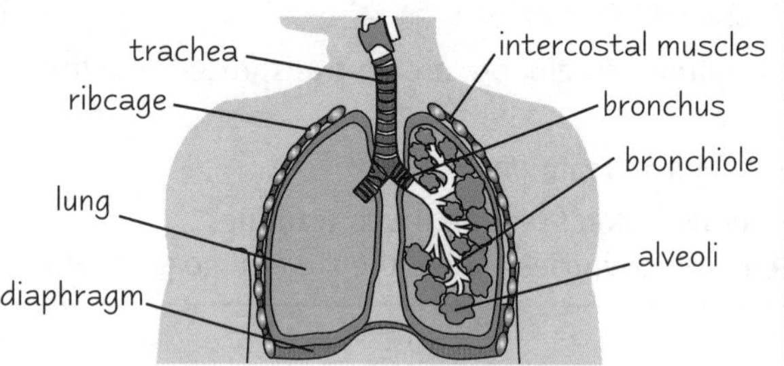
trachea
intercostal muscles
ribcage
bronchus
bronchiole
lung
alveoli
diaphragm

1) As you breathe in, air enters the **trachea** (windpipe).
2) The trachea splits into two **bronchi** — one **bronchus** leading to each lung.
3) Each bronchus then branches off into smaller tubes called **bronchioles**.
4) The bronchioles end in small 'air sacs' called **alveoli** (this is where gases are exchanged — see next page).
5) The **ribcage**, **intercostal muscles** and **diaphragm** all work together to move air in and out (see below).

Ventilation is Breathing In and Breathing Out

Ventilation consists of **inspiration** (breathing in) and **expiration** (breathing out).

Inspiration

1) The **intercostal** and **diaphragm muscles contract**.
2) This causes the **ribcage to move upwards and outwards** and the **diaphragm to flatten**, **increasing the volume** of the thorax (the space where the lungs are).
3) As the volume of the thorax increases the lung pressure **decreases** (to below atmospheric pressure).
4) This causes air to flow **into the lungs**.
5) Inspiration is an **active process** — it requires **energy**.

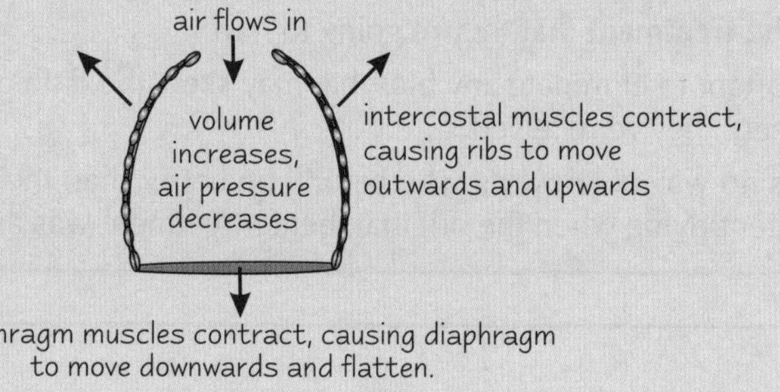
air flows in
volume increases, air pressure decreases
intercostal muscles contract, causing ribs to move outwards and upwards
diaphragm muscles contract, causing diaphragm to move downwards and flatten.

Thankfully, evolution came up with a better ventilation system.

Expiration

1) The **intercostal** and **diaphragm muscles relax**.
2) The **ribcage** moves **downwards and inwards** and the **diaphragm** becomes **curved** again.
3) The thorax volume **decreases**, causing the air pressure to **increase** (to above atmospheric pressure).
4) Air is forced **out of the lungs**.
5) Expiration is a **passive process** — it **doesn't** require energy.

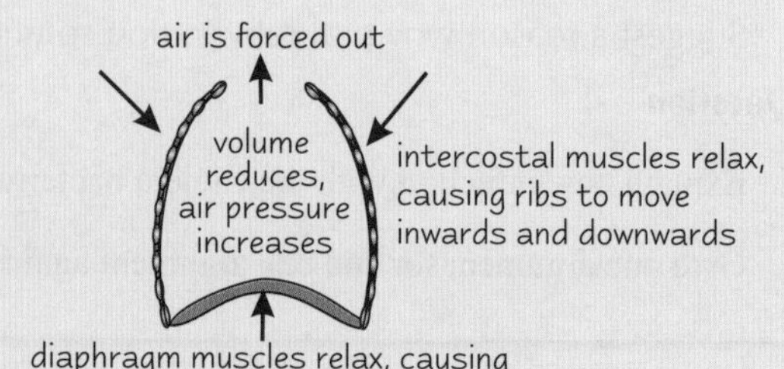
air is forced out
volume reduces, air pressure increases
intercostal muscles relax, causing ribs to move inwards and downwards
diaphragm muscles relax, causing diaphragm to become curved again.

Lung Function

In Humans Gaseous Exchange Happens in the Alveoli

Lungs contain millions of microscopic air sacs where gas exchange occurs — called **alveoli**. Each alveolus is made from a single layer of thin, flat cells called alveolar epithelium.

1) There's a huge number of alveoli in the lungs, which means there's a **big surface area** for exchanging oxygen (O_2) and carbon dioxide (CO_2).
2) The alveoli are surrounded by a network of **capillaries**.

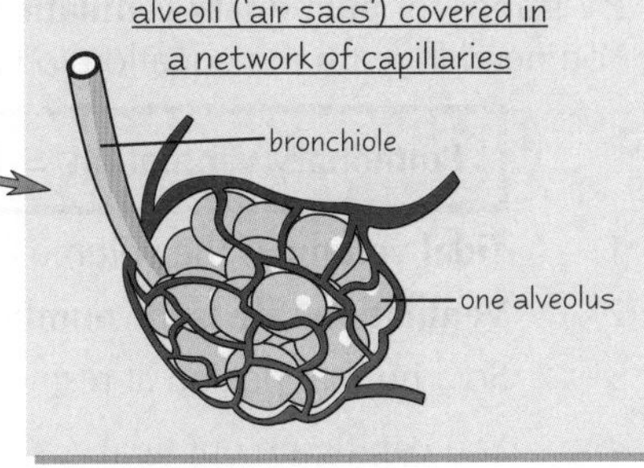

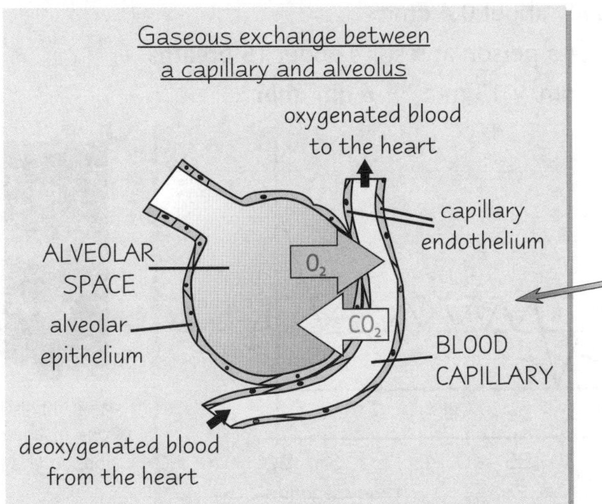

3) O_2 diffuses **out of** the alveoli, across the **alveolar epithelium** and the **capillary endothelium** (a type of epithelium that forms the capillary wall), and into **haemoglobin** (see p. 58) in the **blood**.
4) CO_2 diffuses **into** the alveoli from the blood, and is breathed out.

Epithelial tissue is pretty common in the body. It's usually found on exchange surfaces.

The Alveoli are Adapted for Gas Exchange

Alveoli have features that **speed up** the **rate of diffusion** so gases can be exchanged quickly:

1) **A thin exchange surface** — the **alveolar epithelium** is only **one cell thick**. This means there's a **short diffusion pathway** (which speeds up diffusion).
2) **A large surface area** — the **large number** of alveoli means there's a large surface area for gas exchange.

There's also a **steep concentration gradient** of oxygen and carbon dioxide between the alveoli and the capillaries, which increases the rate of diffusion. This is constantly maintained by the **flow of blood** and **ventilation**.

Practice Questions

Q1 Describe the structure of the human gas exchange system.

Q2 Describe the process of inspiration and expiration.

Q3 Describe the movement of carbon dioxide and oxygen across the alveolar epithelium.

Exam Question

Q1 Explain why there is a fast rate of gas exchange in the alveoli. [6 marks]

Alveoli — useful things... always make me think about pasta...

Just like the digestive system, a mammal's lungs act as an interface with the environment — they take in air and give out waste gases. Ventilation moves these gases into and out of the lungs, but the alveoli have the task of getting them in and out of the bloodstream. Luckily, like many other biological structures, they're well adapted for doing their job.

How Lung Disease Affects Function

It's all very well when your lungs are working perfectly, but some pathogens (and even your lifestyle) can muck them up good and proper, reducing the rate of gas exchange. Not good.

PV is the *Volume of Air* Taken into the *Lungs* in *One Minute*

PV stands for **pulmonary ventilation** — it's measured in $dm^3\ min^{-1}$.
You need to learn the equation to calculate it:

$dm^3\ min^{-1}$ is cubic decimetres per minute — a decimetre is 10 centimetres, and 1 dm^3 is the same as a litre.

Pulmonary Ventilation = Tidal volume × Ventilation rate

1) **Tidal volume** is the volume of air in **each breath** — usually about **0.4 dm^3**.
2) **Ventilation rate** is the **number of breaths per minute**. For a person at rest it's about **15 breaths**.
3) So a normal person at rest would have a PV of about $0.4\ dm^3 \times 15\ min^{-1}$ = **6 $dm^3\ min^{-1}$**.
4) You can figure out tidal volume and ventilation rate from the graph produced from a spirometer (a fancy machine that scientists and doctors use to measure the volume of air breathed in and out):

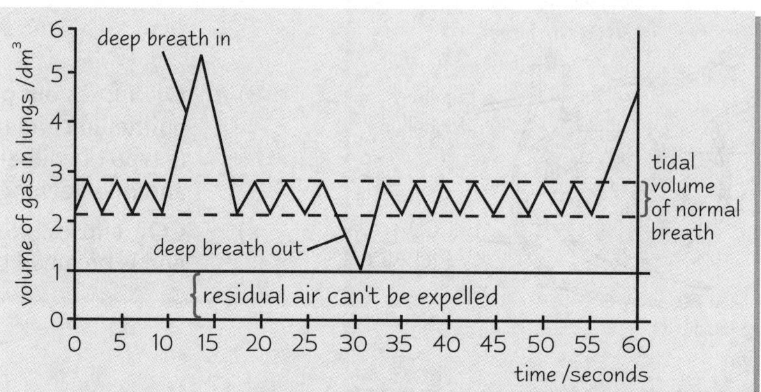

Measuring tidal volume is one of the hardest jobs in the world.

Pulmonary Tuberculosis (TB) is a *Lung Disease* Caused by *Bacteria*

Pulmonary tuberculosis is caused by the bacterium ***Mycobacterium tuberculosis***.

Infection

1) When someone becomes infected with tuberculosis bacteria, immune system cells build a **wall** around the bacteria in the lungs. This forms small, hard lumps known as **tubercles**.
2) Infected tissue within the tubercles **dies**, the gaseous exchange surface is **damaged** so **tidal volume** is **decreased**.
3) Tuberculosis also causes **fibrosis** (see next page), which further reduces the tidal volume.
4) If the bacteria enter the **bloodstream**, they can **spread** to other parts of the body.

Symptoms

1) Common symptoms include a persistent **cough**, coughing up **blood** and **mucus**, **chest pains**, **shortness of breath** and **fatigue**.
2) Sufferers may also have a **fever**.
3) Many **lose weight** due to a reduction in appetite.

Transmission

1) TB is transmitted by **droplet infection** — when an infected person **coughs** or **sneezes**, tiny **droplets** of **saliva** and **mucus** containing the bacteria are released from their mouth and nose. If an uninfected person breathes in these droplets, the bacteria are **passed on**.
2) Tuberculosis tends to be much more widespread in areas where **hygiene** levels are **poor** and where people live in **crowded** conditions.
3) TB can be prevented with the BCG vaccine, and can be treated with antibiotics.

Many people with tuberculosis are **asymptomatic** — they're infected but they **don't show** any symptoms, because the infection is in an **inactive form**. People who are asymptomatic are unable to pass the infection on. But if they become **weakened**, for example by another disease or malnutrition, then the infection can become **active**. They'll show the symptoms and be able to pass on the infection.

How Lung Disease Affects Function

Fibrosis, Asthma and Emphysema all Affect Lung Function

Fibrosis, asthma and emphysema all **reduce the rate of gas exchange** in the alveoli. Less oxygen is able to diffuse into the bloodstream, the body cells **receive less oxygen** and the rate of **aerobic respiration** is **reduced**. This means **less energy is released** and sufferers often feel **tired** and **weak**.

Fibrosis

1) Fibrosis is the formation of **scar tissue** in the lungs. This can be the result of an **infection** or exposure to substances like **asbestos** or **dust**.
2) Scar tissue is **thicker** and **less elastic** than normal lung tissue.
3) This means that the lungs are **less able to expand** and so **can't hold as much air** as normal — the tidal volume is **reduced**. It's also harder to **force air out** of the lungs due to the loss of elasticity.
4) There's a **reduction** in the rate of **gaseous exchange** — **diffusion** is **slower** across a **thicker** scarred membrane.
5) Symptoms of fibrosis include **shortness of breath**, a **dry cough**, **chest pain**, **fatigue** and **weakness**.
6) Fibrosis sufferers have a **faster breathing rate** than normal — to get enough air into their lungs to **oxygenate** their blood.

Asthma

1) Asthma is a respiratory condition where the airways become **inflamed** and **irritated**. The causes vary from case to case but it's usually because of an **allergic reaction** to substances such as **pollen** and **dust**.
2) During an asthma attack, the **smooth muscle** lining the **bronchioles contracts** and a large amount of **mucus** is produced.
3) This causes **constriction** of the airways, making it difficult for the sufferer to **breathe properly**. Air flow in and out of the lungs is **severely reduced**, so less oxygen enters the alveoli and moves into the blood.
4) Symptoms include **wheezing**, a **tight chest** and **shortness of breath**. During an attack the symptoms come on very suddenly. They can be relieved by **drugs** (often in **inhalers**) which cause the muscle in the bronchioles to **relax**, opening up the airways.

Emphysema

1) Emphysema is a lung disease caused by **smoking** or long-term exposure to **air pollution** — foreign particles in the smoke (or air) become **trapped** in the alveoli.
2) This causes **inflammation**, which attracts **phagocytes** to the area. The phagocytes produce an **enzyme** that breaks down **elastin** (a protein found in the **walls** of the **alveoli**).

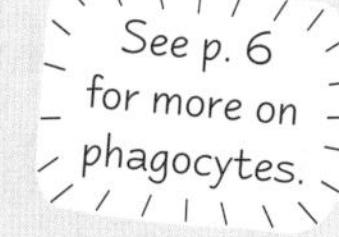

3) Elastin is **elastic** — it helps the alveoli to **return** to their **normal shape** after inhaling and exhaling air.
4) Loss of elastin means the alveoli **can't recoil** to expel air as well (it remains **trapped** in the alveoli).
5) It also leads to **destruction** of the **alveoli walls**, which **reduces** the **surface area** of the alveoli so the rate of **gaseous exchange** decreases.
6) Symptoms of emphysema include **shortness of breath** and **wheezing**. People with emphysema have an **increased breathing rate** as they try to increase the amount of air (containing oxygen) reaching their lungs.

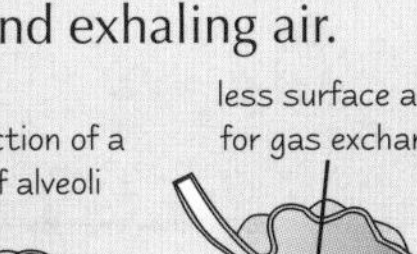

Practice Questions

Q1 How is pulmonary ventilation calculated?

Q2 What happens to the alveoli of someone who suffers from emphysema?

Exam Question

Q1 Explain why a person suffering from fibrosis may feel tired and weak. [7 marks]

Asthma, emphysema and fibrosis — what a wheeze...

Tuberculosis is a pretty grim disease — all that coughing up blood and mucus. In total, around one third of the world's population (that's about two billion people) are infected. It's not just humans who are infected though — similar bacteria cause TB in loads of other animals, like cows, pigs, monkeys, goats, badgers, cats, dogs and children.

Interpreting Lung Disease Data

It's very possible that you could be asked to interpret some data on lung disease in the exam. So being my usual nice self, I've given you some examples to show you how to do it. I know it looks a bit dull but believe me, it'll really help.

You Need to be Able to Interpret Data on Risk Factors and Lung Disease

1) All diseases have factors that will **increase** a person's **chance** of getting that disease. These are called **risk factors**. For example, it's widely known that if you **smoke** you're more likely to get **lung cancer** (smoking is a risk factor for lung cancer).
2) This is an example of a **correlation** — a link between two things (see page 90). However, a correlation doesn't always mean that one thing **causes** the other. Smokers have an **increased risk** of getting cancer but that doesn't necessarily mean smoking **causes** the disease — there are lots of other factors to take into consideration.
3) You need to be able to describe and analyse data given to you in your exams. Here are two examples of the kind of thing you might get:

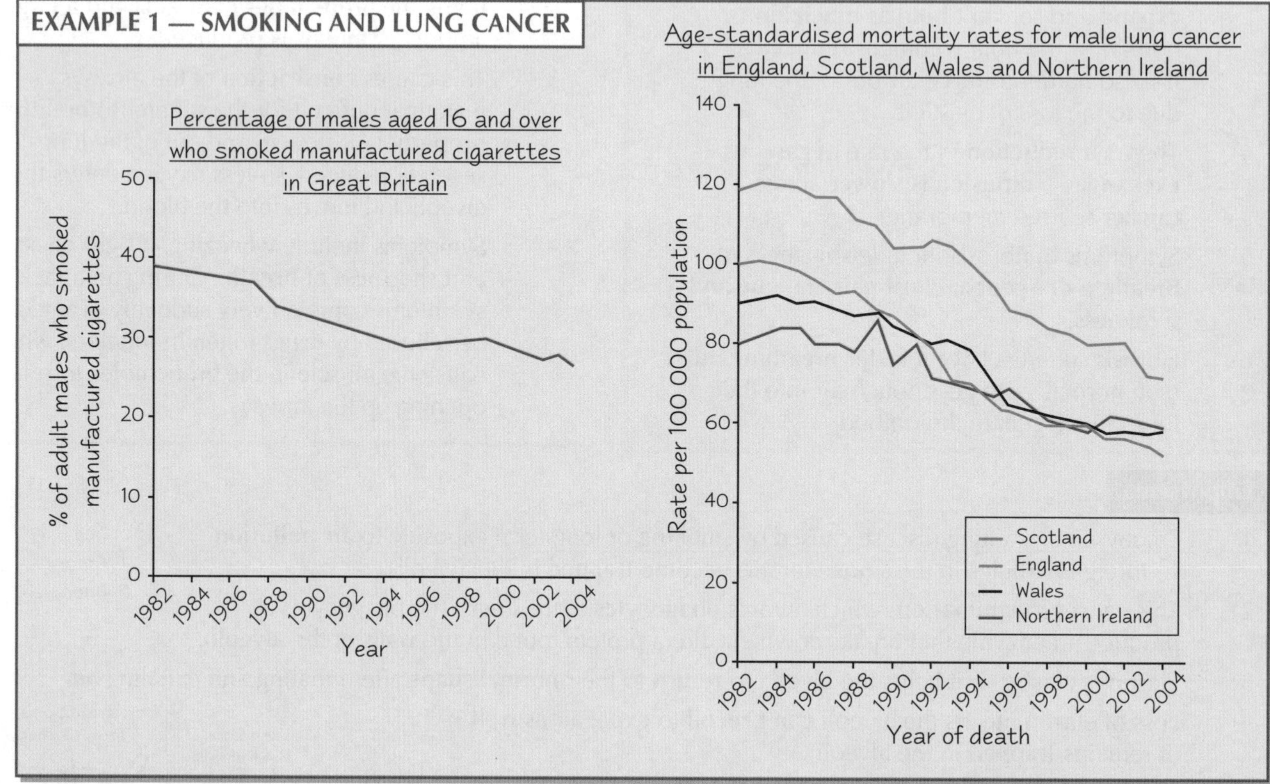

You might be asked to:

1) **Explain the data** — The graph on the left shows that the **number** of adult males in Great Britain (England, Wales and Scotland) who **smoke decreased** between 1982 and 2004. The graph on the right shows that the male lung cancer **mortality (death) rate decreased** between 1982 and 2004 for each of the countries shown. Easy enough so far.

See pages 90-92 and 196-198 for more on interpreting data.

2) **Draw conclusions** — You need to be careful what you say here. There's a **correlation** (link) between the **number** of males **who smoked** and the **mortality rate** for male lung cancer. But you **can't** say that one **caused** the other. There could be **other reasons** for the trend, e.g. deaths due to lung cancer may have decreased because less asbestos was being used in homes (not because fewer people were smoking).

Other points to consider — The graph on the right shows mortality (**death**) rates. The rate of **cases** of lung cancer **may have been increasing** but medical advances may mean more people were **surviving** (so only mortality was decreasing). Some information about the **people involved** in the studies would be helpful. For example, we don't know whether both studies used similar groups — e.g. similar diet, occupation, alcohol consumption etc. If they didn't then the results might not be reliable.

Interpreting Lung Disease Data

EXAMPLE 2 — AIR POLLUTION AND ASTHMA

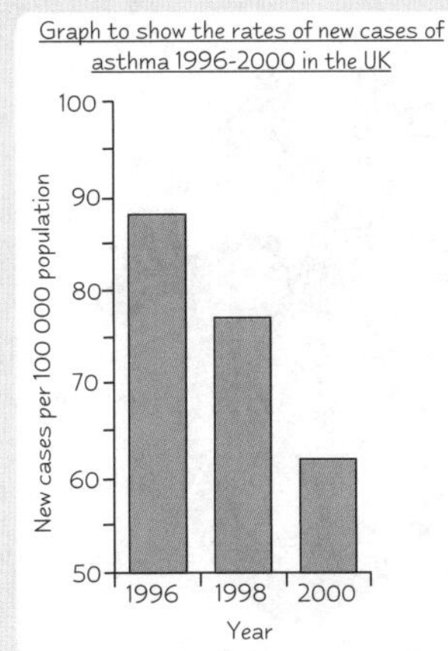

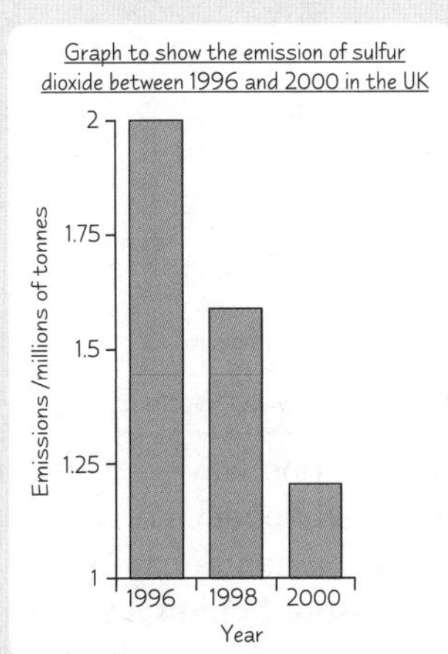

The **top graph** shows the number of **new cases of asthma** per 100 000 of the population diagnosed in the UK from 1996 to 2000. The **bottom graph** shows the **emissions** (in millions of tonnes) of **sulfur dioxide** (an **air pollutant**) from 1996 to 2000 in the UK.

You might be asked to explain the data...

1) The **top graph** shows that the number of **new cases of asthma** in the UK **fell** between 1996 and 2000, from 87 to 62 per 100 000 people.
2) The **bottom graph** shows that the **emissions of sulfur dioxide** in the UK **fell** between 1996 and 2000, from 2 to 1.2 million tonnes.

... **or draw conclusions**

1) Be careful what you say when drawing conclusions. Here there's a **link** between the **number** of new cases of **asthma** and **emissions** of **sulfur dioxide** in the **UK** — the rate of new cases of asthma has **fallen** as sulfur dioxide emissions have **fallen**. You **can't** say that one **causes** the other though because there could be **other reasons** for the trend, e.g. the number of new cases of asthma could be falling due to the **decrease** in the number of people **smoking**.
2) You can't say the **reduction** in asthma cases is **linked** to a **reduction in air pollution** (in general) either as **only** sulfur dioxide levels were studied.

Other points to consider:

1) The top graph shows **new cases** of asthma. The rate of new cases may be **decreasing** but existing cases may be becoming **more severe**.
2) The emissions were for the whole of the UK but air pollution **varies from area to area**, e.g. **cities** tend to be **more polluted**.
3) The asthma data doesn't take into account any **other factors** that may **increase** the risk of developing asthma, e.g. allergies, smoking, etc.

Practice Exam Question

Q1 In early December 1952, a dense layer of cold air trapped pollutants close to ground level in London. The graph opposite shows daily deaths and levels of sulfur dioxide and smoke between 1st and 15th December.

a) Describe the changes in the daily death rate and the levels of pollutants over the days shown. [3 marks]

b) What conclusions can be drawn from this graph? [1 mark]

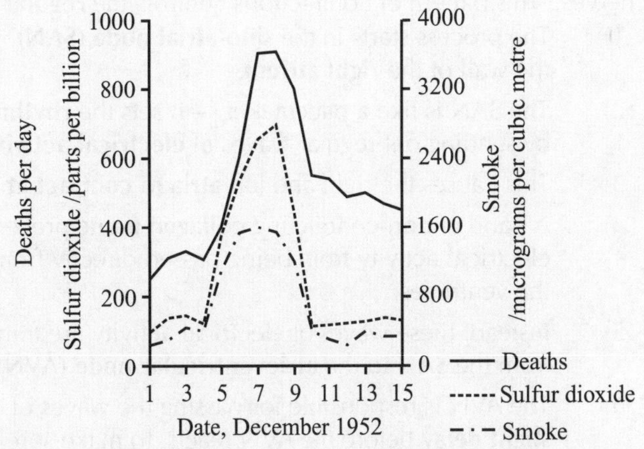

Drawing conclusions — you'll need your wax crayons and some paper...

These pages give examples to help you deal with what the examiners are sure to hurl at you — and boy, do they like throwing data around. There's some important advice here (even if I say so myself) — it's easy to leap to a conclusion that isn't really there — stick to your guns about the difference between correlation and cause and you'll blow 'em away.

The Heart

The circulatory system is made up of the heart and blood vessels. Your heart is THE major player when it comes to circulating blood around your body — it's the 'pump' that gets oxygenated blood to your cells. So... unsurprisingly, you need to know how it works. You'll find that these pages definitely get to the heart of it... groan...

The **Heart** Consists of **Two Muscular Pumps**

The diagram on the right shows the **internal structure** of the heart. The **right side** pumps **deoxygenated blood** to the **lungs** and the **left side** pumps **oxygenated blood** to the **whole body**. Note — the **left and right sides** are **reversed** on the diagram, cos it's the left and right of the person that the heart belongs to.

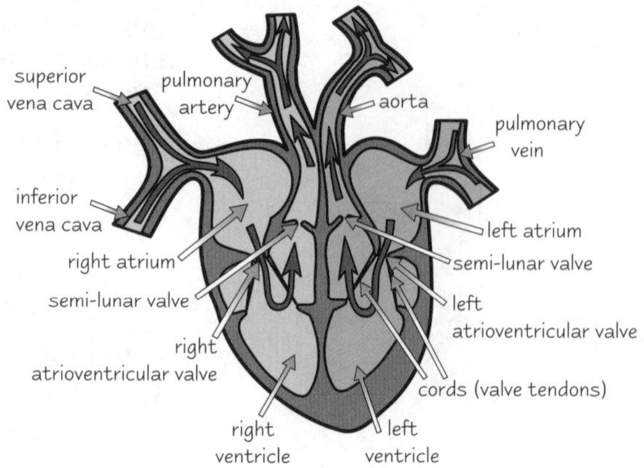

Each bit of the heart is adapted to do its job effectively.

1) The **left ventricle** of the heart has **thicker**, more muscular walls than the **right ventricle**, because it needs to contract powerfully to pump blood all the way round the body. The right side only needs to get blood to the lungs, which are nearby.
2) The **ventricles** have **thicker walls** than the **atria**, because they have to push blood out of the heart whereas the atria just need to push blood a short distance into the ventricles.
3) The **atrioventricular (AV) valves** link the atria to the ventricles and **stop blood flowing back** into the atria when the ventricles contract.
4) The **semi-lunar (SL) valves** link the ventricles to the pulmonary artery and aorta, and **stop blood flowing back** into the heart after the ventricles contract.
5) The **cords** attach the atrioventricular valves to the ventricles to stop them being forced up into the atria when the ventricles contract.

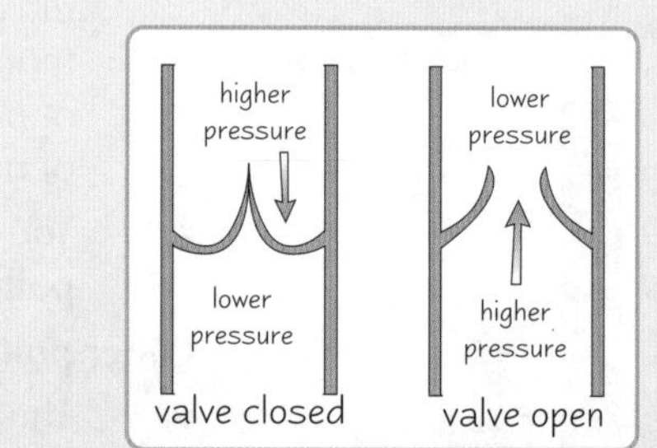

The **valves** only open one way — whether they're open or closed depends on the relative **pressure** of the heart chambers. If there's higher pressure **behind** a valve, it's forced **open**, but if pressure is higher **in front** of the valve it's forced **shut**.

Cardiac Muscle Controls the **Regular Beating** of the Heart

Cardiac (heart) muscle is '**myogenic**' — this means that it can contract and relax without receiving signals from nerves. This pattern of contractions controls the **regular heartbeat**.

1) The process starts in the **sino-atrial node (SAN)**, which is in the wall of the **right atrium**.
2) The SAN is like a pacemaker — it sets the **rhythm** of the heartbeat by sending out regular **waves of electrical activity** to the atrial walls.
3) This causes the right and left **atria** to **contract at the same time**.
4) A band of non-conducting **collagen tissue** prevents the waves of electrical activity from being passed directly from the atria to the ventricles.
5) Instead, these waves of electrical activity are transferred from the SAN to the **atrioventricular node (AVN)**.
6) The AVN is responsible for passing the waves of electrical activity on to the bundle of His. But, there's a **slight delay** before the AVN reacts, to make sure the ventricles contract **after** the atria have emptied.
7) The **bundle of His** is a group of muscle fibres responsible for conducting the waves of electrical activity to the finer muscle fibres in the right and left ventricle walls, called the **Purkyne fibres**.
8) The Purkyne fibres carry the waves of electrical activity into the muscular walls of the right and left ventricles, causing them to **contract simultaneously**, from the bottom up.

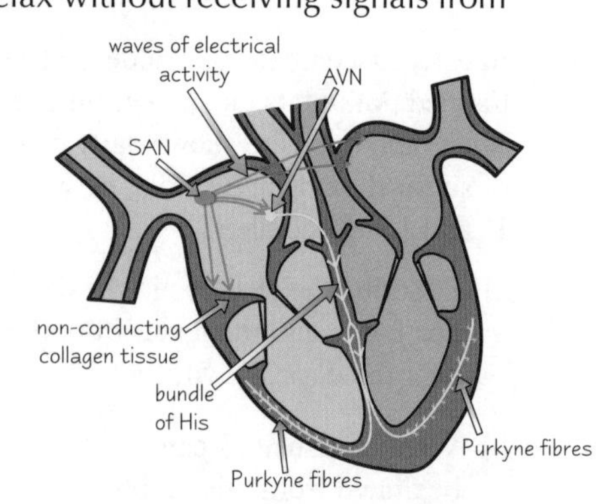

The Heart

Learn the Equation for Cardiac Output

Cardiac output is the **volume** of blood pumped by the **heart per minute** (measured in cm^3 per minute).

It's calculated using this **formula**: **cardiac output = stroke volume × heart rate**

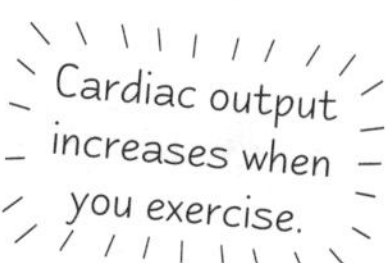
Cardiac output
increases when
you exercise.

1) **Heart rate** — the **number** of **heartbeats** per minute. You can measure your heart rate by feeling your pulse, which is basically surges of blood forced through the arteries by the heart contracting.
2) **Stroke volume** — the **volume** of blood pumped during **each heartbeat**, measured in cm^3.

The Cardiac Cycle Pumps Blood Round the Body

The cardiac cycle is an ongoing sequence of **contraction** and **relaxation** of the atria and ventricles that keeps blood **continuously** circulating round the body. The **volume** of the atria and ventricles **changes** as they contract and relax. **Pressure** changes also occur, due to the changes in chamber volume (e.g. decreasing the volume of a chamber by contraction will increase the pressure of a chamber). The cardiac cycle can be simplified into three stages:

Cardiac contraction is also called systole and relaxation is called diastole.

1 Ventricles relax, atria contract

The **ventricles are relaxed**. The **atria contract**, decreasing the volume of the chamber and **increasing** the **pressure** inside the chamber. This **pushes** the blood into the ventricles. There's a slight **increase** in **ventricular pressure** and **chamber volume** as the **ventricles receive the ejected blood** from the contracting atria.

1 Atria contract
SL valves closed
vena cava
pulmonary vein
AV valves are open

2 Ventricles contract, atria relax

The **atria relax**. The **ventricles contract** (decreasing their volume), **increasing** their **pressure**. The pressure becomes **higher** in the ventricles than the atria, which forces the **AV valves shut** to prevent back-flow. The **pressure** in the **ventricles is also higher than in the aorta and pulmonary artery**, which forces **open** the **SL valves** and blood is forced out into these arteries.

2 Ventricles contract
SL valves forced open
blood leaves via pulmonary artery
blood leaves via aorta
AV valves forced closed

3 Ventricles relax, atria relax

The **ventricles and the atria both relax**. The higher pressure in the pulmonary artery and aorta closes the SL valves to prevent back-flow into the ventricles. Blood returns to the heart and the **atria fill again** due to the higher pressure in the vena cava and pulmonary vein. In turn this starts to **increase** the **pressure** of the atria. As the ventricles continue to **relax**, their **pressure falls below the pressure of the atria** and so the **AV valves open**. This allows blood to flow **passively** (without being pushed by atrial contraction) into the ventricles from the atria. The atria contract, and the whole process begins again.

3 Atria and ventricles relax
SL valves forced closed
blood re-enters via vena cava
blood re-enters via pulmonary vein
AV valves forced open

There's a bit about interpreting cardiac cycle data on the next page. So turn over now — it's well exciting...

The Heart

You Might be Asked to *Interpret Data* on the *Cardiac Cycle*

You may well be asked to analyse or interpret **data** about the changes in **pressure** and **volume** during the cardiac cycle. Here are two examples of the kind of things you might get:

Example 1

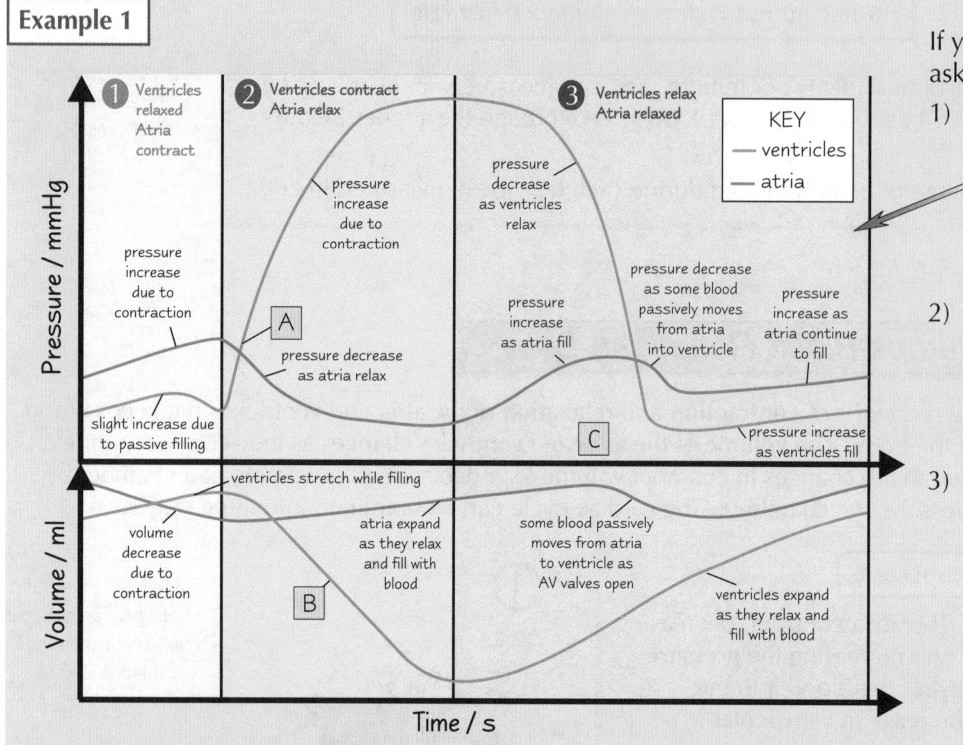

If you get a graph you could be asked **questions** like this:

1) **When** does blood start flowing into the **aorta**?
   At **point A**, the ventricles are **contracting** (and the AV valves are shut), forcing blood into the aorta.
2) Why is **ventricular volume decreasing** at **point B**?
   The ventricles are **contracting**, **reducing** the volume of the chamber.
3) Are the **semi-lunar valves** open or closed at **point C**?
   Closed. The ventricles are **relaxed** and **refilling**, so the pressure is **higher** in the **pulmonary artery** and **aorta**, forcing the SL valves **closed**.

The left ventricle has a thicker wall than the right ventricle and so it contracts more forcefully. This means the pressure is higher in the left ventricle (and in the aorta).

Example 2

You may have to describe the changes in pressure and volume shown by a **diagram**, like the one on the right. In this diagram the **AV valves** are **open**. So you know that the **pressure** in the **atria** is **higher** than in the **ventricles**. So you also know that the **atria are contracting** because that's what causes the **increase** in **pressure**.

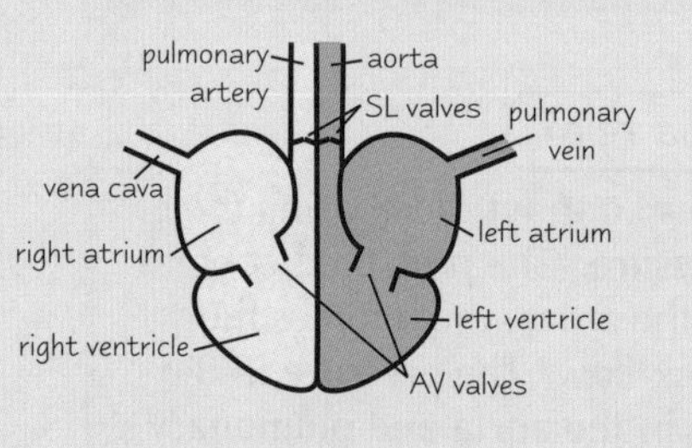

Practice Questions

Q1 Which side of the heart carries oxygenated blood?

Q2 What does "myogenic" mean?

Q3 Describe the roles of the SAN and AVN.

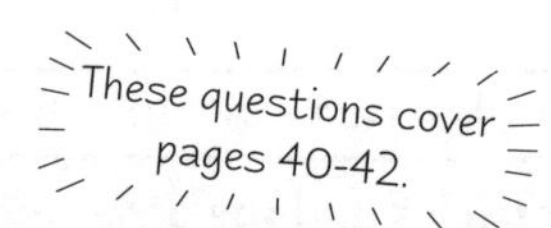

Exam Questions

Q1 Explain how valves in the heart stop blood going back the wrong way. [6 marks]

Q2 The table opposite shows the blood pressure in two heart chambers at different times during part of the cardiac cycle. Between what times:

a) are the AV valves shut? [1 mark]

b) do the ventricles start to relax? [1 mark]

Time / s	Blood pressure / kPa	
	Left atrium	Left ventricle
0.0	0.6	0.5
0.1	1.3	0.8
0.2	0.4	6.9
0.3	0.5	16.5
0.4	0.9	7.0

My heart will go on... — well, I think I sound like Céline Dion anyway...

Three whole pages to learn here, all full of really important stuff. If you understand all the pressure and volume changes then whether you get a diagram or a graph or something entirely different in the exam, you'll be able to interpret it, no probs.

Cardiovascular Disease

No, your heart won't break if HE/SHE (delete as appropriate) doesn't return your call... but there are diseases associated with the heart and blood vessels that you have to learn...

*Most **Cardiovascular Disease** starts with **Atheroma** Formation*

1) The wall of an artery is made up of **several layers** (see p. 72).
2) The **endothelium** (inner lining) is usually smooth and unbroken.
3) If **damage** occurs to the endothelium (e.g. by high blood pressure), **white blood cells** (mostly macrophages) and **lipids** (fat) from the blood, clump together under the lining to form **fatty streaks**.
4) Over time, more white blood cells, lipids and **connective tissue** build up and harden to form a **fibrous plaque** called an **atheroma**.
5) This plaque **partially blocks** the lumen of the **artery** and **restricts blood flow**, which causes **blood pressure** to **increase**.

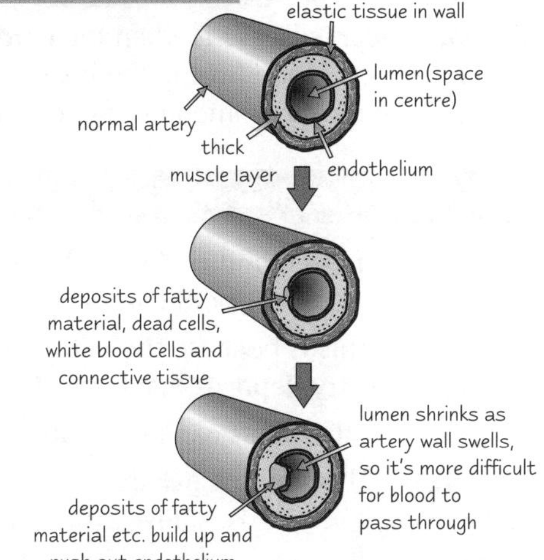

*Atheromas **Increase** the **Risk** of **Aneurysm** and **Thrombosis***

Two types of **disease** that affect the **arteries** are:

Aneurysm — a **balloon-like swelling** of the artery.

1) Atheroma plaques **damage** and **weaken arteries**. They also **narrow** arteries, **increasing blood pressure**.
2) When **blood** travels through a weakened artery at **high pressure**, it may **push** the **inner layers** of the artery **through the outer elastic layer** to form a **balloon-like swelling** — an **aneurysm**.
3) This aneurysm may **burst**, causing a **haemorrhage** (bleeding).

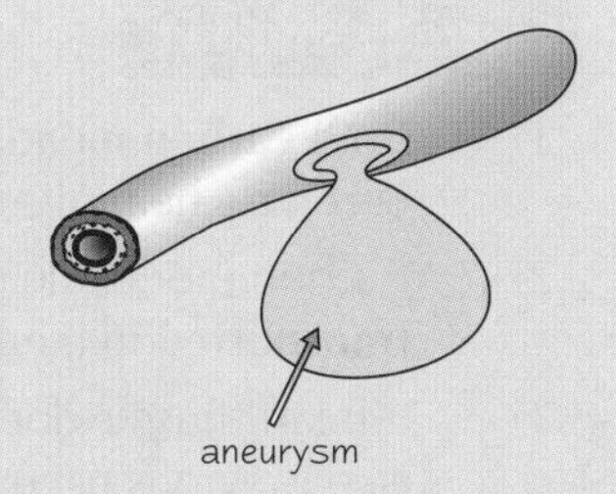

Thrombosis — formation of a **blood clot**.

1) An atheroma plaque can **rupture** (burst through) the **endothelium** (inner lining) of an artery.
2) This **damages** the artery wall and leaves a **rough** surface.
3) **Platelets** and **fibrin** (a protein) accumulate at the site of damage and form a **blood clot** (a thrombus).
4) This blood clot can cause a complete **blockage** of the artery, or it can become **dislodged** and block a blood vessel elsewhere in the body.
5) **Debris** from the rupture can cause another blood clot to form further down the artery.

Interrupted Blood Flow** to the **Heart** can cause a **Myocardial Infarction

1) The **heart muscle** is supplied with **blood** by the **coronary arteries**.
2) This blood contains the **oxygen** needed by heart muscle cells to carry out **respiration**.
3) If a coronary artery becomes **completely blocked** (e.g. by a **blood clot**) an area of the heart muscle will be totally **cut off** from its blood supply, receiving **no oxygen**.
4) This causes a **myocardial infarction** — more commonly known as a **heart attack**.
5) A heart attack can cause **damage** and **death** of the **heart muscle**.
6) **Symptoms** include **pain** in the chest and upper body, **shortness of breath** and **sweating**.
7) If **large areas** of the heart are affected complete **heart failure** can occur, which is often **fatal**.

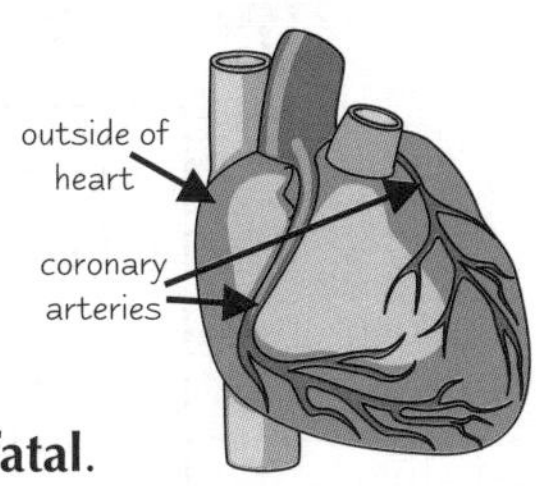

Cardiovascular Disease

There are treatments for cardiovascular disease out there, but it's best to try to avoid these diseases in the first place. Because lifestyle plays a large part, it's pretty easy to make preventive changes.

Some **Factors Increase** the **Risk** of **Coronary Heart Disease (CHD)**

Coronary heart disease is when the **coronary arteries** have lots of **atheromas** in them, which restricts blood flow to the heart. It's a type of **cardiovascular disease**. Some of the most common risk factors are:

1 High blood cholesterol and poor diet

1) If the **blood cholesterol level** is **high** (above 240 mg per 100 cm^3) then the risk of coronary heart disease is increased.
2) This is because **cholesterol** is one of the main constituents of the **fatty deposits** that form **atheromas** (see p. 43).
3) Atheromas can lead to **increased blood pressure** and **blood clots**.
4) This could **block** the flow of blood to **coronary arteries**, which could cause a **myocardial infarction** (see previous page for details).
5) A diet **high in saturated fat** is associated with high blood cholesterol levels.
6) A diet **high in salt** also **increases** the **risk** of cardiovascular disease because it increases the risk of **high blood pressure** (see below).

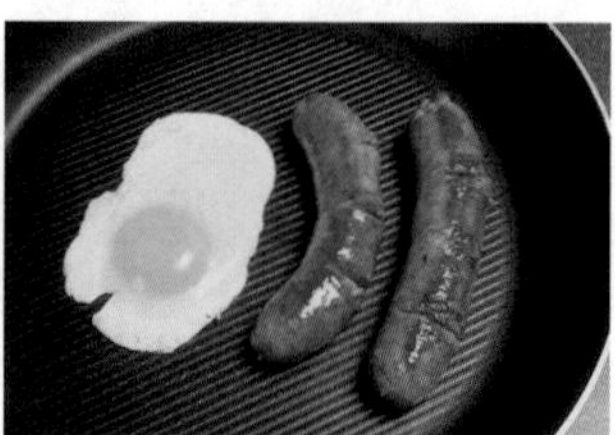

John decided to live on the edge and ordered a fry-up.

2 Cigarette smoking

1) Both **carbon monoxide** and **nicotine**, found in **cigarette smoke**, increase the risk of coronary heart disease.
2) Carbon monoxide combines with **haemoglobin** and **reduces** the amount of **oxygen transported** in the **blood**, and so reduces the amount of oxygen available to tissues.
3) If heart muscle doesn't receive enough oxygen it can lead to a **heart attack** (see previous page).
4) Smoking also **decreases** the **amount** of **antioxidants** in the blood — these are important for **protecting cells** from damage. Fewer antioxidants means **cell damage** in the **coronary artery walls** is more likely, and this can lead to **atheroma formation**.

3 High blood pressure

1) High blood pressure **increases** the **risk** of **damage** to the **artery walls**.
2) Damaged walls have an **increased risk** of **atheroma** formation, causing a further increase in blood pressure.
3) Atheromas can also cause **blood clots** to form (see p. 43).
4) A blood clot could **block flow** of **blood** to the heart muscle, possibly resulting in **myocardial infarction** (see p. 43).
5) So **anything** that **increases** blood pressure also increases the risk of **CHD**, e.g. being **overweight**, **not exercising** and excessive **alcohol** consumption.

Other factors include age (risk increases with age) and sex (men are more at risk than women).

Most of these factors are within our **control** — a person can **choose** to smoke, eat fatty foods, etc. However, some risk factors can't be controlled, such as having a **genetic predisposition** to coronary heart disease or having high blood pressure as a result of another **condition**, e.g. some forms of diabetes. Even so, the risk of developing CHD can be reduced by removing as many **risk factors** as you possibly can.

Cardiovascular Disease

You Might Have to Interpret Data on Risk Factors and CHD

Take a look at the following example of the sort of study you might see in your exam.

The graph shows the results of a study involving **34 439 male British doctors. Questionnaires** were used to find out the smoking habits of the doctors. The number of **deaths** among the participants from ischaemic heart disease (coronary heart disease) was counted, and **adjustments** were made to account for **differences in age.**

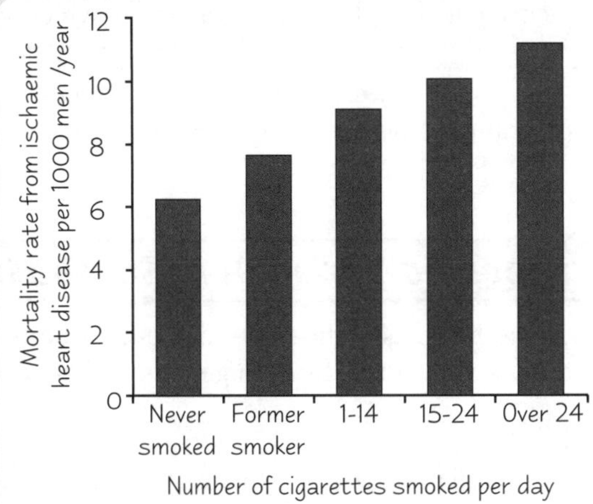

Here are some of the things you might be asked to do:

1) Describe the data — The **number** of deaths from ischaemic heart disease **increased** as the number of cigarettes smoked per day **increased. Fewer former smokers** and **non-smokers** died of ischaemic heart disease than smokers.
2) Draw conclusions — The **graph shows** a **positive correlation** between the number of cigarettes smoked per day by **male doctors** and the **mortality rate** from ischaemic heart disease.
3) Explain the link — You may get asked to **explain** the link between the risk factor under investigation (smoking) and CHD. E.g. **carbon monoxide** in **cigarette smoke** combines with **haemoglobin, reducing** the amount of **oxygen transported** in the **blood.** This **reduces** the amount of oxygen available to tissues, including **heart muscle**, which could lead to a **heart attack.** You would also talk about **antioxidants** (see previous page).
4) Check any conclusions are valid — make sure the conclusions **match** the data, e.g. this study only looked at **male doctors** — no females were involved, so you can't say that this trend is true for **everyone.** Also, you couldn't say smoking more cigarettes causes an increased **risk** of heart disease. The data shows **deaths only** and **specifically** from ischaemic heart disease. It could be that the **morbidity rate** (the number who have heart disease) **decreases** with the number of cigarettes a day. But you can't tell that from this data.
5) Comment on the reliability of the results — For example:
   - A **large sample size** was used — 34 439, which **increases** reliability.
   - People (even doctors) can tell **porkies** on questionnaires, **reducing** the **reliability** of results.

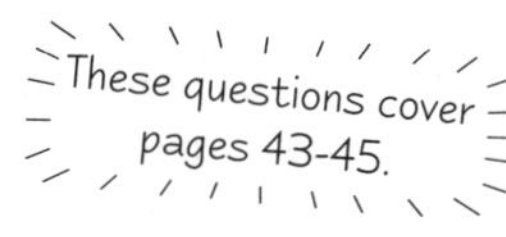

Practice Questions

Q1 What is the name of the arteries that supply the heart muscle with oxygen?

Q2 What is the medical term for a heart attack?

Q3 What does CHD stand for?

Q4 Name three factors that can increase the chance of developing CHD.

These questions cover pages 43-45.

Exam Questions

Q1 Describe how atheroma formation in the coronary arteries can lead to a myocardial infarction. [4 marks]

Q2 Describe how atheromas can increase the risk of a person suffering from an aneurysm. [3 marks]

Revision — increasing my risk of headache, stress, boredom...

I know there's a lot to take in on these pages... but make sure you understand the link between atheromas, thrombosis and heart attacks — basically an atheroma forms, which can cause thrombosis, which can lead to a heart attack. Anything that increases the chance of an atheroma forming (high blood pressure, smoking, fatty diet) is bad news for your heart...

Causes of Variation

Every organism — man, woman, gibbon or platypus — differs in some way from every other. The differences between organisms is called variation. It's all down to genes and the environment.

Variation Exists Between All Individuals

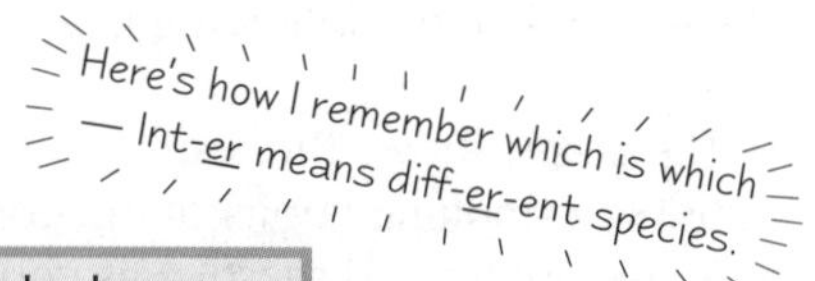
Here's how I remember which is which — Int-er means diff-er-ent species.

Variation is the **differences** that exist between **individuals**. There are two types:

Interspecific — the variation that exists between **different species**. For example, horses vary from ducks, which vary from mice, which vary from the lesser-known spotted teenager.

Intraspecific — the differences that occur **within a species**. For example, the number of 'eyes' on peacocks' feathers, or the length of giraffes' necks.

Intraspecific Variation is Caused by Genetic and Environmental Factors

Although individuals of the same species may **appear similar**, no two individuals are **exactly alike**. **Variation** can be caused by both **genetic** and **environmental** factors.

Genetic

1) All the members of a species have the **same genes** — that's what makes them the **same species**.
2) But **individuals** within a species can have **different versions** of those genes — called **alleles** (see p. 53).
3) The alleles an organism has make up its **genotype**.
4) Different genotypes result in **variation** in **phenotype** — the **characteristics** displayed by an organism.
5) Examples of variation in humans caused by genetic factors include **eye colour** (which can be blue, green, grey, brown) and **blood type** (O, A, B or AB).
6) You **inherit** your genes from your parents. This means genetic variation is **inherited**.

Environmental

The **appearance** (**phenotype**) of an individual is also affected by the **environment**, for example:

1) **Plant growth** is affected by the amount of **minerals**, such as **nitrate** and **phosphate**, available in the soil.
2) **Fur colour** of the **Himalayan rabbit** is affected by **temperature**. Most of the rabbits' fur is **white** except the ears, feet and tail, which are **black**. The black colour only develops in temperatures **below 25 °C**. If a patch of their white fur is **shaved** and a cold pad applied to the shaved area, the hair that grows back will be **black**.
3) **Identical twins** are **genetically** identical — they have the same alleles, so any differences between them will be due to the **environment**. For example, they may have had different **illnesses** that affected their **development** or, if they grew up in **different areas**, they may have different **accents**.

Thumper and Biggles went to any length to not touch the cold floor.

Variation is Often a Combination of Genetic and Environmental Factors

An individual may have the **genetic information** for a **particular characteristic**, but **environmental factors** may affect the **expression** of this characteristic, for example:

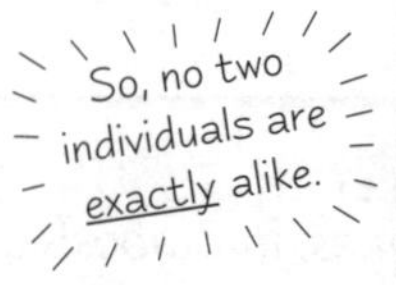
So, no two individuals are exactly alike.

1) A person might have the **genes** to potentially grow to be **six foot tall**. Whether or not they grow to this height will **depend** on **environmental factors** such as their **diet** and **health**.
2) The amount of **melanin** (pigment) people have in their **skin** is partially controlled by their **genes**, but skin colour is influenced by the **amount of sunlight** a person is exposed to.

Causes of Variation

Be Careful When Drawing Conclusions About the Cause of Variation

In any **group of individuals** there's a lot of **variation** — think how different all your friends are. It's not always **clear** whether the variation is caused by **genes**, the **environment** or **both**. Scientists draw conclusions based on the information they have until **new evidence** comes along that **challenges it** — have a look at these two examples:

Example 1 — Overeating

1) **Overeating** was thought to be caused only by environmental factors, like an **increased availability of food** in developed countries.
2) It was later discovered that food consumption **increases** brain **dopamine** levels in animals.
3) Once enough dopamine was released, people would **stop** eating.
4) Researchers discovered that people with one particular **allele** had **30% fewer** dopamine receptors.
5) They found that people with this particular allele were **more likely** to overeat — they wouldn't stop eating when dopamine levels increased.
6) Based on this evidence, scientists now think that overeating has **both genetic** and **environmental** causes.

Example 2 — Antioxidants

1) Many foods in our diet contain **antioxidants** — compounds that are thought to play a role in **preventing chronic diseases.**
2) Foods such as **berries** contain **high levels** of antioxidants.
3) Scientists thought that the berries produced by different **species** of plant contained **different levels** of antioxidants because of **genetic factors.**
4) But experiments that were carried out to see if **environmental** conditions affected antioxidant levels found that environmental conditions caused a great deal of **variation.**
5) Scientists now believe that antioxidant levels in berries are due to **both genetic** and **environmental** factors.

Practice Questions

Q1 What is intraspecific variation?

Q2 What is interspecific variation?

Q3 Which two kinds of factors can cause variation?

Q4 Give one example of a characteristic that varies due to both genetic and environmental factors.

Exam Question

Q1 The graph shows the results of an investigation into the effects of temperature on the length of time it took for ladybird larvae to emerge as adults. Two species of ladybird were investigated, species A and species B.

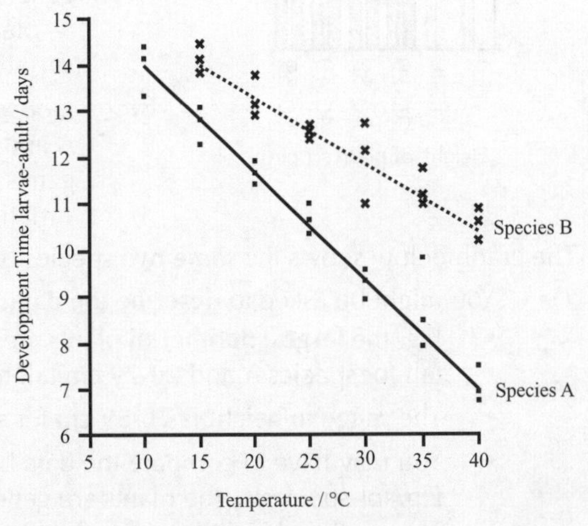

a) Describe the results of the study. [3 marks]

b) Explain what causes the variation between the species and within each species. [4 marks]

Environmental Factor — the search is on for the most talented environment...

Err... Inter... Intra... I don't know who thought that using two words almost exactly the same was a good idea, but it wasn't very helpful. Good thing I've made these pages then I reckon. Copy out the definitions a few times and make sure you understand the examples — they'll use the same ideas but different case studies in the exam.

Investigating Variation

It's a lot of work studying variation in an entire population (imagine studying all the ants in one nest) — so instead you can take a random sample and use this to give you a good idea of what's going on in the entire population.

To **Study** Variation You Have to **Sample** a **Population**

When studying variation you usually only look at a **sample** of the population, **not** the **whole thing**. For most species it would be too **time-consuming** or **impossible** to catch all the individuals in the group. So samples are used as **models** for the **whole population**.

The **Sample** has to be **Random**

Because sample data will be used to **draw conclusions** about the **whole population**, it's important that it **accurately represents** the whole population and that any patterns observed are tested to make sure they're not due to chance.

1) To make sure the sample isn't **biased**, it should be **random**. For example, if you were looking at plant species in a field you could pick random sample sites by dividing the field into a **grid** and using a **random number generator** to select coordinates.
2) To ensure any variation observed in the sample isn't just due to **chance**, it's important to analyse the results **statistically**. This allows you to be more **confident** that the results are true and therefore will reflect what's going on in the **whole population**.

You Need to be Able to **Analyse** and **Interpret Data** Relating to **Variation**

You might be asked to **analyse** and **interpret** data relating to **interspecific and intraspecific variation** in your exam. So here's a big **example** to give you an idea of what you might get:

The graph below shows the growth of two **different** species of plant in the **same environment**. You might be asked to:

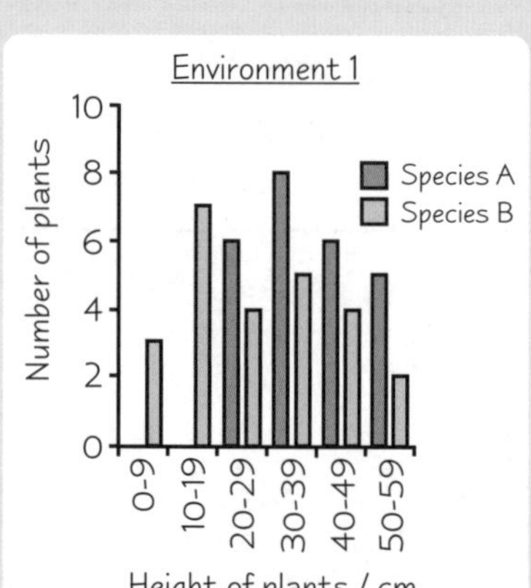

1) <u>Describe the data</u>...
   - The largest number of plants are **30-39 cm** tall for species **A** and **10-19 cm** tall for species **B**.
   - Species **A** plants **range in height** from **20-59 cm** but the range is **larger** for **species B (0-59 cm)**.
2) ...or <u>draw conclusions</u>
   There is **interspecific variation** in plant height — Species **A** plants are **generally taller** than Species B. **Both species** show **intraspecific variation** — plant height varies for both species. There is **more intraspecific variation** in species **B** — the range of heights is bigger.
3) ...or <u>suggest a reason</u> for the differences
   A and B are **separate species**, grown in the same area. This means their **genes are different** but their **environment is the same**. So any **interspecific** variation in height is down to **genetic factors**, **not** the environment.

The graph below shows the **same** two species of plant but grown in a **different environment** than in the first graph.

1) You might be asked to <u>describe the data</u>...
   - E.g. the **largest number** of plants are **40-49 cm** tall for species **A** and **20-29 cm** tall for species **B**.
   - The **range in height** is **20-59 cm** for species **A** and **10-59 cm** for species **B**.
   - You may have to **compare** the data between the two graphs.
     E.g. for species **A**, the plants are generally **taller** in **environment 2**.
     The range in height has **stayed the same**. For species **B**, the plants are also generally **taller** in **environment 2**, but the range in height is **smaller**.
2) ...or <u>draw conclusions and suggest a reason</u> for the differences
   - Both species are **generally taller** in environment **2** than in environment 1.
     So, variation in height is affected by **environmental factors**.
   - Species A shows **similar** height variation in **both environments**.
     The variation in species B **differs** between the two environments.
     This suggests that **environmental factors influence** height **more** in species **B**.

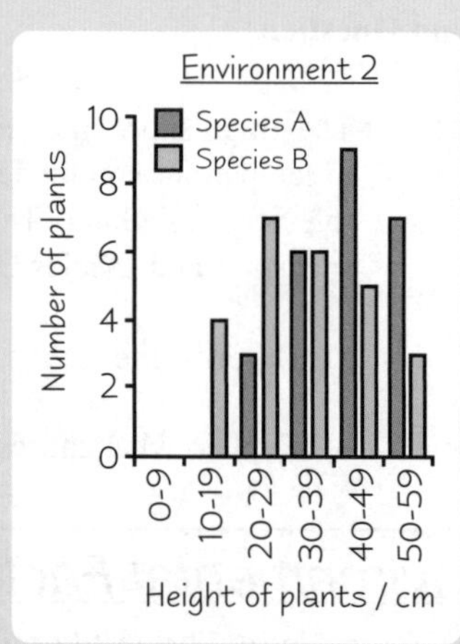

Investigating Variation

Standard Deviation Tells You About the Variation **Within a Sample**

1) The **mean value** tells you the **average** of the values collected in a **sample**.
2) It can be used to tell if there **is variation between samples**, e.g. the mean number of apples produced by species A = 26 and B = 32. So the **number** of apples produced by different tree species **does vary**.
3) Most samples give you a **bell-shaped graph** — this is called a **normal distribution**.
4) The **standard deviation** tells you **how much** the values in a **single sample vary**. It's a **measure** of the **spread of values about the mean**.
5) For example:
   - Species A: mean = 26, standard deviation = 3 — most of the trees in the sample produced between 23 and 29 apples (26 ± 3).
   - Species B: mean = 32, standard deviation = 9 — most of the trees in the sample produced between 23 and 41 apples (32 ± 9).
   - So species B generally produces **more apples** but shows a **greater variation** in the number produced, compared to species A.

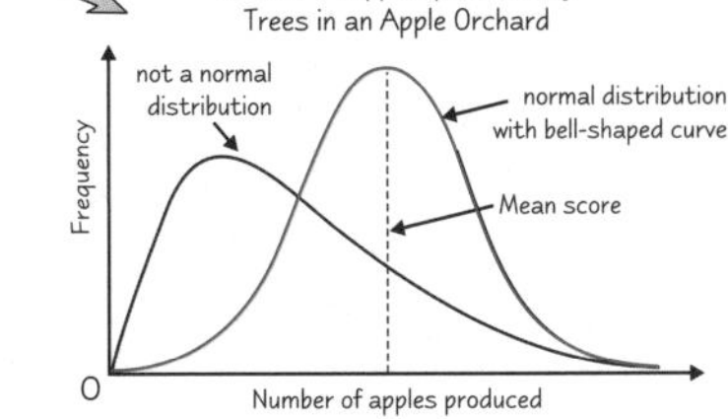

6) A **large standard deviation** means the values in the sample **vary a lot**. A **small standard deviation** tells you most of the sample data is around the mean value, so **varies little**:

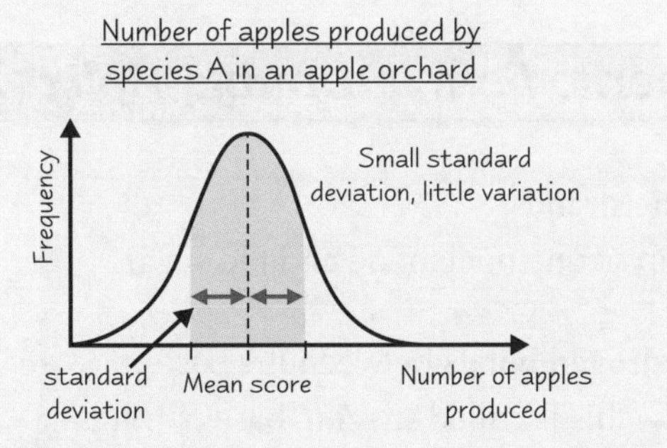

When all the values are **similar**, so vary little, the graph is **steep** and the standard deviation is **small**.

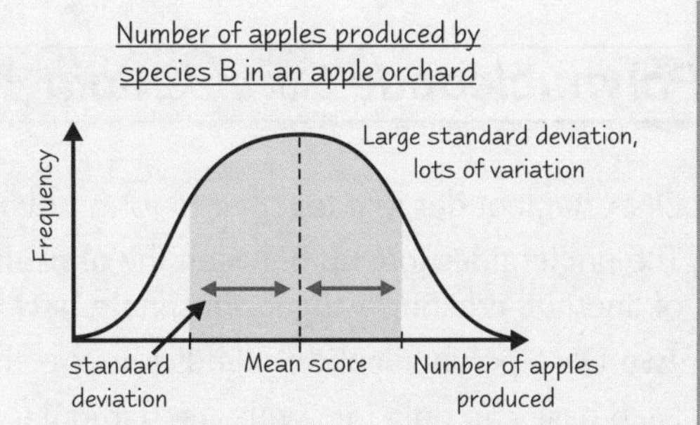

When all the values vary a lot, the graph is **fatter** and the standard deviation is **large**.

Practice Questions

Q1 Why do scientists look at a sample of a population, rather than the whole population?

Q2 Why does a population sample have to be chosen at random?

Q3 What does the standard deviation of a data set tell us?

Exam Question

Q1 A study was conducted into how smoking during pregnancy affects the birth mass of newborn babies, depending on the genotype of the mother. The results showed that women who smoked during the entire pregnancy had babies with a mean reduction in birth mass of 377 grams. But the reduction was as much as 1285 grams among women with certain genotypes.

a) Data on variation in child birth mass was also collected from a group of non-smokers. Suggest why this data was collected. [1 mark]

b) What can be concluded about the influence of genetic factors and environmental factors on birth mass? Give evidence from the study to support your answer. [4 marks]

c) Give two other factors that should be controlled in this experiment. [2 marks]

Investigating standards — say that to your teachers to scare them...

Bet you thought you'd finished with maths — 'fraid not. Thankfully you don't need to know how to work out standard deviation though. But you do need to know how to go about interpreting data — which is a bit more interesting than maths, and I can exclusively reveal it's what all the examiners dream about at night.

DNA

*These pages are about the wonderful world of genetics. But you won't be able to make head or tail of it without learning all about DNA (**d**eoxyribo**n**ucleic **a**cid), so here goes...*

DNA is Made of Nucleotides that Contain a Sugar, a Phosphate and a Base

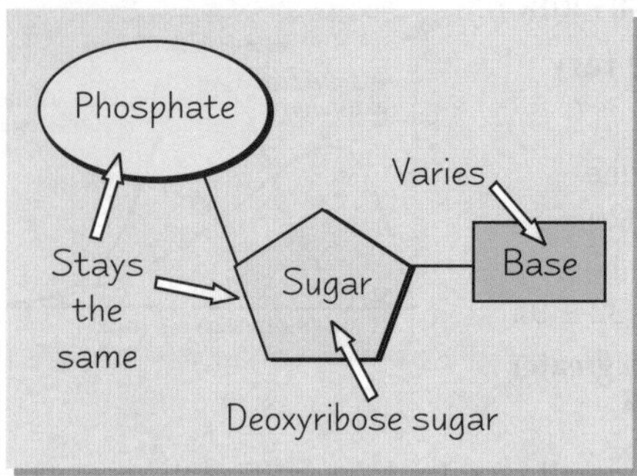

1) DNA is a polynucleotide — it's made up of lots of **nucleotides** joined together.
2) Each nucleotide is made from a **pentose sugar** (with 5 carbon atoms), a **phosphate** group and a **nitrogenous base**.
3) The **sugar** in DNA nucleotides is a **deoxyribose** sugar.
4) Each nucleotide has the **same sugar and phosphate**. The **base** on each nucleotide can **vary** though.
5) There are **four** possible bases — adenine (**A**), thymine (**T**), cytosine (**C**) and guanine (**G**).

Two Polynucleotide Strands Join Together to Form a Double-Helix

1) DNA nucleotides join together to form **polynucleotide strands**.
2) The nucleotides join up between the **phosphate** group of one nucleotide and the **sugar** of another, creating a **sugar-phosphate backbone**.
3) **Two** DNA polynucleotide strands join together by **hydrogen bonds** between the bases.
4) Each base can only join with one particular partner — this is called **specific base pairing**.
5) **Adenine** always pairs with **thymine** (**A - T**) and **guanine** always pairs with **cytosine** (**G - C**).
6) The two strands **wind up** to form the **DNA double-helix**.

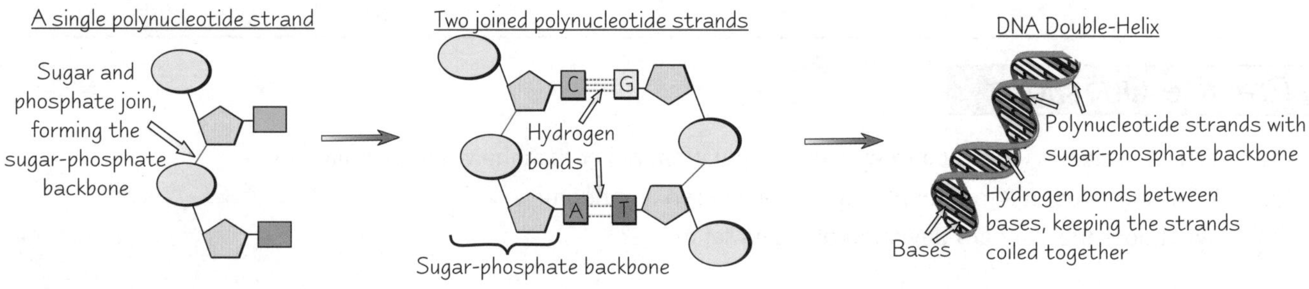

DNA's Structure Makes It Good at Its Job

1) DNA contains your **genetic information** — that's **all the instructions** needed to **grow and develop** from a fertilised egg to a fully grown adult.
2) The DNA molecules are very **long** and are **coiled** up very tightly, so a lot of genetic information can fit into a **small space** in the cell nucleus.
3) DNA molecules have a **paired structure**, which makes it much easier to **copy itself**. This is called **self-replication** (see p. 62). It's important for cell division (see p. 64) and for passing genetic information from **generation to generation** (see p. 54).
4) The double-helix structure means DNA is **very stable** in the cell.

Geoff's helix wasn't as tightly coiled as DNA but it was a lot more fun.

DNA

DNA is Stored Differently in Different Organisms

Although the **structure** of DNA is the same in all organisms, **eukaryotic** and **prokaryotic** cells store DNA in slightly different ways. (For a recap on the differences between prokaryotic and eukaryotic cells see p. 32.)

Eukaryotic DNA is Linear and Associated with Proteins

1) Eukaryotic cells contain **linear** DNA molecules that exist as **chromosomes** — thread-like structures, each made up of **one long molecule** of DNA.
2) The DNA molecule is **really long** so it has to be **wound up** so it can **fit** into the nucleus.
3) The DNA molecule is wound around **proteins** (called **histones**).
4) Histone proteins also help to **support** the DNA.
5) The DNA (and protein) is then coiled up **very tightly** to make a **compact chromosome**.

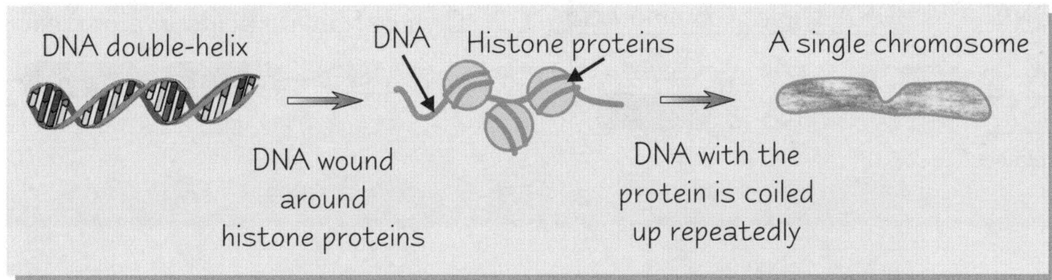

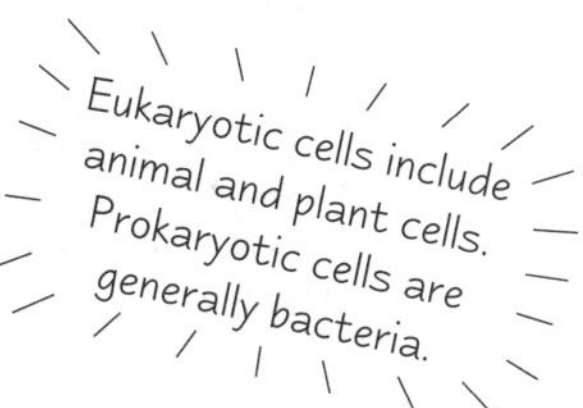

DNA with the protein is coiled up repeatedly

Eukaryotic cells include animal and plant cells. Prokaryotic cells are generally bacteria.

DNA Molecules are Shorter and Circular in Prokaryotes

1) Prokaryotes also carry DNA as **chromosomes** — but the DNA molecules are **shorter** and **circular**.
2) The DNA **isn't** wound around proteins — it condenses to fit in the cell by **supercoiling**.

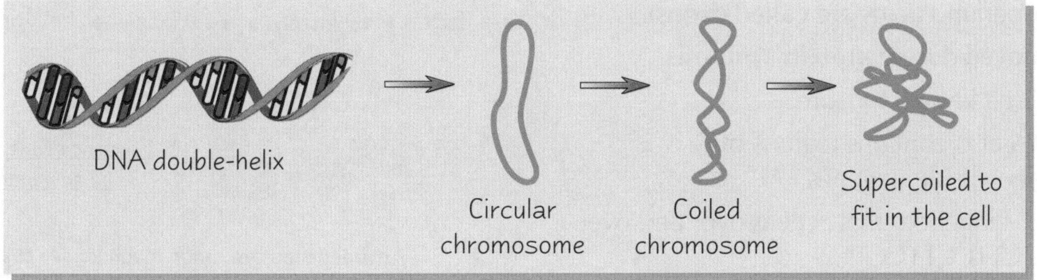

If one more person confused Clifford with supercoiled DNA, he'd have 'em.

Practice Questions

Q1 What are the three main components of nucleotides?

Q2 Which bases join together in a DNA molecule?

Q3 What type of bonds join the bases together?

Q4 Why is DNA so tightly coiled?

Exam Questions

Q1 Describe, using diagrams where appropriate, how nucleotides are joined together in DNA and how two single polynucleotide strands of DNA are joined. [4 marks]

Q2 Describe how DNA is stored in prokaryotic and eukaryotic cells. [5 marks]

Give me a D, give me an N, give me an A! What do you get? — Very confused...

You need to learn the structure of DNA — the sugar-phosphate backbone, the hydrogen bonds, and don't forget the base pairing. Then there's the differences in the way that DNA is stored in eukaryotic and prokaryotic cells. Pheww.

Genes

Now you've got to grips with the structure of DNA, you can learn how DNA is used to carry information. It's all in the sequence of bases you see...

*DNA Contains **Genes** Which are **Instructions** for **Proteins***

1) Genes are **sections of DNA**. They're found on **chromosomes**.
2) Genes **code** for **proteins** (polypeptides) — they contain the **instructions** to make them.
3) Proteins are made from **amino acids**.
4) Different proteins have a **different number** and **order** of amino acids.
5) It's the **order** of **nucleotide bases** in a gene that determines the **order of amino acids** in a particular **protein**.
6) Each amino acid is coded for by a sequence of **three bases** (called a **triplet**) in a gene.
7) Different sequences of **bases** code for different **amino acids**. For example:

Polypeptide is just another word for a protein.

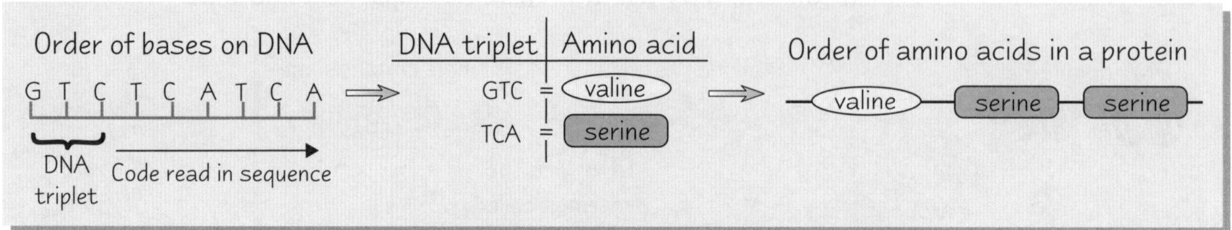

*Not All the **DNA** in **Eukaryotic Cells** Codes for **Proteins***

1) Genes in eukaryotic DNA contain sections that **don't code** for amino acids.
2) These sections of DNA are called **introns** (all the bits that do code for amino acids are called **exons**).
3) Introns are **removed** during **protein synthesis**. Their purpose isn't known for sure.
4) Eukaryotic DNA also contains regions of **multiple repeats** outside of genes.
5) These are DNA sequences that **repeat** over and over. For example: CCTTCCTTCCTT.
6) These areas **don't code** for amino acids either.

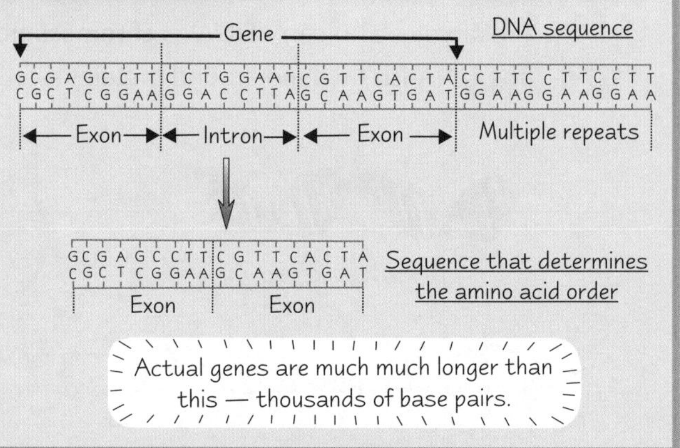

*The **Nature** and **Development** of Organisms is **Determined** by **Genes***

1) **Enzymes** speed up most of our **metabolic pathways** — the chemical reactions that occur in the body. These pathways determine how we **grow and develop**.
2) Because enzymes control the metabolic pathways, they **contribute** to our **development**, and ultimately what we look like (our **phenotype**).
3) All enzymes are **proteins**, which are built using the **instructions** contained within genes. The **order of bases** in the gene decides the order of **amino acids** in the protein and so what type of protein (or enzyme) is made.
4) So, our genes help to **determine** our **nature**, **development** and **phenotype** because they contain the information to **produce** all our proteins and enzymes.

Ken's bad fashion sense was literally down to his genes. (A genes/jeans joke — classic CGP.)

This flowchart shows how **DNA** determines our **nature** and **development**:

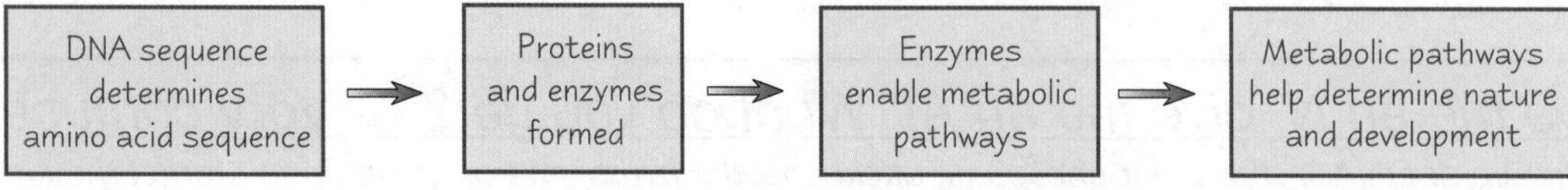

Genes

Genes can Exist in Different Forms Called Alleles

1) A gene can exist in more than one form. These forms are called **alleles**.
2) The order of bases in each allele is slightly different, so they code for **slightly different versions** of the **same characteristic**. For example, the gene that codes for **blood type** exists as one of three alleles — one codes for type O, another for type A and the other for type B.

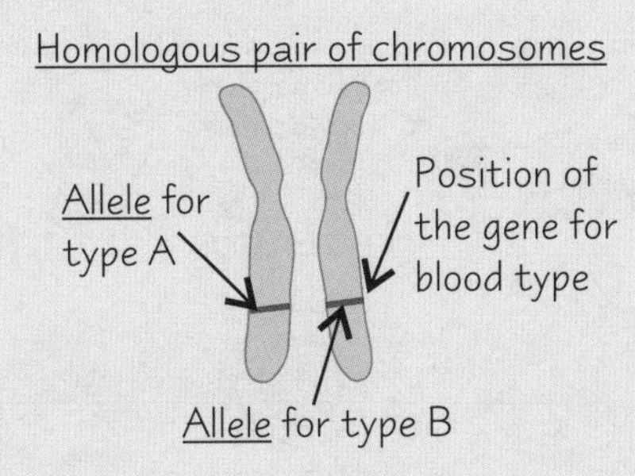

Our DNA is stored as **chromosomes** in the nucleus of cells. Humans have **23 pairs** of chromosomes, 46 in total — two number 1s, two number 2s, two number 3s etc. Pairs of matching chromosomes (e.g. the 1s) are called **homologous pairs**. In a homologous pair both chromosomes are the same size and have the **same genes**, although they could have **different alleles**. Alleles coding for the same characteristic will be found at the **same position** (**locus**) on each chromosome in a homologous pair.

Gene Mutations can Result in Non-functioning Proteins

1) **Mutations** are **changes** in the **base sequence** of an organism's **DNA**.
2) So, mutations can produce **new alleles** of genes.
3) A gene codes for a particular protein, so if the sequence of bases in a gene changes, a **non-functional** or **different protein** could be produced.
4) All **enzymes** are **proteins**. If there's a mutation in a gene that codes for an enzyme, then that enzyme may not **fold up** properly. This may produce an **active site** that's the wrong shape and so a **non-functional enzyme**.

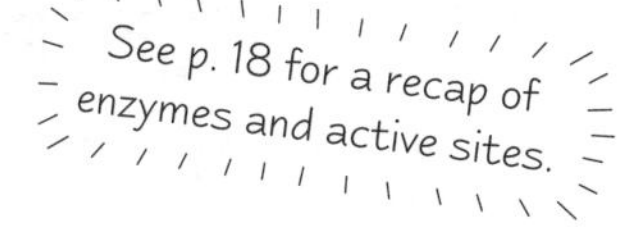
See p. 18 for a recap of enzymes and active sites.

Example

Gene X codes for an enzyme that catalyses the conversion of A to B. A mutation in the gene may result in the formation of a non-functional enzyme. This means that the reaction **can't happen**.

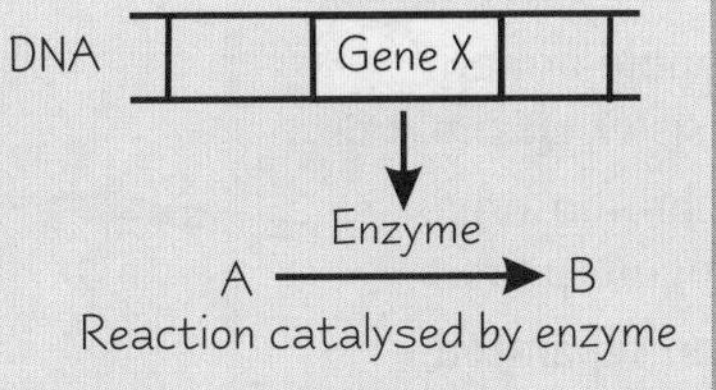

Practice Questions

Q1 What is a DNA triplet?

Q2 What is an intron?

Q3 What is an allele?

Q4 How can a mutation result in a non-functional enzyme?

Amino acid	DNA triplet
Glycine	GGC
Glutamic acid	GAG
Proline	CCG
Tryptophan	TGG

Exam Questions

Q1 a) Write a definition of a gene. [2 marks]

b) Use the table above to write the protein sequence coded for by the DNA sequence TGGCCGCCGGAG. [1 mark]

Q2 Describe how the DNA of an organism helps to determine its nature and development. [4 marks]

Exons stay in, introns go out, in out, in out, and shake it all about...

Quite a few terms to learn here I'm afraid. Some are a bit confusing too. Just try to remember which way round they go. Introns are the non-coding regions but exons are extremely important — they actually code for the protein (polypeptide).

Meiosis and Genetic Variation

Humans are all similar because we have the same genes. But we show genetic variation because we inherit different combinations of alleles from our parents. This is what leads to the differences you can see between you and your siblings, and your friends, and your postman... (unless your mum's not told you something...)

DNA from **One Generation** is **Passed** to the Next by **Gametes**

1) **Gametes** are the **sperm** cells in males and **egg** cells in females. They join together at **fertilisation** to form a **zygote**, which divides and develops into a **new organism**.
2) Normal **body cells** have the **diploid number** (**2n**) of chromosomes — meaning each cell contains **two** of each chromosome, one from the mum and one from the dad.
3) **Gametes** have a **haploid** (**n**) number of chromosomes — there's only one copy of each chromosome.
4) At **fertilisation**, a **haploid sperm** fuses with a **haploid egg**, making a cell with the normal diploid number of chromosomes. Half these chromosomes are from the father (the sperm) and half are from the mother (the egg).

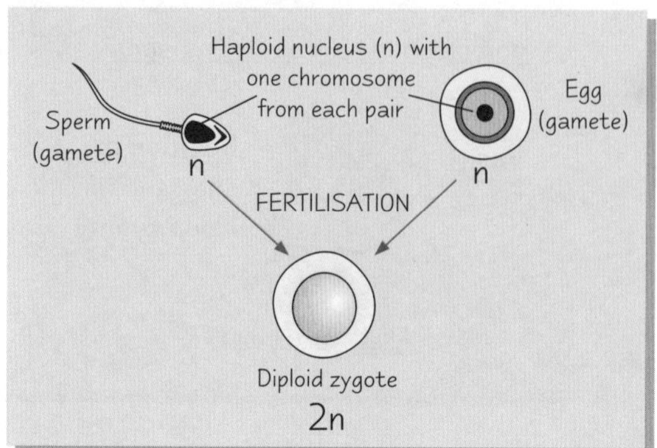

Gametes are Formed by **Meiosis**

Meiosis is a type of cell division. Cells that divide by meiosis are **diploid** to start with, but the cells that are formed from meiosis are **haploid** — the chromosome number **halves**. Without meiosis, you'd get **double** the number of chromosomes when the gametes fused. Not good.

1) The DNA unravels and **replicates** so there are **two** copies of **each** chromosome, called **chromatids**.
2) The DNA condenses to form double-armed chromosomes, made from **two sister chromatids**.
3) **Meiosis I** (first division) — the chromosomes arrange themselves into **homologous pairs**.
4) These homologous **pairs** are then **separated**, **halving** the chromosome number.
5) **Meiosis II** (second division) — the pairs of sister **chromatids** that make up each chromosome are **separated**.
6) **Four haploid cells** (gametes) that are **genetically different** from each other are produced.

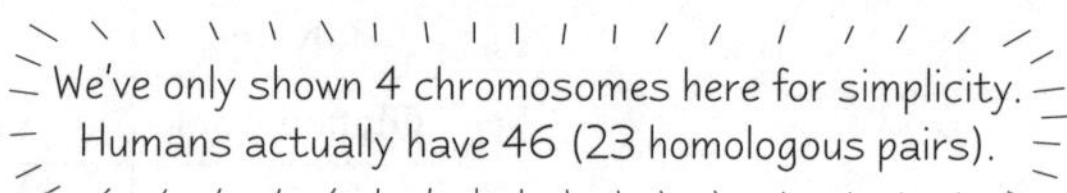

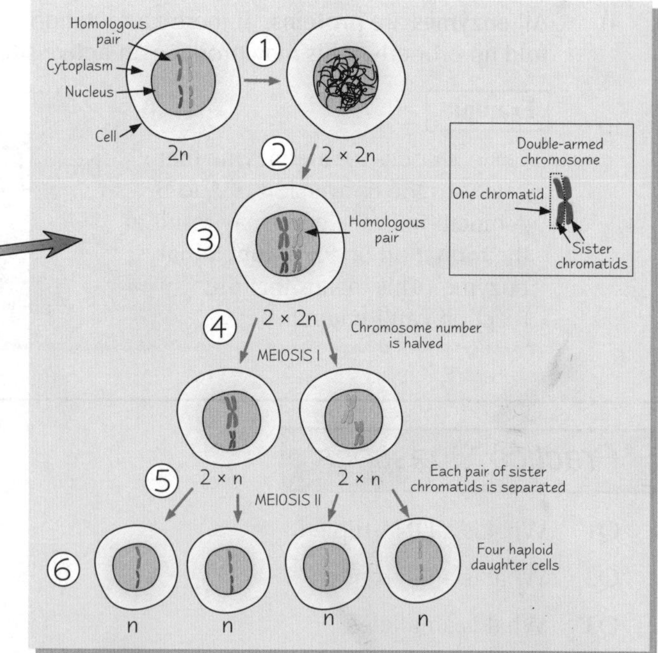

Chromatids **Cross Over** in **Meiosis I**

During meiosis I, **homologous pairs** of chromosomes come together and pair up. The chromatids twist around each other and bits of **chromatids** swap over. The chromatids still contain the **same genes** but now have a different combination of **alleles**.

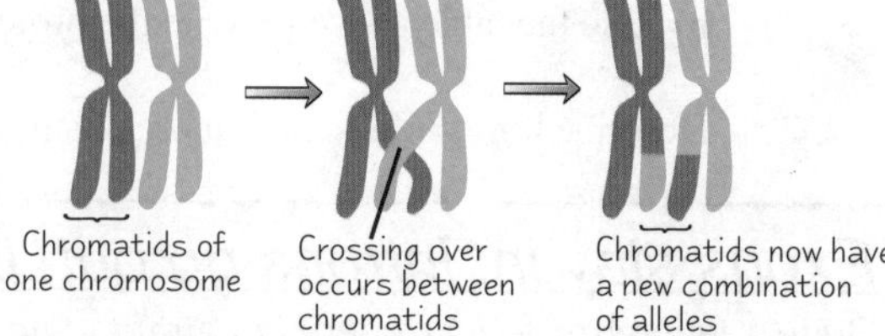

Meiosis and Genetic Variation

Meiosis Produces Cells that are Genetically Different

There are two main events during meiosis that lead to **genetic variation**:

1 Crossing over of chromatids

The **crossing over** of chromatids in meiosis I means that each of the **four daughter cells** formed from meiosis contain chromatids with **different alleles**:

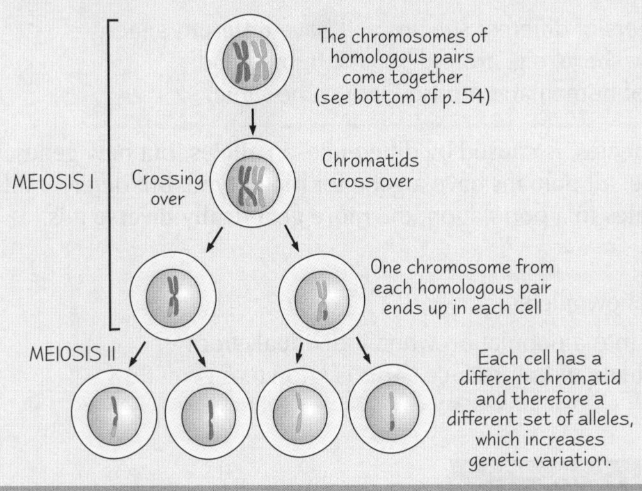

MEIOSIS I
MEIOSIS II
The chromosomes of homologous pairs come together (see bottom of p. 54)
Crossing over
Chromatids cross over
One chromosome from each homologous pair ends up in each cell
Each cell has a different chromatid and therefore a different set of alleles, which increases genetic variation.

2 Independent segregation of chromosomes

1) The four daughter cells formed from meiosis have completely **different combinations** of **chromosomes.**

2) All your cells have a **combination** of chromosomes from your parents, half from your mum (**maternal**) and half from your dad (**paternal**).

3) When the gametes are produced, different **combinations** of those maternal and paternal **chromosomes** go into each cell.

4) This is called **independent segregation** (separation) of the chromosomes.

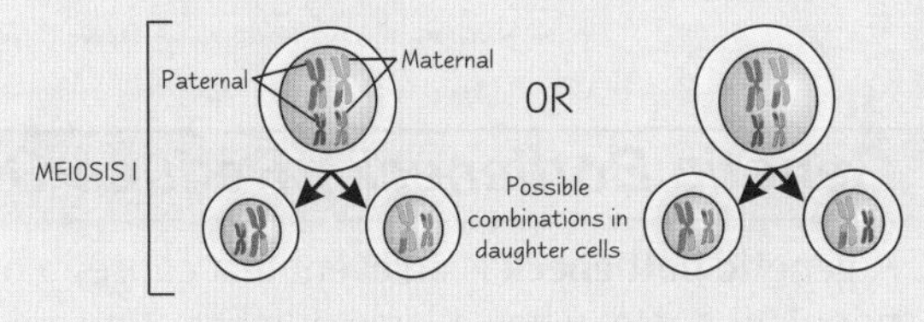

Paternal
Maternal
OR
MEIOSIS I
Possible combinations in daughter cells

Practice Questions

Q1 Explain what is meant by the terms haploid and diploid.

Q2 What happens to the chromosome number at fertilisation?

Q3 What is a chromatid?

Q4 How many divisions are there in meiosis?

Q5 How many cells does meiosis produce?

Exam Questions

Q1 Explain why it's important for gametes to have half the number of chromosomes as normal body cells. [2 marks]

Q2 Describe, using diagrams where appropriate, the process of meiosis. [6 marks]

Q3 a) Explain what crossing over is and how it leads to genetic variation. [4 marks]

b) Explain how independent segregation leads to genetic variation. [2 marks]

Reproduction isn't as exciting as some people would have you believe...

This page is quite tricky, so use the diagrams to help you understand — they might look evil, but they really do help. The key thing to understand is that meiosis produces four genetically different haploid (n) daughter cells. And that the genetic variation in the daughter cells occurs because of two processes — crossing over and independent segregation.

Genetic Diversity

Genetic diversity describes the variety of alleles in a species or population.

Variation in DNA can Lead to Genetic Diversity

Genetic diversity is all about **variety**. The **more variety** in a population's DNA, the **more genetically diverse** it is.

1) Genetic diversity exists **within** a species. The DNA within a species varies **very little** though. **All** the members of the species will have the **same genes** but **different alleles**. For example, approximately 99.5% of DNA is the same in all humans.
2) The DNA of **different species** varies **a lot**. Members of different species will have **different genes**. The more **related** a species is, the more DNA they **share**, e.g. around 94% of human and chimpanzee DNA is the same, and around 85% of human and mouse DNA is the same.

Genetic diversity within a species, or a **population** of a species, is caused by differences in **alleles**, but new genes **don't appear** and old genes **don't disappear**. For example, all humans have a gene for blood type, but different alleles (versions) of blood type may come and go. The **more alleles** in a population, the **more genetically diverse** it is. Genetic diversity within a population is **increased** by:

1) **Mutations** in the DNA — forming new alleles.
2) Different alleles being introduced into a population when individuals from another population **migrate into them** and reproduce. This is known as **gene flow**.

Genetic Bottlenecks Reduce Genetic Diversity

A **genetic bottleneck** is an event that causes a big **reduction** in a population, e.g. when a large number of organisms within a population **die** before reproducing. This reduces the number of **different alleles** in the gene pool and so reduces **genetic diversity**. The survivors **reproduce** and a larger population is created from a few individuals.

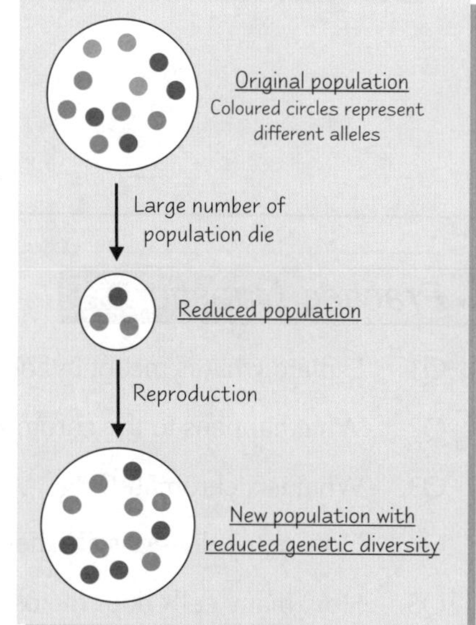

The gene pool is the complete range of alleles in a population.

Example — Northern Elephant Seals

Northern elephant seals were hunted by humans in the late 1800s. Their **original population** was reduced to around **50 seals** who have since produced a population of around 100 000. This new population has **very little** genetic diversity compared to the southern elephant seals who never suffered such a **reduction** in numbers.

Colin's offer to introduce new alleles into the population had yet to be accepted.

The Founder Effect is a Type of Genetic Bottleneck

The **founder effect** describes what happens when just a **few** organisms from a population start a **new colony**. Only a small number of organisms have contributed their **alleles** to the **gene pool**. There's more **inbreeding** in the new population, which can lead to a **higher incidence** of genetic disease.

Example — The Amish

The **Amish population** of North America are all descended from a **small** number of Swiss who **migrated** there. The population shows **little genetic diversity**. They have remained **isolated** from the surrounding population due to their **religious beliefs**, so **few new alleles** have been introduced. The population suffers an unusually high incidence of certain **genetic disorders**.

The founder effect can occur as a result of **migration** leading to geographical **separation** or if a new colony is separated from the original population for **another reason**, such as **religion**.

Genetic Diversity

Selective Breeding Involves Choosing Which Organisms Reproduce

Changes in genetic diversity aren't just brought about by **natural events** like bottlenecks or migration. **Selective breeding** of plants and animals by humans has resulted in **reduced genetic diversity** in some populations. Selective breeding involves humans **selecting** which domesticated animals or strains of plants **reproduce** together in order to produce **high-yielding** breeds. For example:

1) A farmer wants a strain of **corn plant** that is tall and produces lots of ears, so he **breeds** a **tall** corn strain with one that produces **multiple ears**.
2) He selects the **offspring** that are tallest and have most ears, and breeds them **together**.
3) The farmer **continues** this until he produces a **very tall** strain that produces **multiple ears** of corn.

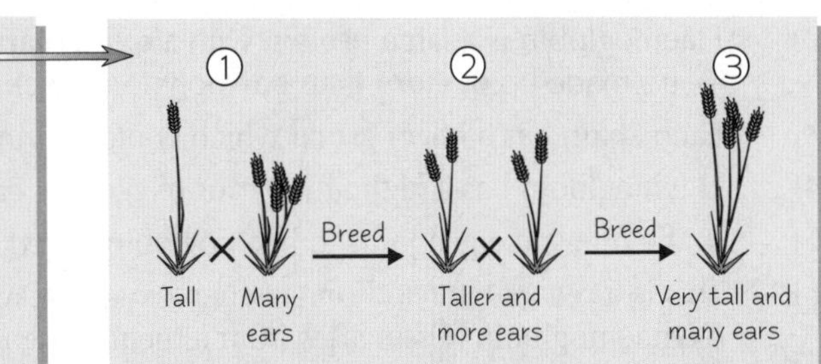

Selective breeding leads to a **reduction** in genetic diversity — once an organism with the **desired characteristics** (e.g. tall with multiple ears) has been produced, only that type of organism will continue being **bred**. So only similar organisms with **similar traits** and therefore **similar alleles** are bred together. It results in a type of **genetic bottleneck** as it reduces the **number of alleles** in the gene pool.

Selective Breeding can Cause Problems for the Organisms Involved

You need to be able to discuss the **ethical issues** involved with selective breeding.

Arguments FOR selective breeding

1) It can produce **high-yielding** animals and plants.
2) It can be used to produce animals and plants that have increased **resistance** to disease. This means farmers have to use **fewer** drugs and pesticides.
3) Animals and plants could be bred to have increased tolerance of bad conditions, e.g. **drought** or **cold**.

Arguments AGAINST selective breeding

1) It can cause **health problems**. E.g. dairy cows are often **lame** and have a **short life expectancy** because of the extra strain making and carrying loads of milk puts on their bodies.
2) It **reduces genetic diversity**, which results in an increased incidence of **genetic disease** and an **increased susceptibility** to new diseases because of the lack of **alleles** in the population.

Practice Questions

Q1 How does the founder effect reduce genetic diversity?

Q2 Describe the process of selective breeding.

Q3 Describe two arguments for and two arguments against selective breeding.

Exam Questions

Q1 Describe what a genetic bottleneck is and explain how it causes reduced genetic diversity within a population. [3 marks]

Q2 Describe what selective breeding is and explain why it leads to reduced genetic diversity within a population. [3 marks]

Sausage dogs didn't come from the wild...

You might think that selective breeding is a relatively new thing that we've developed with our knowledge of genetics... but you'd be wrong. We've been selectively breeding animals for yonks and yonks. All the different breeds of dog are just selectively bred strains which came from a general wolf-type dog back in the day. Even sausage dogs. Amazing...

Variation in Haemoglobin

Haemoglobin's a protein that carries oxygen around the body. Different species have different versions of it depending on where each species lives. All of which adds up to two pages of no-holds-barred fun...

Oxygen is Carried Round the Body by Haemoglobin

1) **Red blood cells** contain **haemoglobin** (Hb).
2) Haemoglobin is a large **protein** with a **quaternary** structure (see p. 14 for more) — it's made up of **more than one** polypeptide chain (**four** of them in fact).
3) Each chain has a **haem group** which contains **iron** and gives haemoglobin its **red** colour.
4) Haemoglobin has a **high affinity for oxygen** — each molecule can carry **four oxygen molecules**.
5) In the lungs, oxygen **joins** to haemoglobin in red blood cells to form **oxyhaemoglobin**.
6) This is a **reversible reaction** — when oxygen leaves oxyhaemoglobin (**dissociates** from it) near the body cells, it turns back to haemoglobin.

$$\mathbf{Hb} + \mathbf{4O_2} \rightleftharpoons \mathbf{HbO_8}$$

haemoglobin + oxygen ⇌ oxyhaemoglobin

'Affinity' for oxygen means tendency to combine with oxygen.

There are many **chemically similar** types of haemoglobin found in many different organisms, all of which carry out the **same function**. As well as being found in all vertebrates, haemoglobin is found in earthworms, starfish, some insects, some plants and even in some bacteria.

Haemoglobin Saturation Depends on the Partial Pressure of Oxygen

1) The **partial pressure** of **oxygen** (**pO_2**) is a measure of **oxygen concentration**. The **greater** the concentration of dissolved oxygen in cells, the **higher** the partial pressure.
2) Similarly, the **partial pressure** of **carbon dioxide** (**pCO_2**) is a measure of the concentration of CO_2 in a cell.
3) Haemoglobin's **affinity** for oxygen **varies** depending on the **partial pressure** of **oxygen**:

> Oxygen **loads onto** haemoglobin to form oxyhaemoglobin where there's a **high pO_2**.
> Oxyhaemoglobin **unloads** its oxygen where there's a **lower pO_2**.

4) Oxygen enters blood capillaries at the **alveoli** in the **lungs**. Alveoli have a **high pO_2** so oxygen **loads onto** haemoglobin to form oxyhaemoglobin.
5) When **cells respire**, they use up oxygen — this **lowers the pO_2**. Red blood cells deliver oxyhaemoglobin to respiring tissues, where it unloads its oxygen.
6) The haemoglobin then returns to the lungs to pick up more oxygen.

There was no use pretending — the pCH_4 had just increased, and Keith knew who was to blame.

Dissociation Curves Show How Affinity for Oxygen Varies

A **dissociation curve** shows how **saturated** the haemoglobin is with oxygen at any given partial pressure.

100% saturation means every haemoglobin molecule is carrying the maximum of 4 molecules of oxygen.

0% saturation means none of the haemoglobin molecules are carrying any oxygen.

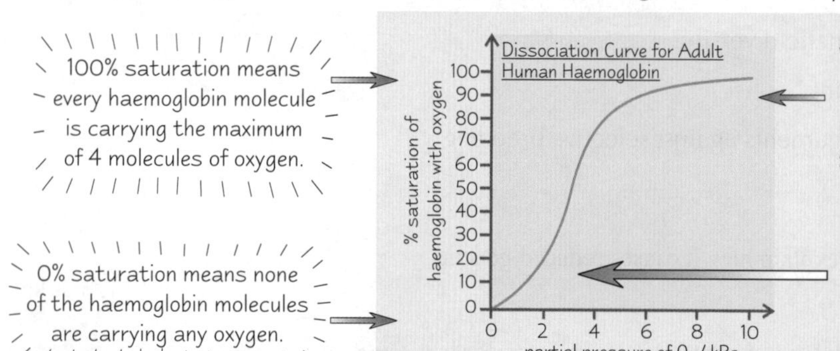

Where **pO_2 is high** (e.g. in the lungs), haemoglobin has a **high affinity** for oxygen (i.e. it will **readily combine** with oxygen), so it has a **high saturation** of oxygen.

Where **pO_2 is low** (e.g. in respiring tissues), haemoglobin has a **low affinity** for oxygen, which means it **releases oxygen** rather than combines with it. That's why it has a **low saturation** of oxygen.

The graph is **'S-shaped'** because when haemoglobin (Hb) combines with the **first O_2 molecule**, its **shape alters** in a way that makes it **easier** for other molecules to join too. But as the Hb starts to become saturated, it gets **harder** for more oxygen molecules to join. As a result, the curve has a **steep** bit in the middle where it's really easy for oxygen molecules to join, and **shallow** bits at each end where it's harder. When the curve is steep, a **small change in pO_2** causes a **big change** in the **amount of oxygen** carried by the Hb.

Variation in Haemoglobin

Carbon Dioxide Concentration Affects Oxygen Unloading

To complicate matters, haemoglobin gives up its oxygen **more readily** at **higher partial pressures of carbon dioxide** (pCO_2). It's a cunning way of getting more oxygen to cells during activity.

1) When cells respire they produce carbon dioxide, which **raises the pCO_2**.
2) This increases the rate of **oxygen unloading** — the dissociation curve '**shifts**' down. The saturation of blood with oxygen is **lower** for a given pO_2, meaning that **more oxygen** is being **released**.
3) This is called the **Bohr effect.**

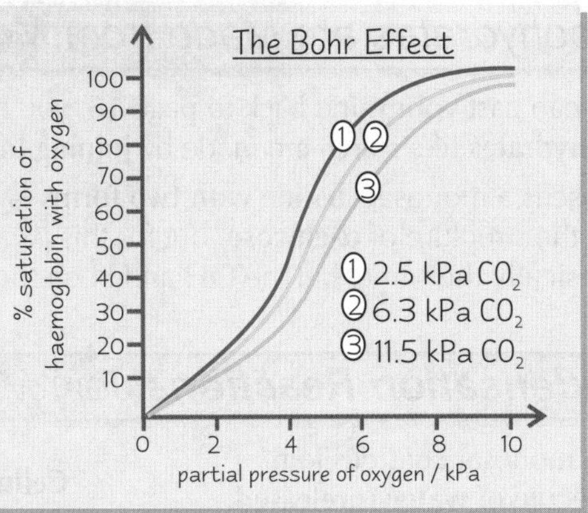

Haemoglobin is Different in Different Organisms

Different organisms have different **types** of haemoglobin with different **oxygen transporting capacities**.

1) Organisms that live in environments with a **low concentration of oxygen** have haemoglobin with a **higher affinity** for oxygen than human haemoglobin — the dissociation curve is to the **left** of ours.
2) Organisms that are very **active** and have a **high oxygen demand** have haemoglobin with a **lower affinity** for oxygen than human haemoglobin — the curve is to the **right** of the human one.

A = animal living in depleted oxygen environment, e.g. a lugworm
B = animal living at high altitude where the partial pressure of oxygen is lower, e.g. a llama in the Andes.
C = human dissociation curve
D = active animal with a high respiratory rate living where there's plenty of available oxygen, e.g. a hawk.

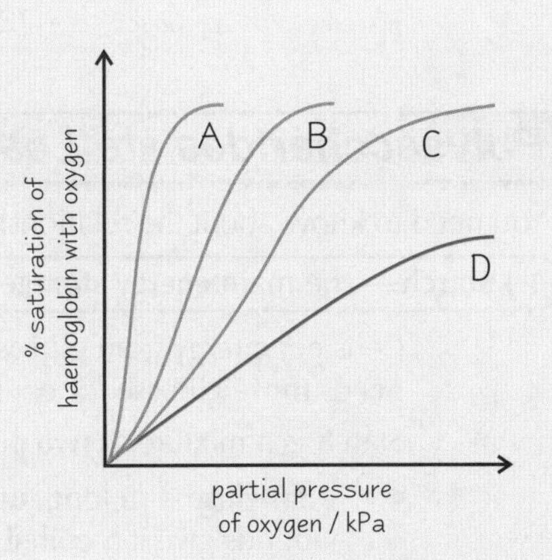

Practice Questions

Q1 How many oxygen molecules can each haemoglobin molecule carry?

Q2 Where in the body would you find a low partial pressure of oxygen?

Q3 Why are oxygen dissociation curves S-shaped?

Q4 What is the Bohr effect?

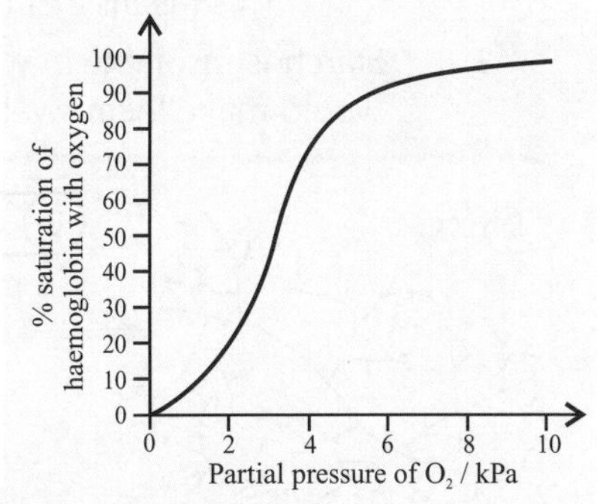

Exam Questions

Q1 The graph shows the oxygen dissociation curve for human haemoglobin. On the graph, sketch the curves you would expect for an earthworm (which lives in a low oxygen environment) and a human in a high carbon dioxide environment. Explain the position of your sketched curves. [6 marks]

Q2 Haemoglobin is a protein with a quaternary structure. Explain what this means. [1 mark]

There's more than partial pressure on you to learn this stuff...

Well, I don't know about you but after these two pages I need a sit-down. Most people get their knickers in a twist over partial pressure — it's not the easiest thing to get your head round. Whenever you see it written down just pretend it says concentration instead — cross it out and write concentration if you have to — and everything should become clearer. Honest.

Variation in Carbohydrates and Cell Structure

Mmmmm, tasty tasty glucose... Unfortunately you have to learn its structure rather than eat it. You also need to cover the structure and function of tasty tasty starch and not-so-tasty-but-equally-important cellulose and glycogen. Enjoy...

Carbohydrates are Made from ***Monosaccharides***

If you can cast your mind back to page 16, you might remember that **complex carbohydrates** like starch are made by **joining** lots of **monosaccharides** together.

Glucose is a monosaccharide with **two forms** — α and β. On page 16 you learnt about the structure of **α-glucose**, but for this page you need to know **β-glucose**... (It's basically the same, but the OH and H on the right are swapped around.)

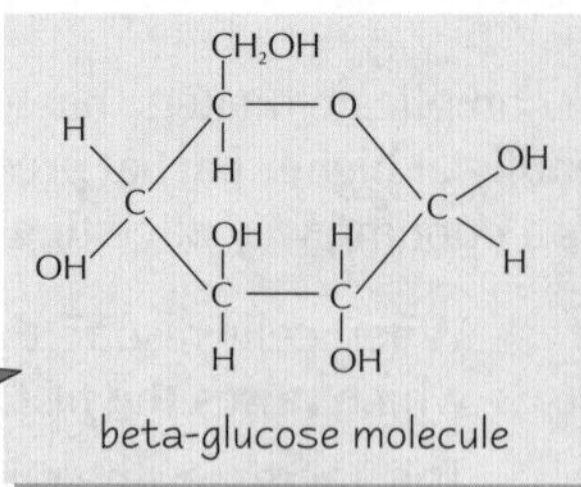

beta-glucose molecule

Condensation Reactions Join Monosaccharides (Sugars) Together

When monosaccharides join, a molecule of **water** is **released**. This is called a **condensation reaction**. The bonds that join sugars together are called **glycosidic bonds**.

If you're asked to show a condensation reaction, don't forget to put the water molecule in as a product.

Cellulose is formed when beta-glucose is linked by condensation.

glycosidic bond

monosaccharide + monosaccharide ⇌ disaccharide

H_2O is removed

Polysaccharides are ***Loads of Sugars*** Joined Together

You need to know about the relationship between the **structure** and **function** of three polysaccharides:

1) Starch — the main **energy storage material** in **plants**

1) Cells get **energy** from **glucose**. Plants **store** excess glucose as **starch** (when a plant **needs more glucose** for energy it **breaks down** starch to release the glucose).

2) Starch is a mixture of **two** polysaccharides of **alpha-glucose** — **amylose** and **amylopectin**:

- **Amylose** — a long, **unbranched chain** of α–glucose. The angles of the glycosidic bonds give it a **coiled structure**, almost like a cylinder. This makes it **compact**, so it's really **good for storage** because you can **fit more in** to a small space.
- **Amylopectin** — a long, **branched chain** of α–glucose. Its **side branches** allow the **enzymes** that break down the molecule to get at the **glycosidic bonds easily**. This means that the glucose can be **released quickly**.

3) Starch is **insoluble** in water so it doesn't cause water to enter cells by **osmosis**, which would make them swell (see p. 28). This makes it good for **storage**.

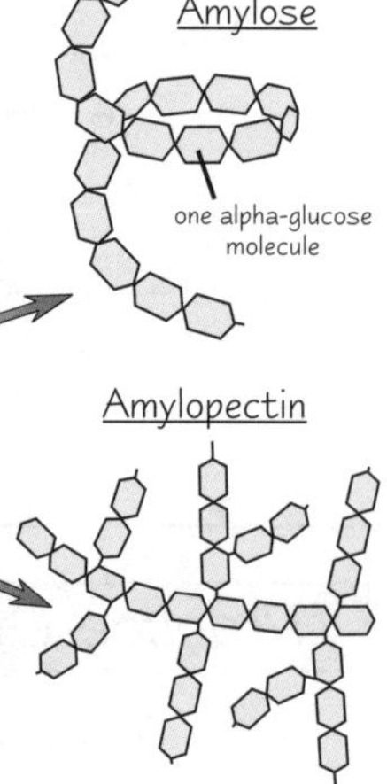

2) Glycogen — the main **energy storage material** in **animals**

1) Animal cells get **energy** from **glucose** too. But animals **store** excess glucose as **glycogen** — another polysaccharide of **alpha-glucose**.

2) Its structure is very similar to amylopectin, except that it has **loads** more **side branches** coming off it. Loads of branches means that stored glucose can be **released quickly**, which is **important for energy release** in animals.

3) It's also a very **compact** molecule, so it's good for storage.

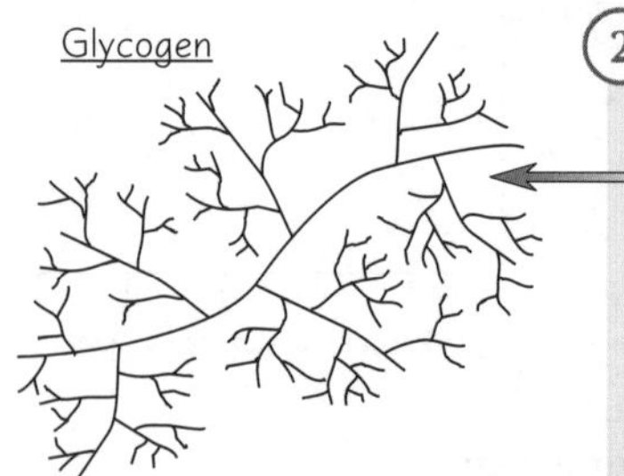

3) Cellulose — the major component of **cell walls in plants**

1) Cellulose is made of **long**, **unbranched** chains of **beta-glucose**.

2) The **bonds** between the sugars are **straight**, so the cellulose chains are straight.

3) The cellulose chains are linked together by **hydrogen bonds** to form strong fibres called **microfibrils**. The strong fibres mean cellulose provides **structural support** for cells (e.g. in plant cell walls).

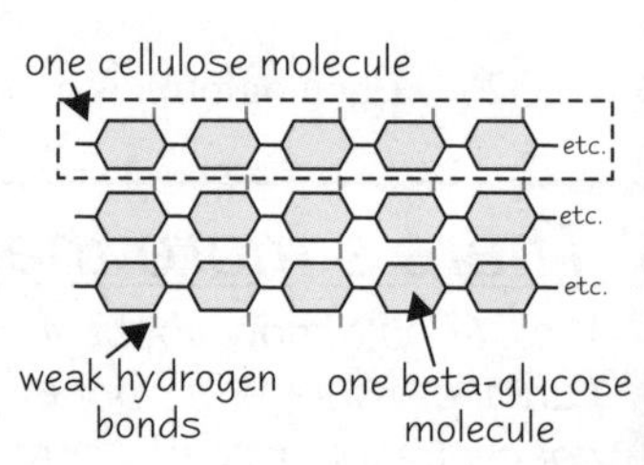

Variation in Carbohydrates and Cell Structure

Plant and *Animal Cells* have *Similarities* and *Differences*

Way, way back in Unit 1 Section Three you learnt about animal cell structure and organelles.
Well, you need to know about their structure for this section too so you can see how plant cell structure differs.

Most **animal** cells have the following parts — make sure you know them all:

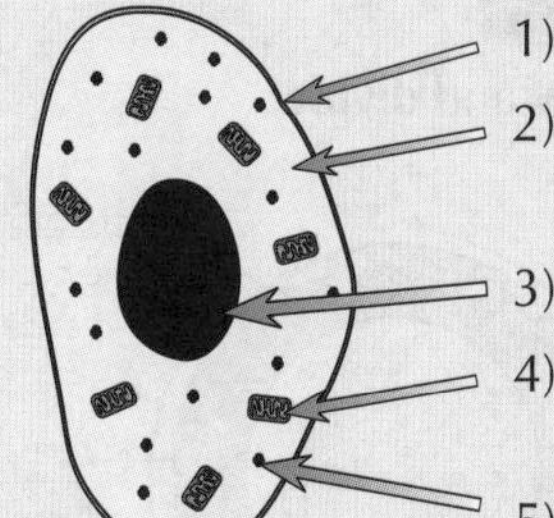

1) **Plasma membrane** — holds the cell together and controls what goes **in** and **out**.
2) **Cytoplasm** — gel-like substance where most of the **chemical reactions** happen. It contains **enzymes** (see page 18) that control these chemical reactions.
3) **Nucleus** — contains **genetic material** that controls the activities of the cell.
4) **Mitochondria** — where most of the reactions for **respiration** take place. Respiration releases **energy** that the cell needs to work.
5) **Ribosomes** — where **proteins** are made in the cell.

Plant cells usually have **all the bits** that **animal** cells have, plus a few **extra** things that animal cells **don't** have:

1) Rigid **cell wall** — made of **cellulose**. It **supports** and **strengthens** the cell.
2) **Permanent vacuole** — contains **cell sap**, a weak solution of sugar and salts.
3) **Chloroplasts** — where **photosynthesis** occurs, which makes food for the plant. They contain a **green** substance called **chlorophyll**.

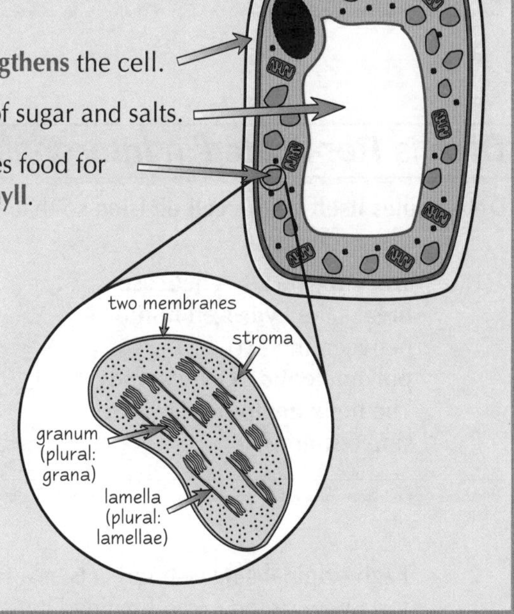

- Chloroplasts are surrounded by a **double membrane**, and also have membranes inside called **thylakoid membranes**. These membranes are stacked up in the chloroplast to form **grana**.
- Grana are linked together by **lamellae** — thin, flat pieces of thylakoid membrane.
- Some parts of photosynthesis happen in the **grana**, and other parts happen in the **stroma** (a thick fluid found in chloroplasts).

In the exam, you might get a question where you need to apply your **knowledge** of the **organelles** in a plant cell to explain why it's particularly **suited** to its **function**. Here are some tips:

- Think about **where** the cell's **located** in the plant — e.g. if it's exposed to **light**, then it'll have **lots** of **chloroplasts** to maximise **photosynthesis**.
- Think about **what** the cell **needs** to do its **job** — e.g. if the cell uses a lot of **energy**, it'll need lots of **mitochondria**. If it makes a lot of **proteins**, it'll need a lot of **ribosomes**.

Practice Questions

Q1 Draw a diagram to illustrate a condensation reaction between two molecules of β-glucose.

Q2 Give three structures found in plant cells but not in animal cells.

Exam Question

Q1 Describe how the structure of starch makes it suited to its function. [6 marks]

Starch — I thought that was just for shirt collars...

It's important to understand that every cell in an organism is adapted to perform a function — you can always trace some of its features back to its function. Different cells even use the exact same molecules to do completely different things. Take glucose, for example — all plant cells use it to make cellulose, but they can also make starch from it if they need to store energy.

The Cell Cycle and DNA Replication

Ever wondered how you grow from one tiny cell to a complete whole person? Or how that big cut you got in that horrific guitar strumming incident healed? No, oh... well you grow bigger and heal because your cells replicate and you need to learn the processes involved.

The Cell Cycle is the Process of Cell Growth and Division

The **cell cycle** is the process that all body cells from **multicellular organisms** use to **grow** and **divide**.

1) The cell cycle **starts** when a cell has been produced by cell division and **ends** with the cell dividing to produce two identical cells.
2) The cell cycle consists of a period of **cell growth** and **DNA replication**, called **interphase**, and a period of **cell division**, called **mitosis**.
3) Interphase (cell growth) is subdivided into three separate growth stages. These are called G_1, **S** and G_2.

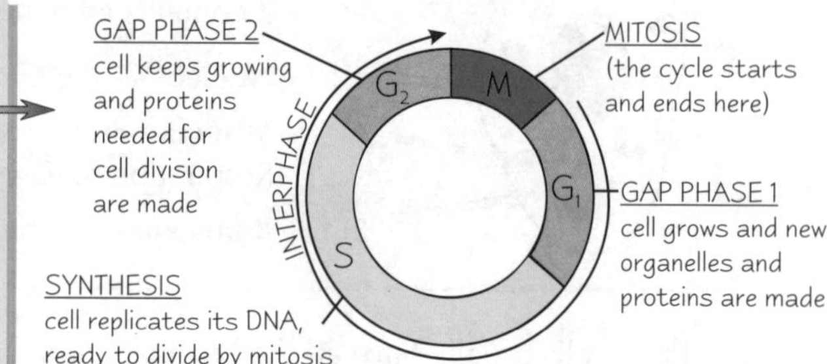
GAP PHASE 2
cell keeps growing and proteins needed for cell division are made
MITOSIS
(the cycle starts and ends here)
G2
M
G1
INTERPHASE
S
GAP PHASE 1
cell grows and new organelles and proteins are made
SYNTHESIS
cell replicates its DNA, ready to divide by mitosis

DNA is Replicated in Interphase

DNA copies itself before **cell division** so that each new cell has the full amount of DNA.

1) The enzyme **DNA helicase breaks the hydrogen bonds** between the two **polynucleotide** DNA strands. The helix **unzips** to form two single strands.

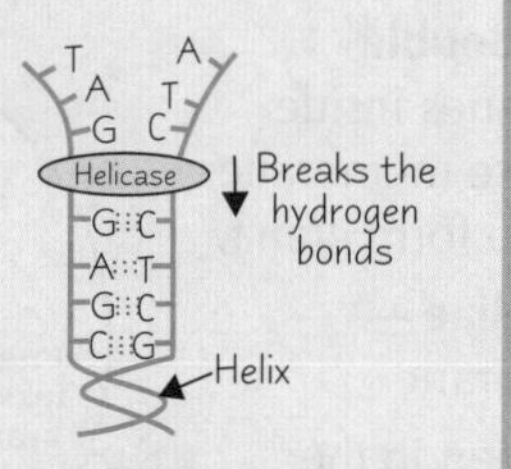
Helicase
Breaks the hydrogen bonds
Helix

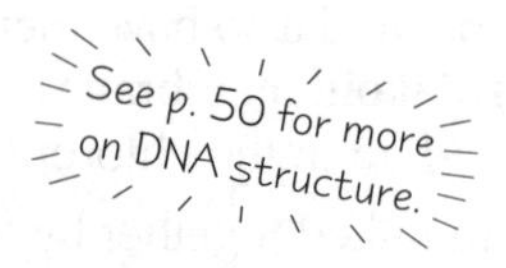
See p. 50 for more on DNA structure.

Mandy took her cells for a cycle.

2) Each **original** single strand acts as a **template** for a new strand. Free-floating DNA nucleotides join to the **exposed bases** on each original template strand by **specific base pairing** — A with T and C with G.

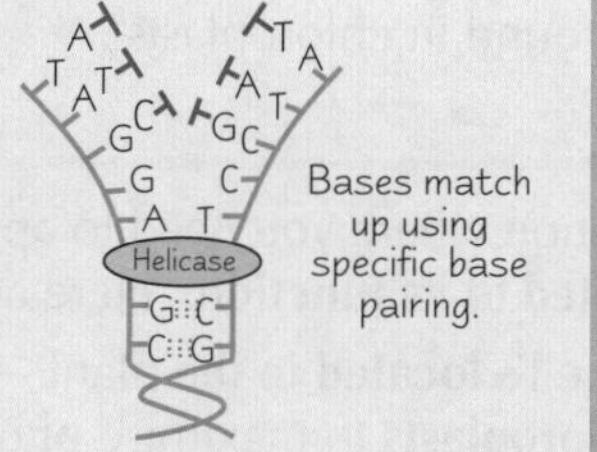
Bases match up using specific base pairing.
Helicase

3) The nucleotides on the new strand are joined together by the enzyme **DNA polymerase**. Hydrogen bonds **form** between the bases on the original and new strand.

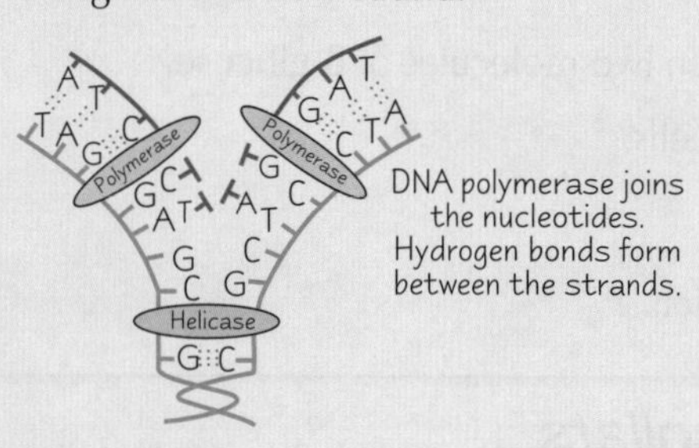
Polymerase
Polymerase
DNA polymerase joins the nucleotides. Hydrogen bonds form between the strands.
Helicase

4) Each new DNA molecule contains **one strand** from the **original** DNA molecule and one **new strand**.

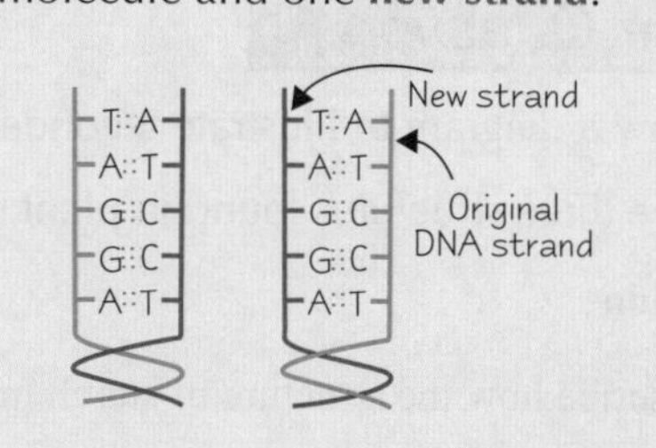
New strand
Original DNA strand

This type of copying is called **semi-conservative replication** because **half** of the new strands of DNA are from the **original** piece of DNA.

The Cell Cycle and DNA Replication

You May Have to *Interpret* Early *Experimental Work* About *DNA*

In the exam you might have to interpret **experimental evidence** that shows the **role** and **importance** of DNA. Here are some examples of the **early experiments** that were carried out. You **don't** need to learn them, just **understand** how the results show the role and importance of DNA.

Evidence of hereditary molecules

An experiment with **mice** and two kinds of **pneumonia**, a **disease-causing** strain (**D**) and a **non-disease-causing** strain (**N**), showed there's a **hereditary molecule** (genetic material).

1) Mice injected with **strain D died** and with **strain N survived**.
2) **Killed D** was injected into mice — they **survived**.
3) **Killed D** and **live N** were injected together — they **died**.

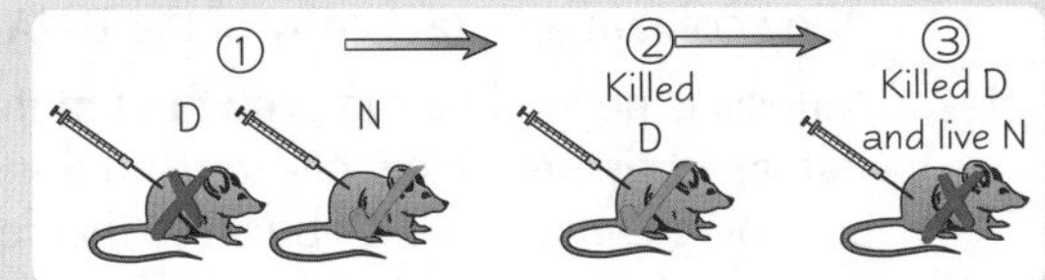

Killed D had **passed on** an inheritance molecule to the live N strain, making it **capable** of causing **disease**.

Evidence that DNA is the genetic material

Scientists were unsure if the hereditary molecule was **DNA, RNA** or **protein**. They investigated it by treating the killed D strain with **protease** (destroys protein), **RNase** (destroys RNA) or **DNase** (destroys DNA) and then **injecting** it along with live N strain into mice. The strains that had been treated with DNase **didn't** kill the mice, so DNA was shown to be the **hereditary molecule** (genetic material).

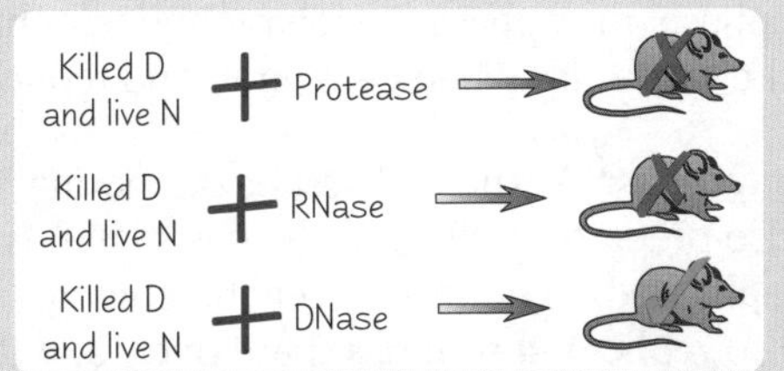

More evidence that DNA is the genetic material

When viruses infect bacteria they **inject** their genetic material into the **cell**. So whatever viral material is found **inside** the bacterial cell must be the genetic material.

1) Scientists labelled the **DNA** of some viruses with radioactive **phosphate**, ^{32}**P** (**blue**), and the **protein** of some more viruses with radioactive **sulfur**, ^{35}**S** (**red**).
2) They then let the viruses **infect** some bacteria.
3) When they **separated** the bacteria and viruses they found ^{32}**P** (**blue**) inside the bacteria and ^{35}**S** (**red**) on the outside, providing **evidence** that DNA was the **genetic material**.

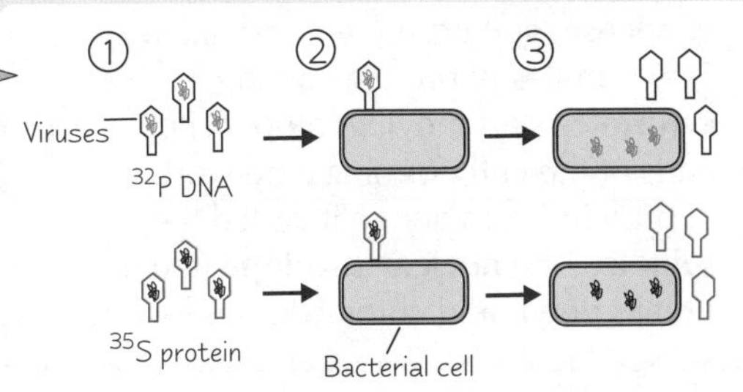

Practice Questions

Q1 Name the two main stages of the cell cycle.

Q2 Why is DNA replication described as semi-conservative?

Q3 Name two enzymes involved in DNA replication.

Exam Question

Q1 a) Fill in the missing base pairs on the diagram opposite. [1 mark]

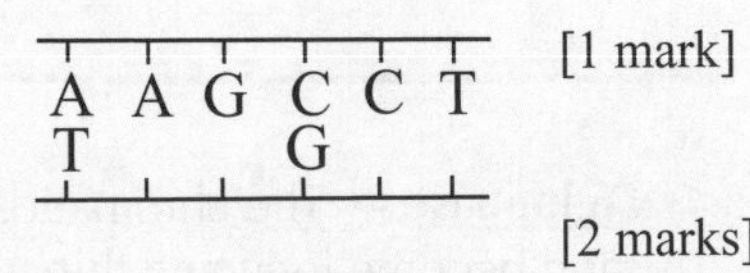

b) Draw a diagram to show the DNA molecule on the right after it has replicated. Label the original and new strands. [2 marks]

I went through a gap phase — I just love their denim...

DNA and its self-replication is important — so make sure you understand what's going on. Diagrams are handy for learning stuff like this. I don't just put them in to keep myself amused you know — so get drawing and learning.

Cell Division — Mitosis

I don't like cell division. There, I've said it. It's unfair of me, because if it wasn't for cell division I'd still only be one cell big. It's all those diagrams that look like worms nailed to bits of string that put me off.

Mitosis is *Cell Division* that Produces *Genetically Identical Cells*

1) There are two types of cell division — **mitosis** and **meiosis** (see p. 54 for more on meiosis).
2) Mitosis is the form of cell division that occurs during the **cell cycle**.
3) In **mitosis** a **parent cell** divides to produce **two genetically identical daughter cells** (they contain an **exact copy** of the **DNA** of the parent cell).
4) Mitosis is needed for the **growth** of multicellular organisms (like us) and for **repairing damaged tissues**. How else do you think you get from being a baby to being a big, strapping teenager — it's because the cells in our bodies grow and divide.

Mitosis has *Four Division Stages*

Mitosis is really one **continuous process**, but it's described as a series of **division stages** — prophase, metaphase, anaphase and telophase. **Interphase** comes **before** mitosis in the cell cycle — it's when cells grow and replicate their DNA ready for division (see p. 62).

Interphase — The cell carries out normal functions, but also prepares to divide. The cell's **DNA** is unravelled and **replicated**, to double its genetic content. The **organelles** are also **replicated** so it has spare ones, and its ATP content is increased (ATP provides the energy needed for cell division).

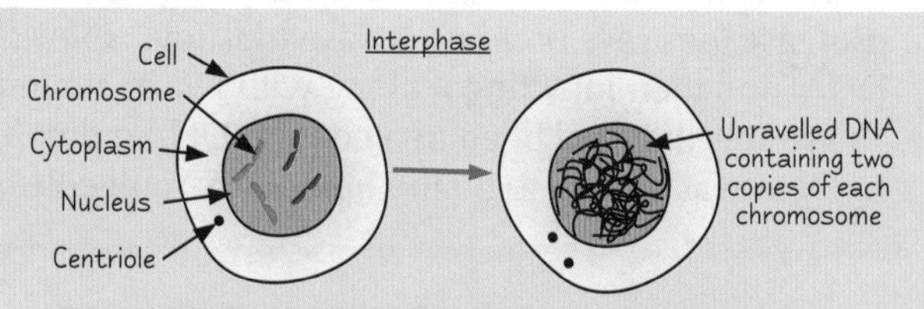

1) **Prophase** — The **chromosomes condense**, getting shorter and fatter. Tiny bundles of protein called **centrioles** start moving to opposite ends of the cell, forming a network of protein fibres across it called the **spindle**. The **nuclear envelope** (the membrane around the nucleus) **breaks down** and chromosomes lie free in the cytoplasm.

Centrioles move to opposite ends of the cell
Nuclear envelope starts to break down
Centromere

As mitosis begins, the chromosomes are made of two strands joined in the middle by a centromere.
The separate strands are called chromatids.

One chromatid
Centromere
Sister chromatids

There are two strands because each chromosome has already made an identical copy of itself during interphase. When mitosis is over, the chromatids end up as one-strand chromosomes in the new daughter cells.

2) **Metaphase** — The chromosomes (each with two chromatids) **line up** along the middle of the cell and become **attached** to the **spindle** by their **centromere**.

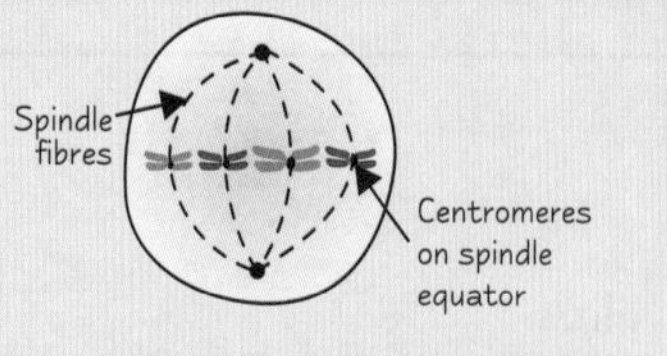

You need to be able to recognise each stage in mitosis from photographs. But don't worry — they'll look pretty much like the diagrams here.

3) **Anaphase** — The centromeres divide, **separating** each pair of sister **chromatids**. The spindles contract, pulling chromatids to opposite ends of the cell, centromere first.

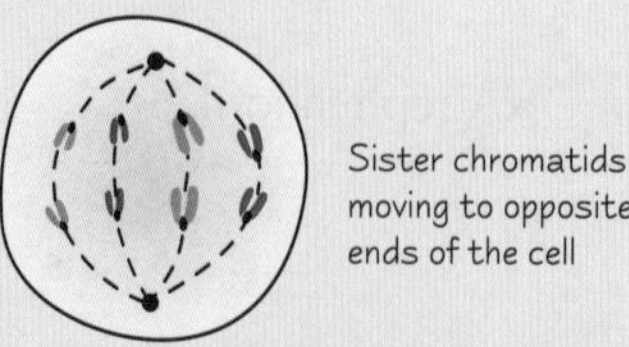

4) **Telophase** — The chromatids reach the **opposite poles** on the spindle. They uncoil and become long and thin again. They're now called **chromosomes** again. A **nuclear envelope** forms around each group of chromosomes, so there are now **two nuclei**. The **cytoplasm divides** and there are now **two daughter cells** that are **genetically identical** to the original cell and to each other. Mitosis is finished and each daughter cell starts the **interphase** part of the cell cycle to get ready for the next round of mitosis.

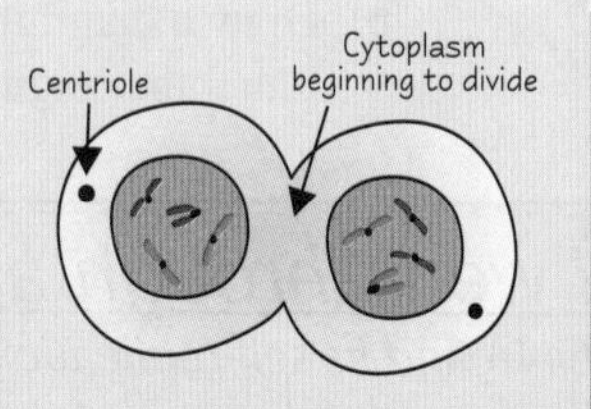

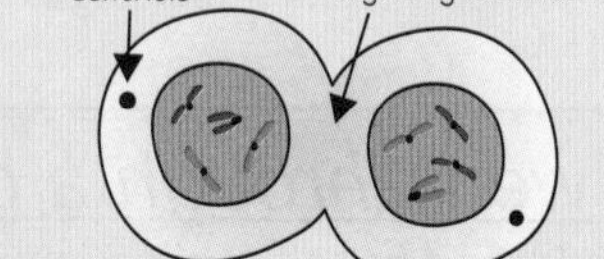

Cell Division — Mitosis

Cancer is the Result of *Uncontrolled Cell Division*

Mutations are changes in the base sequence of an organism's DNA (see p. 53).

1) Cell growth and cell division are **controlled by genes**.
2) Normally, when cells have divided enough times to make **enough new cells**, they stop. But if there's a **mutation** in a gene that controls cell division, the cells can **grow out of control**.
3) The cells **keep on dividing** to make more and more cells, which form a **tumour**.
4) **Cancer** is a tumour that **invades** surrounding tissue.

Some *Cancer Treatments* Target the *Cell Cycle*

Some treatments for cancer are designed to **disrupt** the cell cycle.
These treatments don't **distinguish** tumour cells from normal cells though — they also **kill normal body cells** that are dividing. However, tumour cells **divide much more frequently** than normal cells, so the treatments are **more likely** to kill tumour cells. Some cell cycle **targets** of cancer treatments include:

1) **G1 (cell growth and protein production)** — Some chemical drugs (chemotherapy) prevent the **synthesis of enzymes** needed for DNA replication. If these aren't produced, the cell is unable to enter the **synthesis phase** (S), disrupting the cell cycle and forcing the cell to **kill itself**.
2) **S phase (DNA replication)** — **Radiation** and some drugs **damage DNA**. When the cell gets to S phase it checks for **damaged DNA** and if any is detected it **kills itself**, preventing **further** tumour growth.

Because cancer treatments **kill normal cells** too certain steps are taken to **reduce the impact** on normal body cells:

1) A **chunk** of tumour is often removed first using **surgery**. This removes a lot of tumour cells and increases the access of any left to nutrients and oxygen, which triggers them to enter the **cell cycle**, making them **more susceptible** to treatment.
2) **Repeated treatments** are given with periods of **non-treatment** (breaks) in between. A **large dose** could kill **all the tumour** but also so many normal cells that the patient could **die**. Repeated treatments with **breaks** allows the body to **recover** and produce new cells. The treatment is **repeated** as any tumour cells **not killed** by the treatment will keep **dividing and growing** during the breaks too. The break period is kept short so the body can **recover** but the **cancer** can't grow back to the same size as before.

Practice Questions

Q1 Give the two main uses of mitosis.

Q2 List the four stages of mitosis.

Q3 Describe how tumours are formed.

Q4 Describe how repeated cancer treatment with breaks can help to reduce the impact on normal body cells.

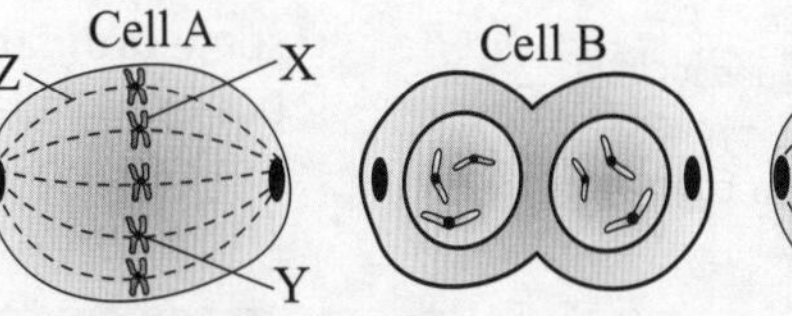

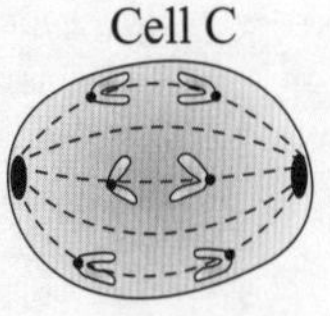

Exam Question

Q1 The diagrams show cells at different stages of mitosis.

a) For each of the cells A, B and C state the stage of mitosis, giving a reason for your answer. [6 marks]

b) Name the structures labelled X, Y and Z in cell A. [3 marks]

Doctor, I'm getting short and fat — don't worry, it's just a phase...

Quite a lot to learn on these pages — but it's all important stuff, so no slacking. Mitosis is vital — it's how cells multiply and how organisms like us grow. Don't forget — the best way to learn is to get drawing those diagrams.

Cell Differentiation and Organisation

In complex multicellular organisms like me (well, I wouldn't say I'm complex, but multicellular at least), cells are adapted for different jobs. And all these cells are organised to work together.

*Cells of **Multicellular** Organisms Can **Differentiate***

1) **Multicellular organisms** are made up from many **different** cell types, e.g. nerve cells, muscle cells, white blood cells.
2) **All** these cell types are **specialised** — they're designed to carry out **specific functions** (see below).
3) The **structure** of each specialised cell type is **adapted** to suit its particular job.
4) The **process** of **becoming specialised** is called **differentiation.**

Joe knew his cells were specialised — specialised to look good.

Differentiated Cells** are **Adapted** for **Specific Functions

Here are two examples of differentiated cells to show you how they're adapted for their function:

1) **Squamous epithelium cells** are found in many places. They're **thin**, with not much cytoplasm. In the lungs they line the **alveoli** and are thin to allow gases to pass through them easily.

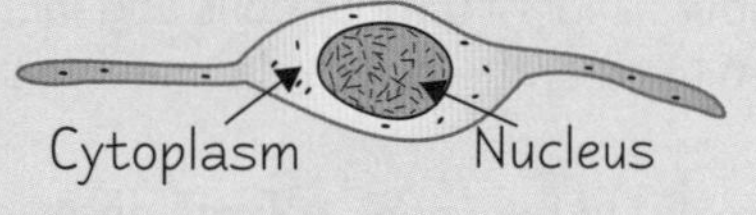

2) **Palisade mesophyll cells** in leaves are where **photosynthesis** occurs. They contain **many chloroplasts**, so they can absorb as much sunlight as possible. The walls are **thin**, so carbon dioxide can **easily enter**.

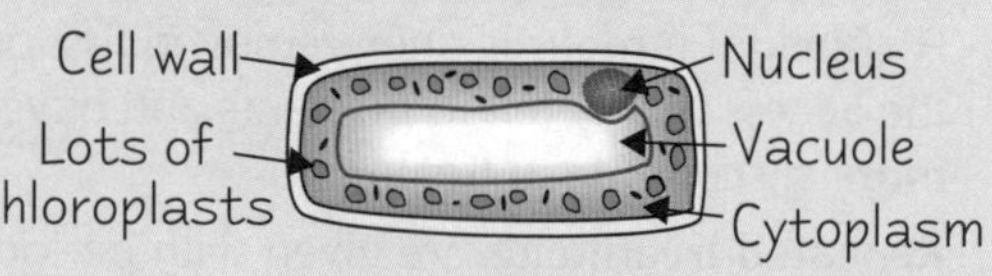

Similar Cells** are Organised into **Tissues

Similar cells are grouped together into **tissues**. Here are some examples:

1) **Squamous epithelium tissue** is a **single layer** of **flat cells** lining a surface. Squamous epithelium tissue is found in many places including the alveoli in the lungs.

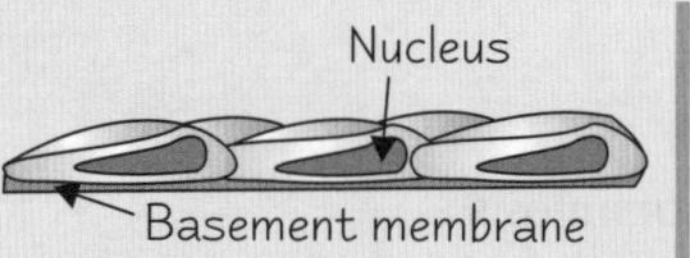

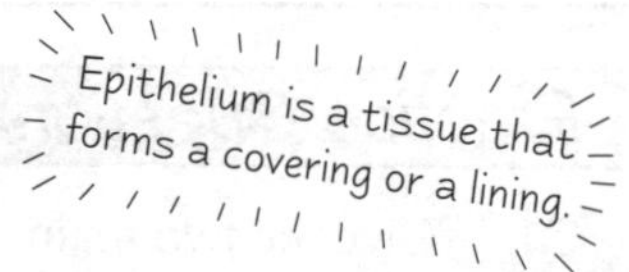

Epithelium is a tissue that forms a covering or a lining.

Tissues aren't always made up of one type of cell. Some tissues include different types of cell working together.

2) **Phloem tissue transports sugars** around the plant. It's arranged in **tubes** and is made up of **sieve cells**, **companion cells**, and some **ordinary** plant cells. Each sieve cell has end walls with **holes** in them, so that sap can move easily through them. These end walls are called **sieve plates**.

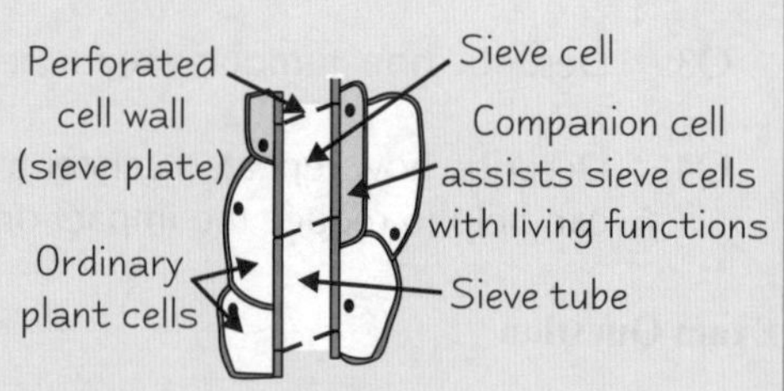

3) **Xylem tissue** is a plant tissue with two jobs — it **transports water** around the plant, and it **supports** the plant. It contains **xylem vessel cells** and **parenchyma cells**.

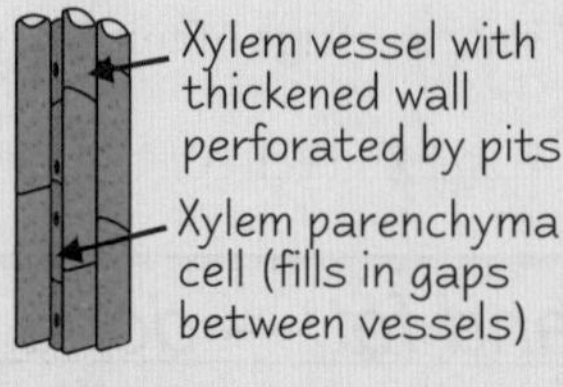

Cell Differentiation and Organisation

Tissues are Organised into Organs

An **organ** is a group of different tissues that **work together** to perform a particular function. Here are two examples:

The leaf is an example of a plant organ. It's made up of the following **tissues**:

1) **Lower epidermis** — contains stomata (pores) to let air in and out for gas exchange.
2) **Spongy mesophyll** — full of spaces to let gases circulate.
3) **Palisade mesophyll** — most photosynthesis occurs here.
4) **Xylem** — carries water to the leaf.
5) **Phloem** — carries sugars away from the leaf.
6) **Upper epidermis** — covered in a waterproof waxy cuticle to reduce water loss.

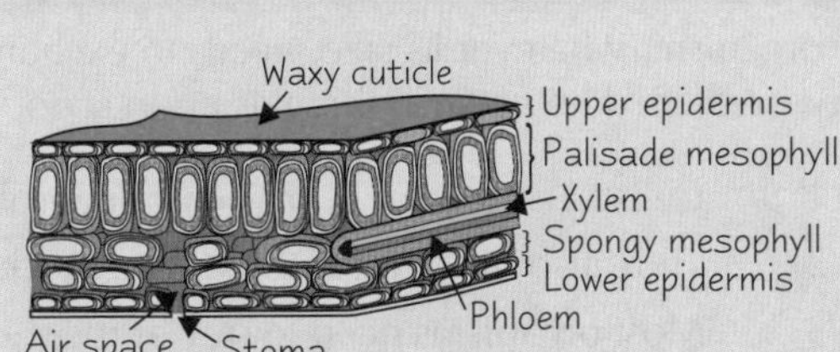

The lungs are an example of an animal organ. They're made up of the following **tissues**:

1) **Squamous epithelium tissue** — surrounds the alveoli (where gas exchange occurs).
2) **Fibrous connective tissue** — forms a continuous mesh around the lungs and contains fibres that help to force the air back out of the lungs when exhaling.
3) **Blood vessels** — capillaries surround the alveoli.

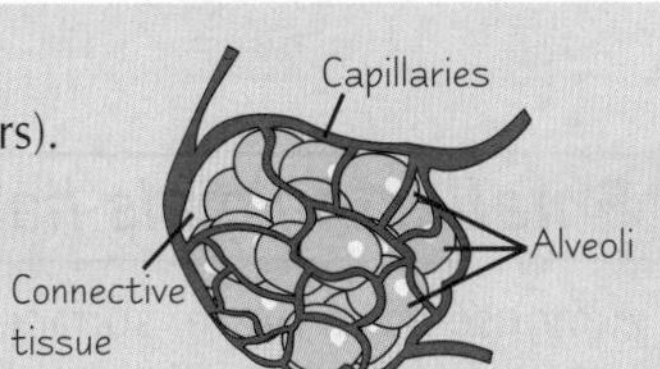

Organs are Organised into Systems

Organs work together to form **organ systems** — each system has a **particular function**. Here are some examples:

The Circulatory System allows the transport of gases and other substances around the body. It includes:

1) **The heart** — pumps the blood around the body.
2) **Blood vessels** — carries the blood to the tissues.

The Shoot System in plants includes:
leaves (site of photosynthesis), **buds** (growing regions), **stems** (for support) and **flowers** (for sexual reproduction).

The Respiratory System brings oxygen into the body and removes carbon dioxide. It includes:

1) **The lungs** — where gas exchange occurs.
2) **The trachea** — allows air flow.
3) **The bronchi** — carry the air into the lungs.

Practice Questions

Q1 What is meant by cell differentiation?

Q2 Define what is meant by a tissue.

Q3 Give one animal and one plant example of an organ system.

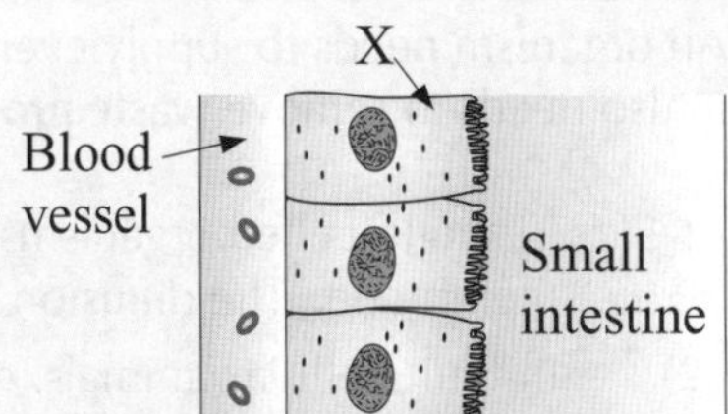

Exam Questions

Q1 Tissue X above is found lining the small intestine, where nutrients are absorbed into the bloodstream. Outline how it is adapted for its function. [4 marks]

Q2 The liver is made of hepatocyte cells that form the main tissue, blood vessels to provide nutrients and oxygen, and connective tissue that holds the organ together. Discuss whether the liver is best described as a tissue or an organ. [2 marks]

Soft and quilted — the best kind of tissue...

The important thing to remember from these pages is that in multicellular organisms, cells are adapted for their function. You don't need to learn the examples off by heart — just understand how the adaptations are related to the function.

Size and Surface Area

Exchanging things with the environment is pretty easy if you're a single-celled organism, but if you're multicellular it all gets a bit more complicated... and it's all down to this 'surface area to volume ratio' malarkey.

Organisms Need to **Exchange Substances** with their **Environment**

Every organism, whatever its size, needs to exchange things with its environment. Otherwise there'd be no such thing as poop scoops...

1) Cells need to take in **oxygen** (for aerobic respiration) and **nutrients**.
2) They also need to excrete **waste products** like **carbon dioxide** and **urea**.
3) Most organisms need to stay at roughly the **same temperature**, so **heat** needs to be exchanged too.

Raj was glad he'd exchanged his canoe for a bigger boat.

How easy the exchange of substances is depends on the organism's **surface area to volume ratio**.

Smaller Animals have **Higher Surface Area : Volume Ratios**

A mouse has a bigger surface area **relative to its volume** than a hippo. This can be hard to imagine, but you can prove it mathematically. Imagine these animals as cubes:

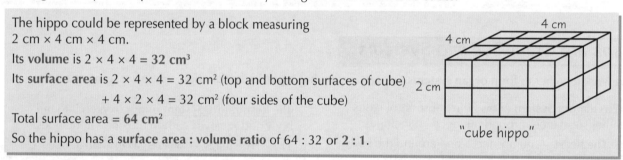

The hippo could be represented by a block measuring 2 cm × 4 cm × 4 cm.

Its **volume** is 2 × 4 × 4 = **32 cm^3**

Its **surface area** is 2 × 4 × 4 = 32 cm^2 (top and bottom surfaces of cube)

+ 4 × 2 × 4 = 32 cm^2 (four sides of the cube)

Total surface area = **64 cm^2**

So the hippo has a **surface area : volume ratio** of 64 : 32 or **2 : 1**.

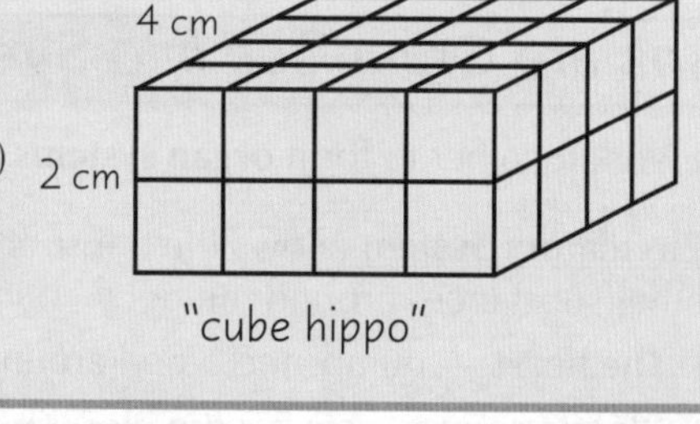

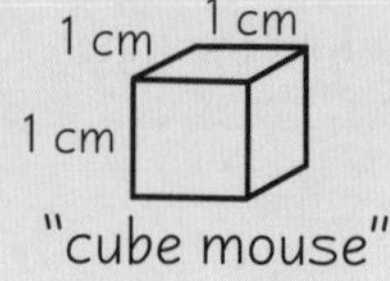

Compare this to a cube mouse measuring 1 cm × 1 cm × 1 cm.

Its **volume** is 1 x 1 x 1 = **1 cm^3**

Its **surface area** is 6 x 1 x 1 = **6 cm^2**

So the mouse has a **surface area : volume ratio** of 6 : 1

The cube mouse's surface area is six times its volume, but the cube hippo's surface area is only twice its volume. Smaller animals have a bigger surface area compared to their volume.

Multicellular Organisms need **Exchange Organs** and **Mass Transport Systems**

An organism needs to supply **every one of its cells** with substances like **glucose** and **oxygen** (for respiration). It also needs to **remove waste products** from every cell to avoid damaging itself.

1) In **single-celled** organisms, these substances can **diffuse directly** into (or out of) the cell across the cell surface membrane. The diffusion rate is quick because of the small distances the substances have to travel (see p. 28).
2) In **multicellular** animals, diffusion across the outer membrane is **too slow**, for two reasons:
   - Some cells are **deep within the body** — there's a big distance between them and the **outside environment**.
   - Larger animals have a **low surface area to volume ratio** — it's difficult to exchange **enough** substances to supply a **large volume of animal** through a relatively **small outer surface**.

So rather than using straightforward diffusion to absorb and excrete substances, multicellular animals need specialised **exchange organs** (like lungs — see p. 35).

They also need an efficient system to carry substances to and from their individual cells — this is **mass transport**. In mammals, 'mass transport' normally refers to the **circulatory system** (see p. 40), which uses **blood** to carry glucose and oxygen around the body. It also carries **hormones**, **antibodies** (p. 6) and **waste** like CO_2.

Size and Surface Area

Body Size and Shape Affect Heat Exchange

As well as creating **waste products** that need to be transported away, the metabolic activity inside cells creates **heat**. Staying at the right temperature is difficult, and it's pretty heavily influenced by your **size** and **shape**...

Size

The **rate of heat loss** from an organism depends on its **surface area**. As you saw on the previous page, if an organism has a large volume, e.g. a hippo, its surface area is relatively **small**. This makes it **harder** for it to lose heat from its body. If an organism is small, e.g. a mouse, its relative surface area is **large**, so heat is lost more **easily**.

Shape

1) Animals with a **compact** shape have a **small surface area** relative to their volume — **minimising heat loss** from their surface.
2) Animals with a **less compact** shape (those that are a bit **gangly** or have **sticky outy** bits) have a **larger surface area** relative to their volume — this **increases heat loss** from their surface.
3) Whether an animal is compact or not depends on the **temperature** of its **environment**. Here's an example:

Arctic fox
Body temperature 37 °C
Average outside temperature 0 °C

The Arctic fox has **small ears** and a **round head** to **reduce** its SA : V ratio and heat loss.

African bat-eared fox
Body temperature 37 °C
Average outside temperature 25 °C

The African bat-eared fox has **large ears** and a more **pointed nose** to **increase** its SA : V ratio and heat loss.

European fox
Body temperature 37 °C
Average outside temperature 12 °C

The European fox is **intermediate** between the two, matching the temperature of its environment.

Organisms have Behavioural and Physiological Adaptations to Aid Exchange

Not all organisms have a body size or shape to suit their climate — some have **other adaptations** instead...

1) Animals with a high SA : volume ratio tend to **lose more water** as it evaporates from their surface. Some **small desert mammals** have **kidney structure adaptations** so that they produce **less urine** to compensate.
2) **Smaller animals** living in **colder regions** often have a much **higher metabolic rate** to compensate for their high SA : volume ratio — this helps to keep them warm by creating **more heat**. To do this they need to eat large amounts of **high energy foods** such as seeds and nuts.
3) Smaller mammals may have thick layers of **fur** or **hibernate** when the weather gets really cold.
4) **Larger organisms** living in **hot regions**, such as elephants and hippos, find it hard to keep cool as their heat loss is relatively slow. **Elephants** have developed **large flat ears** which **increase** their **surface area**, allowing them to lose more heat. **Hippos** spend much of the day in the **water** — a **behavioural adaptation** to help them lose heat.

Practice Questions

Q1 Give four things that organisms need to exchange with their environment.

Q2 Describe how body shape affects heat exchange.

Exam Question

Q1 Explain why diffusion is not an efficient transport system for large mammals. [3 marks]

Cube animals indeed — it's all gone a bit Picasso...

You need to understand why single-celled organisms and large multicellular organisms use different methods for exchange. Most multicellular organisms couldn't survive using diffusion alone — that's why they have exchange organs.

Gas Exchange

Lots of organisms have developed adaptations to improve their rate of gas exchange. It's a tricky business if you're an insect or a plant though — you've got to exchange enough gas but avoid losing all your water and drying to a crisp...

Gas Exchange Surfaces have Two Major Adaptations

Most gas exchange surfaces have two things in common:

1) They have a **large surface area**.
2) They're **thin** (often just one layer of epithelial cells) — this provides a **short diffusion pathway** across the gas exchange surface.

The organism also maintains a **steep concentration gradient** of gases across the exchange surface.

All these features **increase** the **rate of diffusion**.

Single-celled Organisms Exchange Gases across their Body Surface

1) Single-celled organisms absorb and release gases by **diffusion** through their **outer surface**.
2) They have a relatively **large surface area**, a **thin surface** and a **short diffusion pathway** (oxygen can take part in **biochemical reactions** as soon as it **diffuses** into the cell) — so there's **no need** for a gas exchange system.

Fish Use a Counter-Current System for Gas Exchange

There's a **lower concentration** of oxygen in water than in air. So **fish** have special **adaptations** to get enough of it.

1) Water, containing oxygen, enters the fish through its **mouth** and passes out through the gills.
2) Each gill is made of lots of **thin plates** called **gill filaments**, which give a **big surface area** for **exchange** of **gases**.
3) The gill filaments are covered in lots of tiny structures called **lamellae**, which **increase** the **surface area** even more.
4) The lamellae have lots of **blood capillaries** and a thin surface layer of cells to speed up diffusion.

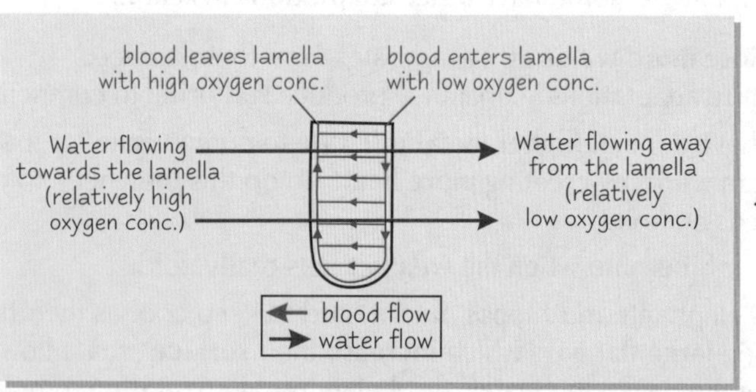

5) **Blood** flows through the lamellae in one direction and **water** flows over in the opposite direction. This is called a **counter-current system**. It maintains a **large concentration gradient** between the water and the blood — so as much oxygen as possible diffuses from the water into the blood.

Insects use Tracheae to Exchange Gases

1) Insects have microscopic air-filled pipes called **tracheae** which they use for gas exchange.
2) Air moves into the tracheae through pores on the surface called **spiracles**.
3) **Oxygen** travels down the **concentration gradient** towards the **cells**. **Carbon dioxide** from the cells moves down its own concentration gradient towards the **spiracles** to be **released** into the atmosphere.
4) The tracheae branch off into smaller **tracheoles** which have **thin, permeable walls** and go to individual cells. This means that oxygen diffuses directly into the respiring cells — the insect's circulatory system doesn't transport O_2.
5) Insects use **rhythmic abdominal movements** to move air in and out of the spiracles.

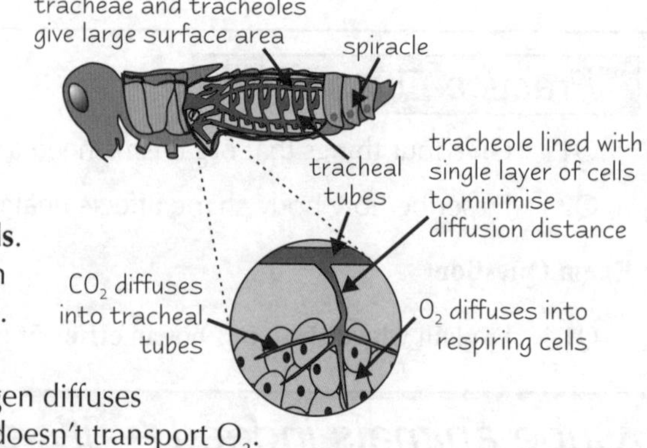

Gas Exchange

Dicotyledonous Plants Exchange Gases at the Surface of the Mesophyll Cells

1) Plants need CO_2 for **photosynthesis**, which produces O_2 as a waste gas. They need O_2 for **respiration**, which produces CO_2 as a waste gas.
2) The main gas exchange surface is the **surface of the mesophyll cells** in the leaf. They're well adapted for their function — they have a **large surface area.**
3) The mesophyll cells are inside the leaf. Gases move in and out through special pores in the **epidermis** called **stomata** (singular = stoma).
4) The stomata can **open** to allow exchange of gases, and **close** if the plant is losing too much water. **Guard cells** control the opening and closing of stomata.

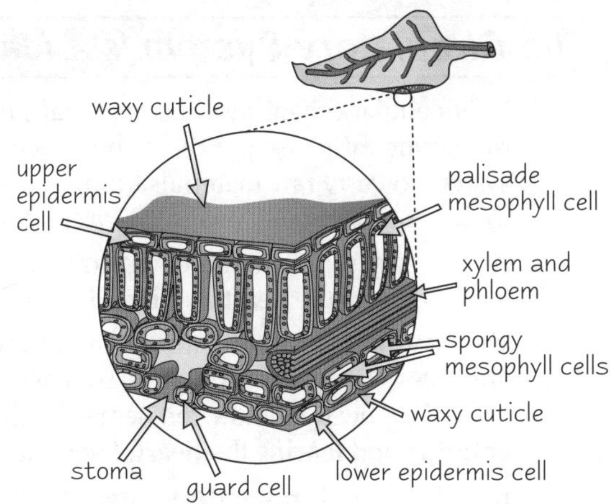

Insects and Plants can Control Water Loss

Exchanging gases tends to make you **lose water** — there's a sort of **trade-off** between the two. Luckily for plants and insects though, they've evolved **adaptations** to **minimise water loss** without reducing gas exchange too much.

1) If **insects** are losing too much water, they **close** their **spiracles** using muscles. They also have a **waterproof, waxy cuticle** all over their body and **tiny hairs** around their spiracles, both of which **reduce evaporation**.
2) Plants' stomata are usually kept **open** during the day to allow **gaseous exchange**. Water enters the guard cells, making them **turgid**, which **opens** the stomatal pore. If the plant starts to get **dehydrated**, the guard cells lose water and become **flaccid**, which **closes** the pore.
3) Some plants are specially adapted for life in **warm**, **dry** or **windy** habitats, where **water loss** is a problem. These plants are called **xerophytes**.

See p. 75 for more on water loss in plants.

Examples of xerophytic adaptations include:

- Stomata sunk in **pits** which trap moist air, reducing evaporation.
- **Curled** leaves with the stomata inside, protecting them from wind.
- A layer of '**hairs**' on the epidermis to trap moist air round the stomata, reducing the concentration gradient of water.
- A **reduced number of stomata**, so there are fewer places for water to escape.
- **Waxy, waterproof cuticles** on leaves and stems to reduce evaporation.

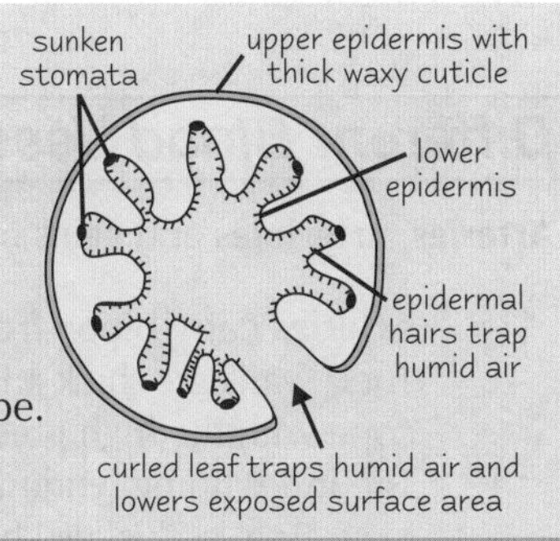

Practice Questions

Q1 How are single-celled organisms adapted for efficient gas exchange?

Q2 What is the advantage to fish of having a counter-current system in their gills?

Q3 What are an insect's spiracles?

Q4 Through which pores are gases exchanged in plants?

Exam Questions

Q1 Give three ways gas exchange organs are adapted to their function. Give a different example for each one. [6 marks]

Q2 Explain why plants that live in the desert often have sunken stomata or stomata surrounded by hairs. [2 marks]

Keep revising and you'll be on the right trachea...

There's a pretty strong theme on these pages — whatever organism it is, to exchange gases efficiently it needs exchange organs with a large surface area, a thin exchange surface and a high concentration gradient. Don't forget that (or I'll hit you with a big stick).

The Circulatory System

As the name suggests, the circulatory system is responsible for circulating stuff around the body— blood, to be specific. Most multicellular organisms (mammals, insects, fish, even French people) have a circulatory system of some type.

The **Circulatory System** is a **Mass Transport System**

1) Multicellular organisms, like **mammals**, have a **low surface area to volume ratio** (see p. 68), so they need a specialised **transport system** to carry raw materials from specialised **exchange organs** to their body cells — this is the **circulatory system**.
2) As you already know, the circulatory system is made up of the **heart** and **blood vessels**.
3) The heart **pumps blood** through blood vessels (arteries, arterioles, veins and capillaries) to reach different parts of the body. You need to **know** the names of **all** the blood vessels **entering** and **leaving** the **heart**, **liver** and **kidneys**.
4) Blood transports **respiratory gases**, products of **digestion**, **metabolic wastes** and **hormones** round the body.

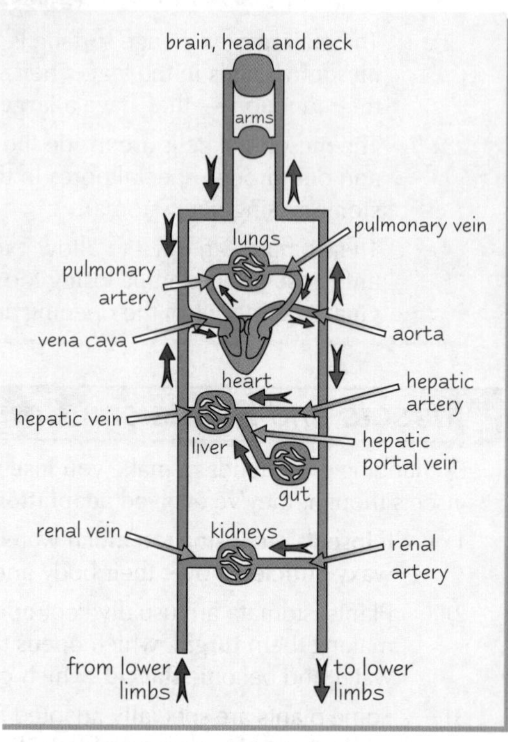

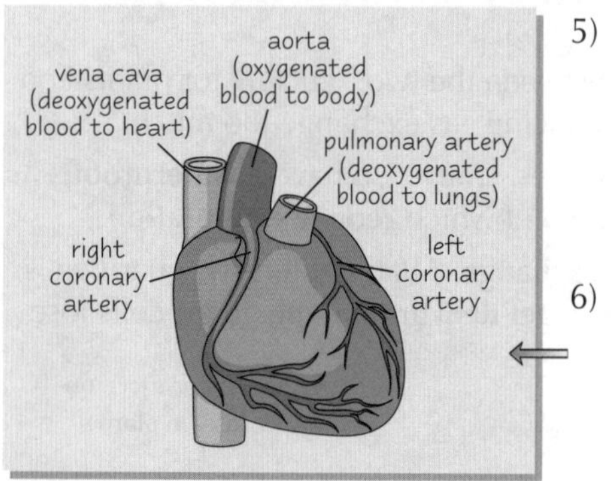

5) There are **two circuits**. One circuit takes blood from the **heart** to the **lungs**, then **back to the heart**. The other loop takes blood around the **rest of the body**.
6) The heart has its own blood supply — the left and right **coronary arteries**.

Different Blood Vessels are Adapted for **Different Functions**

Arteries, **arterioles** and **veins** have different **characteristics**, and you need to know **why**...

1) **Arteries** carry blood **from** the heart **to** the rest of the body. Their walls are thick and **muscular** and have elastic tissue to cope with the **high pressure** produced by the heartbeat. The inner lining (**endothelium**) is **folded**, allowing the artery to **stretch** — this also helps it to cope with high pressure. All arteries carry **oxygenated** blood except for the **pulmonary arteries**, which take deoxygenated blood to the lungs.

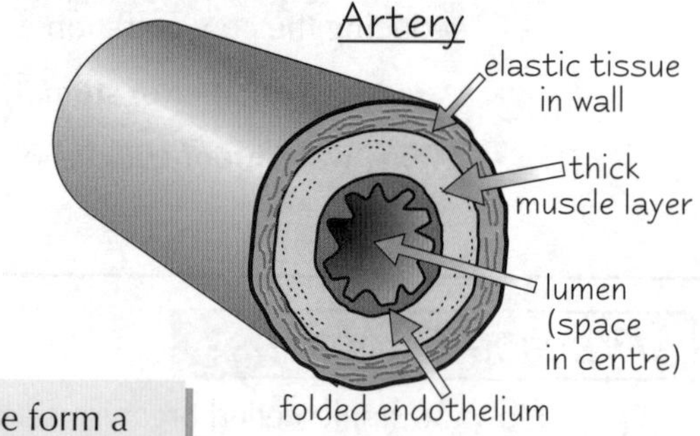

2) Arteries divide into smaller vessels called **arterioles**. These form a network throughout the body. Blood is directed to different **areas of demand** in the body by **muscles** inside the arterioles, which contract to restrict the blood flow or relax to allow full blood flow.

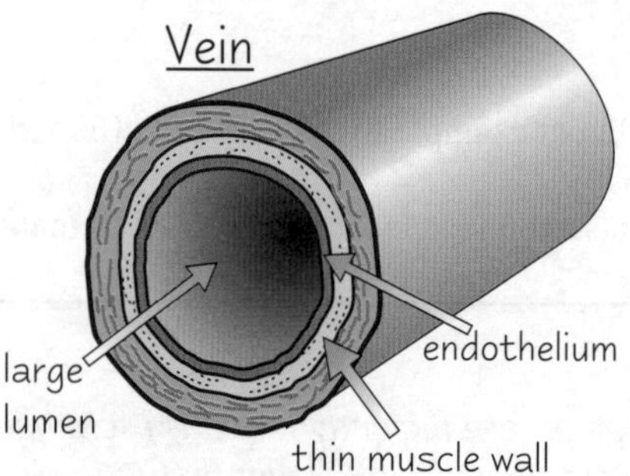

3) **Veins** take blood back **to the heart** under **low pressure**. They have a **wider** lumen than equivalent arteries, with very little elastic or muscle tissue. Veins contain **valves** to stop the blood flowing backwards. Blood flow through the veins is helped by contraction of the **body muscles** surrounding them. All veins carry **deoxygenated** blood (because oxygen has been used up by body cells), except for the **pulmonary veins**, which carry oxygenated blood to the heart from the lungs.

The Circulatory System

Substances are Exchanged between Blood and Body Tissues at Capillaries

Arterioles branch into **capillaries**, which are the **smallest** of the blood vessels. Substances (e.g. glucose and oxygen) are **exchanged** between cells and capillaries, so they're adapted for **efficient diffusion**.

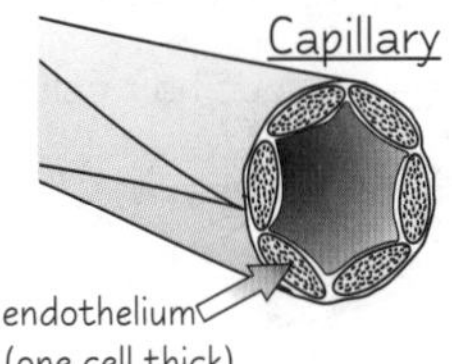

1) They're always found very **near cells in exchange tissues** (e.g. alveoli in the lungs), so there's a **short diffusion pathway**.
2) Their walls are only **one cell thick**, which also shortens the diffusion pathway.
3) There are a large number of capillaries, to **increase surface area** for exchange. Networks of capillaries in tissue are called **capillary beds**.

Tissue Fluid is Formed from Blood

Tissue fluid is the fluid that **surrounds cells** in tissues. It's made from substances that leave the blood, e.g. oxygen, water and nutrients. Cells take in oxygen and nutrients from the tissue fluid, and release metabolic waste into it. Substances move out of blood capillaries, into the tissue fluid, by **pressure filtration**:

1) At the **start** of the capillary bed, nearest the arteries, the pressure inside the capillaries is **greater** than the pressure in the tissue fluid. This difference in pressure **forces fluid out** of the **capillaries** and into the **spaces** around the cells, forming tissue fluid.
2) As fluid leaves, the pressure reduces in the capillaries — so the pressure is much **lower** at the **end** of the capillary bed that's nearest to the veins.
3) Due to the fluid loss, the **water potential** at the end of the capillaries nearest the veins is **lower** than the water potential in the **tissue fluid** — so some **water re-enters** the capillaries from the tissue fluid at the vein end by **osmosis** (see p. 28 for more on osmosis).

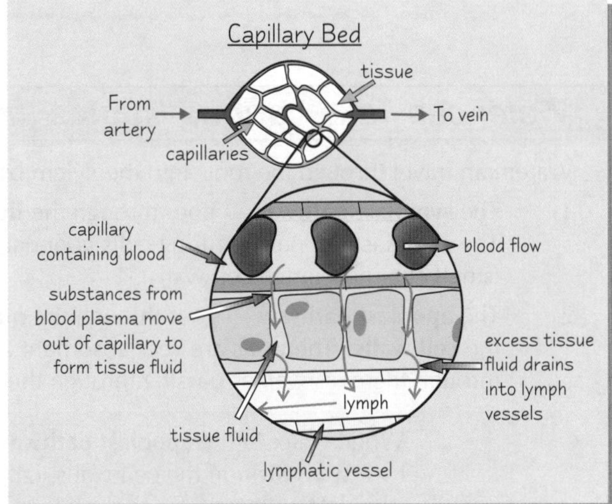

Unlike blood, tissue fluid **doesn't** contain **red blood cells** or **big proteins**, because they're **too large** to be pushed out through the capillary walls. Any **excess** tissue fluid is drained into the **lymphatic system** (a network of tubes that acts a bit like a drain), which transports this excess fluid from the tissues and dumps it back into the circulatory system.

Practice Questions

Q1 Name all the blood vessels entering and leaving the heart.

Q2 Name all the blood vessels entering and leaving the kidney.

Q3 List four types of blood vessel.

Q4 Explain why capillaries are important in metabolic exchange.

Exam Questions

Q1 Describe the structure of an artery and explain how it relates to its function. [6 marks]

Q2 Explain how tissue fluid is formed and how it is returned to the circulation. [4 marks]

If blood can handle transport this efficiently, the trains have no excuse...

Four hours I was waiting at the train station this weekend. Four hours! Anyway, you may have noticed that biologists are obsessed with the relationship between structure and function, so whenever you're learning the structure of something, make sure you know how this relates to its function. Like veins, arteries and capillaries on these pages, for example.

Water Transport in Plants

Water enters a plant through its roots and eventually, if it's not used, exits via the leaves. "Ah-ha," I hear you cry, "but how does it flow upwards, against gravity?" Well that, my friend, is a mystery that's about to be explained.

Water Enters a Plant through its Root Hair Cells

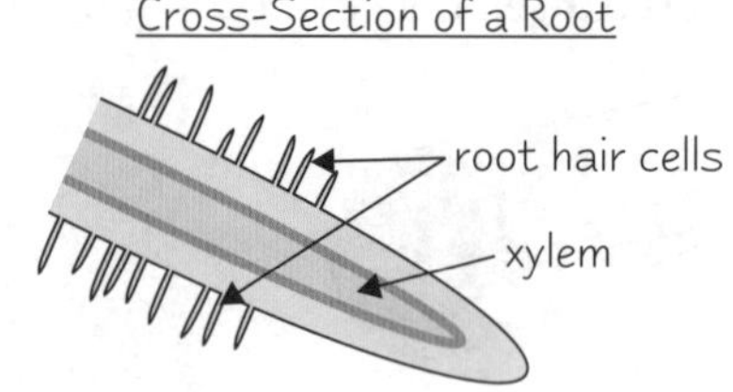

1) Water has to get from the **soil**, through the **root** and into the **xylem** — the system of vessels that **transports** water throughout the plant.
2) The bit of the root that absorbs water is covered in **root hairs**. These **increase** the root's **surface area**, speeding up water uptake.
3) Once it's absorbed, the water has to get through the **cortex**, including the **endodermis**, before it can reach the xylem.

> Water always moves from areas of **higher water potential** to areas of **lower water potential** — it goes down a **water potential gradient**. The **soil** around roots generally has a **high water potential** (i.e. there's lots of water there) and **leaves** have a **lower water potential** (because water constantly **evaporates** from them). This creates a water potential gradient that keeps water moving through the plant in the right direction, **from roots to leaves**.

Water can Take Various Routes through the Root

The prison had been strangely quiet ever since plasmodesmata were installed.

Water can travel through the roots into the xylem by two different paths:

1) The **symplast pathway** — goes through the **living** parts of cells — the **cytoplasm**. The cytoplasm of neighbouring cells connect through **plasmodesmata** (small channels in the cell walls).
2) The **apoplast pathway** — goes through the **non-living** parts of the root — the **cell walls**. The walls are very absorbent and water can simply **diffuse** through them, as well as passing through the spaces between them.

> - When water in the **apoplast pathway** gets to the **endodermis** cells though, its path is blocked by a **waxy strip** in the cell walls, called the **Casparian strip**. Now the water has to take the **symplast pathway**.
> - This is useful, because it means the water has to go through a **cell membrane**. Cell membranes are able to control whether or not substances in the water get through (see p. 26).
> - Once past this barrier, the water moves into the **xylem**.

3) Both pathways are used, but the main one is the **apoplast pathway** because it provides the **least resistance**.

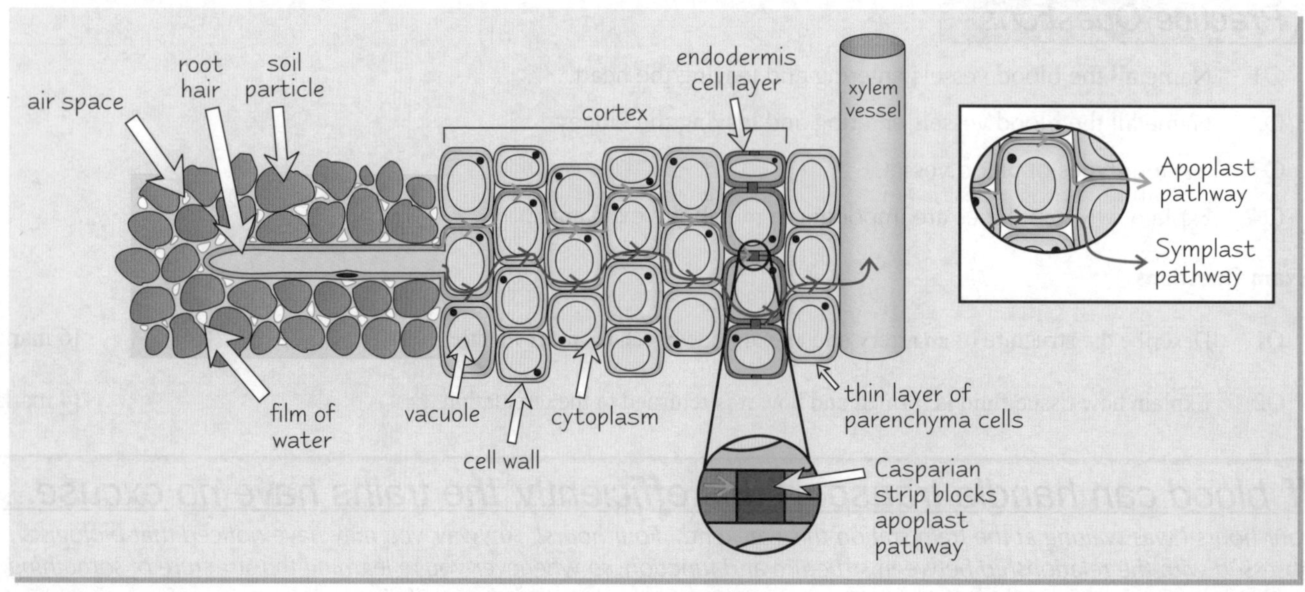

Water Transport in Plants

Water Moves **Up** a Plant **Against** the Force of **Gravity**

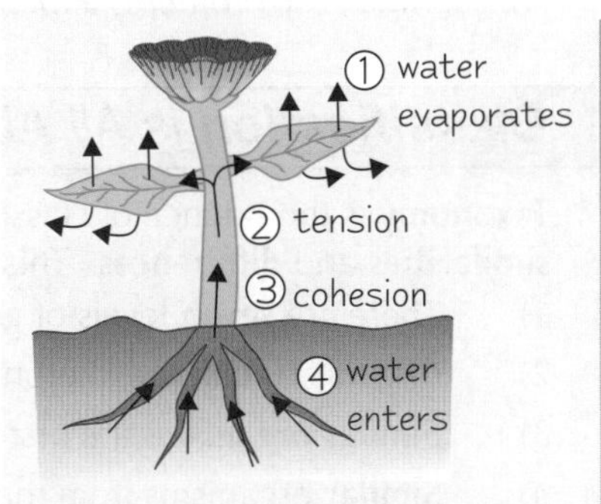

Water can move up a plant in two ways:

① **Cohesion** and **tension** help water move up plants, from roots to leaves, against the force of gravity.

1) Water **evaporates** from the **leaves** at the 'top' of the xylem. →
2) This creates **tension** (**suction**), which pulls more water into the leaf.
3) Water molecules are **cohesive** (they **stick together**) so when some are pulled into the leaf others follow. This means the whole **column** of water in the **xylem**, from the leaves down to the roots, **moves upwards**.
4) **Water** enters the stem through the **roots**.

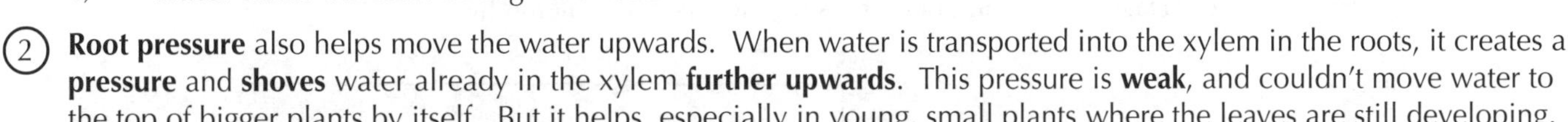

② **Root pressure** also helps move the water upwards. When water is transported into the xylem in the roots, it creates a **pressure** and **shoves** water already in the xylem **further upwards**. This pressure is **weak**, and couldn't move water to the top of bigger plants by itself. But it helps, especially in young, small plants where the leaves are still developing.

Transpiration is **Loss of Water** from a Plant's Surface

Transpiration is the **evaporation** of **water** from a plant's surface, especially the **leaves**.

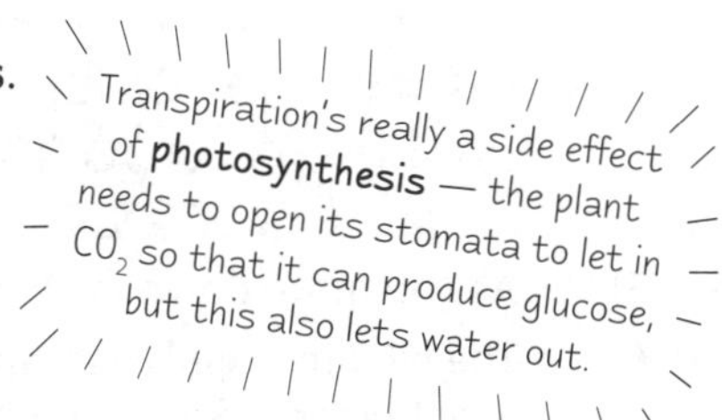

Transpiration's really a side effect of **photosynthesis** — the plant needs to open its stomata to let in CO_2 so that it can produce glucose, but this also lets water out.

1) Water **evaporates** from the moist cell walls and accumulates in the spaces between cells in the leaf.
2) When the **stomata** open, it moves out of the leaf down the **concentration gradient** (there's more water inside the leaf than in the air outside).

Four Main Factors Affect **Transpiration Rate**

1) **Light** — the **lighter** it is the **faster** the **transpiration rate**. This is because the **stomata open** when it gets **light**. When it's **dark** the stomata are usually **closed**, so there's little transpiration.
2) **Temperature** — the **higher the temperature** the **faster** the **transpiration rate**. Warmer water molecules have more energy so they **evaporate** from the cells inside the leaf **faster**. This **increases** the **concentration gradient** between the inside and outside of the leaf, making water **diffuse out** of the leaf **faster**.
3) **Humidity** — the **lower** the **humidity**, the **faster** the **transpiration rate**. If the air around the plant is **dry**, the **concentration gradient** between the leaf and the air is **increased**, which increases transpiration.
4) **Wind** — the **windier** it is, the **faster** the **transpiration rate**. Lots of air movement **blows away** water molecules from around the stomata. This **increases** the **concentration gradient**, which increases the rate of transpiration.

Practice Questions

Q1 How does water enter a plant?

Q2 Explain how root pressure helps to move water through plants.

Q3 Give four factors that affect transpiration rate.

Exam Questions

Q1 Describe the two routes water can take through the roots of a plant. [4 marks]

Q2 Explain why movement of water in the xylem stops if the leaves of a plant are removed. [4 marks]

So many routes through the roots...

Lots of impressive biological words on this page to amaze your friends and confound your enemies. Go through all this stuff again, and whenever you come across a ridiculous word like 'plasmodesmata', just stop and check you know exactly what it means. (Personally, I think they should've just called them 'cell wall gaps', but nobody ever listens to me.)

Principles of Classification

For hundreds of years people have been putting organisms into groups to make it easier to recognise and name them. For example, my brother is a member of the species Idioto bigearian (Latin for idiots with big ears).

Classification is All About Grouping Together Similar Organisms

Taxonomy is the science of classification. It involves **naming** organisms and **organising them** into **groups** based on their **similarities** and **differences**. This makes it **easier** for scientists to **identify** and **study** them.

1) There are **seven** levels of groups (called taxonomic groups) used to classify organisms.
2) Organisms can only belong to **one group** at **each level** in the taxonomic hierarchy — there's **no overlap**.
3) **Similar organisms** are first sorted into **large groups** called **kingdoms**, e.g. all animals are in the animal kingdom.
4) **Similar** organisms from that kingdom are then grouped into a **phylum**. **Similar** organisms from each phylum are then grouped into a **class**, and **so on** down the seven levels of the hierarchy.

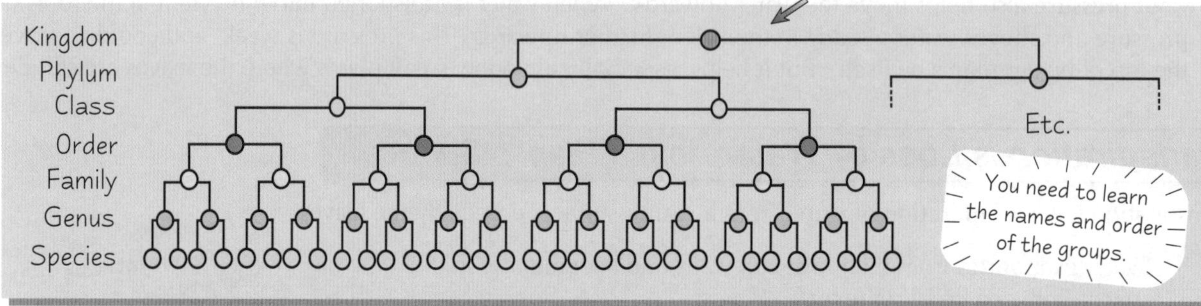

5) As you move **down** the hierarchy, there are **more groups** at each level but **fewer organisms** in each group.
6) The hierarchy **ends** with **species** — the groups that contain only **one type** of organism (e.g. humans, dogs, *E. coli*). You need to **learn** the definition of a **species**:

A species is a group of similar organisms able to reproduce to give fertile offspring.

Species are given a **scientific name** to **distinguish** them from similar organisms. This is a **two-word** name in **Latin**. The **first** word is the **genus** name and the **second** word is the **species** name — e.g. humans are *Homo sapiens*. Giving organisms a scientific name enables scientists to **communicate** about organisms in a standard way that minimises confusion. E.g. Americans call a type of bird **cockatoos** and Australians call them **flaming galahs** (best said with an Australian accent), but it's the **same bird**. If the correct **scientific name** is used — *Eolophus roseicapillus* — there's no confusion.

7) Scientists constantly **update** classification systems because of **discoveries** about new species and new **evidence** about known organisms (e.g. **DNA sequence** data — see p. 78).

Phylogenetics Tells Us About an Organism's Evolutionary History

1) **Phylogenetics** is the study of the **evolutionary history** of groups of **organisms**.
2) All organisms have **evolved** from shared common ancestors (**relatives**). E.g. members of the Hominidae family (great apes and humans) evolved from a common ancestor. First orangutans **diverged** (evolved to become a **different species**) from this common ancestor. Next gorillas diverged, then humans, closely followed by bonobos and chimpanzees.
3) Phylogenetics tells us **who's related** to whom and how **closely related** they are.
4) Closely related species **diverged** away from each other **most recently**. E.g. the phylogenetic tree opposite shows the **Hominidae tree**. Humans and **chimpanzees** are **closely** related, as they diverged very **recently**. You can see this because their branches are **close** together. Humans and orangutans are more **distantly** related, as they diverged longer ago, so their branches are **further** apart.

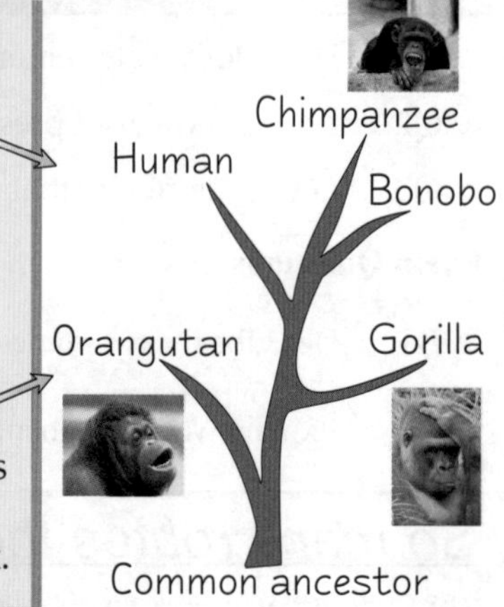

Principles of Classification

Defining Organisms as *Distinct Species* Can be Quite *Tricky*

1) Scientists can have problems when using the **definition** of a species on the previous page to decide which species an organism belongs to or if it's a new, **distinct species**.
2) This is because you can't always see their **reproductive behaviour** (you can't always tell if different organisms can reproduce to give **fertile offspring**).
3) Here are some of the **reasons** why you can't always see their reproductive behaviour:
    1) They're **extinct**, so obviously you **can't** study their reproductive behaviour.
    2) They **reproduce asexually** — they never **reproduce together** even if they belong to the same species, e.g. bacteria.
    3) There are **practical** and **ethical issues** involved — you can't see if some organisms reproduce successfully in the wild (due to geography) and you can't study them in a lab (because it's unethical), e.g. humans and chimps are classed as separate species but has anyone ever tried mating them?

Evidence has been found of human/parrot reproduction.

4) Because of these problems some organisms are **classified** as one species or another using other **techniques**.
5) Scientists can now compare the **DNA** (see p. 78) of organisms to see **how related** they are, e.g. the **more** DNA they have in common the more **closely related** they are. But there's no strict cut-off to say **how much** shared DNA can be used to define a **species**. For example, **only** about 6% of human DNA **differs** from chimpanzee DNA but we are separate species.

Practice Questions

Q1 What is taxonomy?

Q2 Why is taxonomy and the classification system important?

Q3 What name is given to the groups in the taxonomic hierarchy that contain the largest number of species?

Q4 Why is the scientific naming system important?

Q5 What is phylogenetics?

Deer
Sheep
Buffalo
Rhesus Monkey
Llama
Human
Camel
Pig
Green Monkey
Rabbit
Dog
Cat

Exam Questions

Q1 The phylogenetic tree on the right shows the evolutionary history of some mammalian species.

a) Which species shown on the tree is most closely related to humans? Explain how you know this. [2 marks]

b) Which animal is a close relative of both camels and deer? [1 mark]

Q2 Define a species. [2 marks]

Q3 Complete the table for the classification of humans. [5 marks]

	Phylum			Family		
Animalia	Chordata	Mammalia	Primates	Hominidae	Homo	sapiens

Q4 Give three reasons why it can be hard for scientists to define organisms as members of a distinct species. [3 marks]

Phylum — I thought that was the snot you get with a cold...

If you don't understand what a species is, you'll find yourself struggling on the next few pages. So, all together now... a species is a group of similar organisms able to reproduce to give fertile offspring... Now, don't forget that.

Classifying Species

Early classification systems only used observable features to place organisms into groups, e.g. six heads, four toes, a love of Take That. But now a variety of evidence is used to classify organisms...

Species Can be **Classified** by **Their DNA** or **Proteins**

1) Species can be **classified** into different groups in the **taxonomic hierarchy** (see p. 76) based on **similarities** and **differences** in their **genes**.
2) This can be done by comparing their **DNA sequence** or by looking at their **proteins** (which are coded for by their DNA).
3) Organisms that are **more closely** related will have **more similar** DNA and proteins than distantly related organisms.

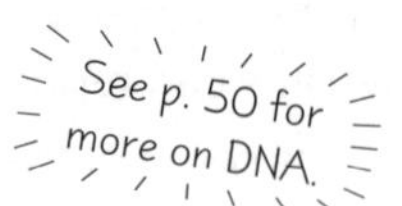

DNA Can be Compared **Directly** or by Using **Hybridisation**

DNA similarity can be measured by looking at the **sequence of bases** or by **DNA hybridisation**:

DNA sequencing

The **DNA** of organisms can be directly compared by looking at the **order of the bases** (As, Ts, Gs and Cs) in each. Closely related species will have a **higher percentage** of similarity in their DNA **base order**, e.g. humans and chimps share around 94%, humans and mice share about 85%.
DNA sequence comparison has led to **new classification systems** for **plants**, e.g. the classification system for flowering plants is based almost entirely on **similarities** between DNA sequences.

DNA Hybridisation

DNA Hybridisation is used to see how similar DNA is **without** sequencing it. Here's how it's done:

1) DNA from **two** different species is collected, separated into **single strands** and **mixed** together.
2) Where the **base sequences** of the DNA are the same on both strands, **hydrogen bonds** form between the base pairs by **specific base pairing**.
The more DNA bases that **hybridise** (bond) together, the more **alike** the DNA is.
3) The DNA is then **heated** to separate the strands again. **Similar DNA** will have **more hydrogen bonds** holding the two strands together so a **higher temperature** (i.e. **more energy**) will be needed to separate the strands.

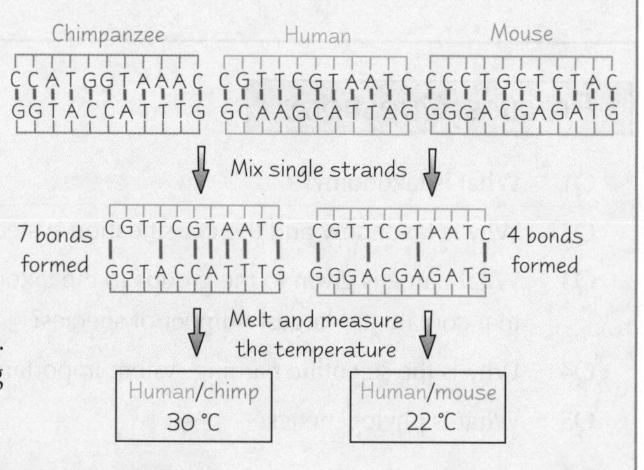

Proteins Can be Compared **Directly** or by Using **Immunology**

Similar organisms will have **similar proteins** in their cells. Proteins can be compared in **two** ways:

1) **Comparing amino acid sequence**
Proteins are made of **amino acids**. The **sequence** of amino acids in a protein is coded for by the **base sequence** in DNA (see p. 52). **Related organisms** have similar DNA sequences and so **similar amino acid sequences** in their proteins.

2) **Immunological comparisons** — Similar proteins will bind the same **antibodies** (see p. 6). E.g. if antibodies to a **human version** of a protein are added to isolated samples from some other **species**, any protein that's like the human version will also be **recognised** (bound) by that antibody.

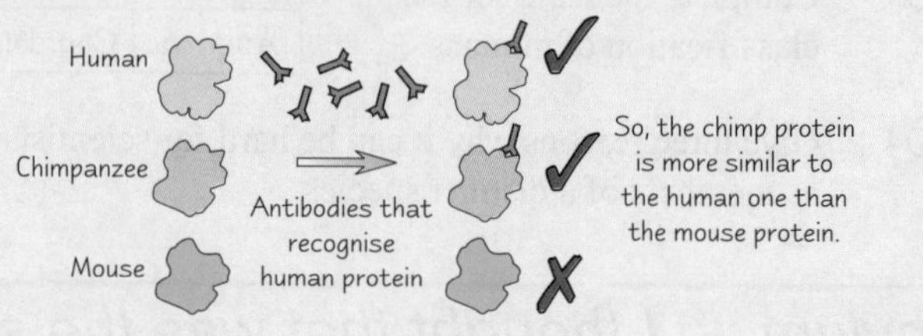

Classifying Species

You Need to be Able to *Interpret Data* on DNA and Protein *Similarities*

Here are two examples of the kind of thing you might get:

	Species A	Species B	Species C	Species D
Species A	100%	86%	42%	44%
Species B	86%	100%	51%	53%
Species C	42%	51%	100%	91%
Species D	44%	53%	91%	100%

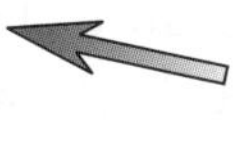

The table on the left shows the **% similarity of DNA** using DNA sequence analysis between several species of bacteria.

The data shows that species **A** and **B** are **more closely related** to each other than they are to either C or D. Species **C** and **D** are also **more closely related** to each other than they are to either A or B.

The diagram on the right shows the **amino acid sequences** of a certain protein from three different species.

You can see that the amino acid sequences from species **A** and **B** are **very similar**. The sequence from species **C** is **very different** to any of the other sequences. This would suggest that species **A** and **B** are **more closely related**.

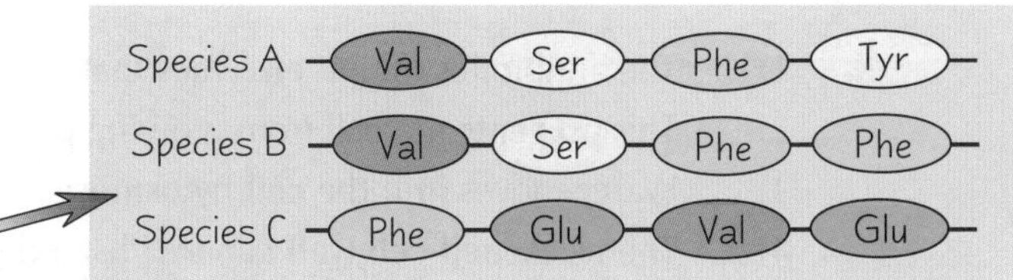

Courtship Behaviour can Also be Used to *Classify Species*

Geoff's jive never failed to attract a mate.

1) **Courtship behaviour** is carried out by organisms to **attract** a mate of the **right species**.
2) It can be fairly simple, e.g. **releasing chemicals**, or quite complex, e.g. a series of **displays**.
3) Courtship behaviour is **species specific** — only members of the same species will do and respond to that courtship behaviour. This prevents **interbreeding** and so makes reproduction **more successful** (as mating with the wrong species won't produce **fertile** offspring).
4) Because of this specificity, courtship behaviour can be used to **classify organisms**.
5) The more **closely related** species are, the **more similar** their courtship behaviour.

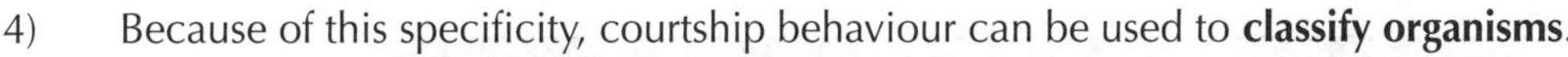

Some examples of courtship behaviour include:

1) **Fireflies** give off **pulses of light**. The pattern of flashes is specific to each species.
2) **Crickets** make **sounds** that are similar to Morse code, the code being different for different species.
3) **Male peacocks** show off their **colourful tails**. This tail pattern is only found in peacocks.
4) **Male butterflies** use **chemicals** to attract females. Only those of the correct species respond.

Practice Questions

Q1 Suggest two ways that DNA from two different species could be compared.

Q2 Suggest two ways that proteins from two different species could be compared.

Exam Questions

Q1 Explain how DNA hybridisation is used to analyse similarities between the DNA of two species. [5 marks]

Q2 The amino acid sequence of a specific protein was used to make comparisons between four species of animal. The results are shown on the right.

Species	Amino acid 1	Amino acid 2	Amino acid 3	Amino acid 4
Rabbit	His	Ala	Asp	Lys
Mouse	Thr	Ala	Asp	Val
Chicken	Ala	Thr	Arg	Arg
Rat	Thr	Ala	Asp	Phy

a) Which two species are the most closely related? [1 mark]

b) Which species is the most distantly related to the other three? [1 mark]

School discos — the perfect place to observe courtship behaviour...

It's important that you understand that the more similar the DNA and proteins, the more closely related (and hence the more recently diverged) two species are. This is because relatives have similar DNA, which codes for similar proteins, made of a similar sequence of amino acids. Just like you and your family — you're all alike because your DNA's similar.

Antibiotic Action and Resistance

These pages are all about antibiotics and how they kill (or inhibit) bacteria. But don't feel sorry for the bacteria — they're getting their own back by evolving antibiotic resistance. Sneaky...

Antibiotics Are Used to Treat Bacterial Diseases

1) Antibiotics are **chemicals** that either **kill** or **inhibit** the **growth** of bacteria.
2) **Different types** of antibiotics kill or inhibit the growth of bacteria in **different ways**.
3) Some **prevent growing** bacterial cells from **forming** the bacterial **cell wall**, which usually gives the cell structure and support (see p. 32).
4) This can lead to **osmotic lysis**:

> 1) The antibiotics **inhibit enzymes** that are needed to make the **chemical bonds** in the cell wall.
> 2) This **prevents** the cell from growing properly and **weakens** the cell wall.
> 3) **Water** moves **into the cell** by **osmosis**.
> 4) The **weakened cell wall** can't withstand the increase in **pressure** and **bursts** (**lyses**).

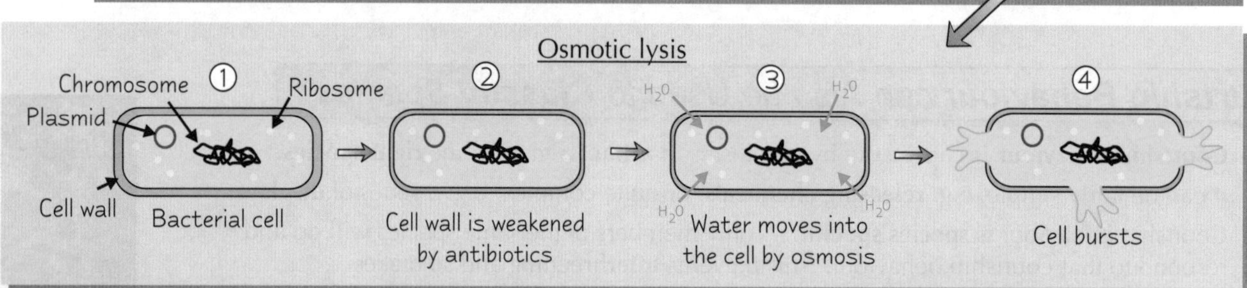

Mutations in Bacterial DNA can Cause Antibiotic Resistance

See p. 53 for more on DNA and mutations.

1) The **genetic material** in bacteria is the same as in most other organisms — **DNA**.
2) The DNA of an organism contains **genes** that carry the instructions for different **proteins**. These proteins determine the organism's **characteristics**.
3) **Mutations** are **changes** in the **base sequence** of an organism's DNA.
4) If a mutation occurs in the DNA of a gene it could change the protein and cause a **different characteristic**.
5) Some mutations in bacterial DNA mean that the bacteria are **not affected** by a particular antibiotic any more — they've developed **antibiotic resistance**.

Example

Methicillin is an antibiotic that inhibits an enzyme involved in **cell wall formation** (see above).
Some bacteria have developed resistance to methicillin, e.g. methicillin-resistant *Staphylococcus aureus* (**MRSA**).
Usually, **resistance** to methicillin occurs because the **gene** for the **target enzyme** of methicillin has **mutated**.
The mutated gene produces an **altered enzyme** that methicillin no longer **recognises**, and so **can't inhibit**.

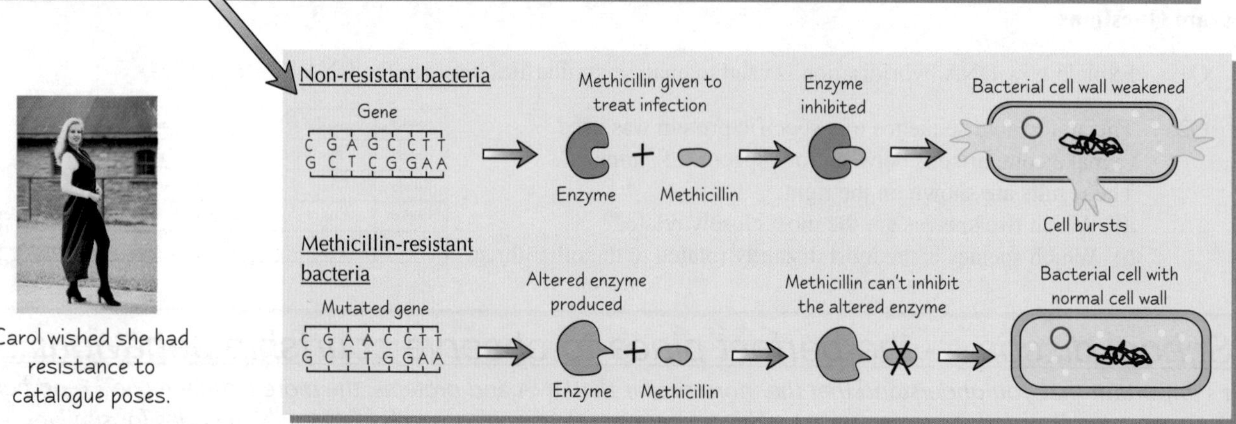

Carol wished she had resistance to catalogue poses.

Antibiotic Action and Resistance

Antibiotic Resistance can be Passed On Vertically...

Vertical gene transmission is where genes are passed on during **reproduction**.

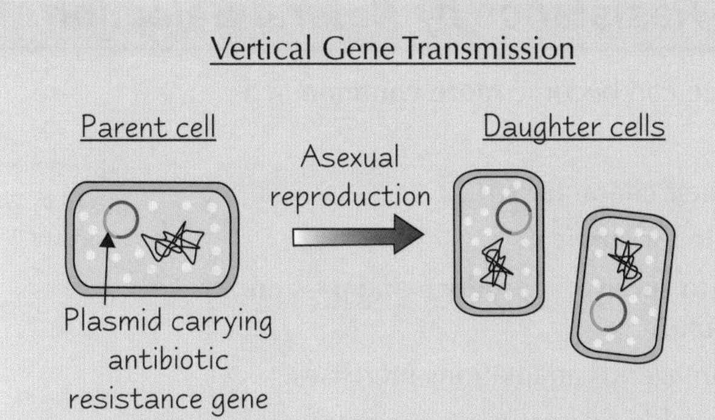

1) Bacteria reproduce **asexually**, so each daughter cell is an **exact copy** of the parent.
2) This means that **each** daughter cell has an exact copy of the parent cell's **genes**, including any that give it **antibiotic resistance.**
3) Genes for antibiotic resistance can be found in the bacterial **chromosome** or in **plasmids** (small **rings of DNA** found in bacterial cells, see p. 32).
4) The chromosome and any plasmids are passed on to the daughter cells during reproduction.

...or Horizontally

1) Genes for resistance can also be passed on **horizontally**.
2) Two bacteria **join together** in a process called **conjugation** and a **copy of a plasmid** is passed from one cell to the other.
3) Plasmids can be passed on to a member of the **same species** or a totally **different species**.

Horizontal Gene Transmission

Bacterial cell with antibiotic resistance gene in plasmid

Bacteria without plasmid

Bacterial conjugation — copy of plasmid transferred through stalk called a pilus

Both bacterial cells have a copy of the plasmid containing the antibiotic resistance gene

Practice Questions

Q1 What is osmotic lysis?

Q2 What is the genetic material in bacteria?

Q3 What is the difference between vertical and horizontal gene transmission?

Exam Question

Q1 More and more bacteria are becoming resistant to antibiotics such as penicillin.

a) Describe how resistance to an antibiotic arises in bacteria. [3 marks]

b) Describe how resistance to antibiotics is spread between two bacteria. [3 marks]

c) Penicillin is a cell wall inhibitor antibiotic. Explain how penicillin kills bacteria. [4 marks]

Horizontal gene transmission — that's not what it was called in my day...

There are lots of nice, colourful pictures on this page, but they're not just here to make the place look pretty you know. They're here to help you learn the different processes you need to understand for your exam — osmotic lysis, vertical transmission and horizontal transmission — so get scribblin' and learnin'. Go on, go on, go on...

Antibiotic Resistance

Mutations arise by accident but if they're useful, e.g. give antibiotic resistance, then natural selection will make sure they're passed on and on and on and on and on and on and on (and on and on and on)...

Bacterial Populations Evolve Antibiotic Resistance by Natural Selection

An **adaptation** (a useful characteristic) like antibiotic resistance can become **more common** in a **population** because of **natural selection**:

1) Individuals within a population **show variation** in their **characteristics**.
2) **Predation**, **disease** and **competition** create a **struggle for survival**.
3) Individuals with **better adaptations** are **more likely** to **survive**, **reproduce** and **pass on** the **alleles** that cause the adaptations to their **offspring**.
4) Over time, the **number** of individuals with the advantageous adaptations **increases**.
5) Over generations this leads to **evolution** as the favourable adaptations become **more common** in the population.

Adaptations are caused by gene mutations.

Here's how populations of antibiotic-resistant bacteria evolve by natural selection:

1) Some individuals in a population have alleles that give them **resistance** to an **antibiotic**.
2) The population is **exposed** to that antibiotic, **killing** bacteria **without** the antibiotic resistance allele.
3) The **resistant bacteria survive** and **reproduce** without competition, passing on the allele that gives antibiotic resistance to their offspring.
4) After some time **most** organisms in the population will carry the antibiotic resistance allele.

Natural Selection also Occurs in Other Organisms

1) Natural selection happens in **all populations** — not just in bacterial populations.
2) There are loads of examples, but they all follow the same basic principle — the organism has a **characteristic** that makes it more likely to **survive**, **reproduce** and pass on the **alleles** for the better characteristic.
3) In your **exam** you might be asked to explain why certain characteristics are common (or have increased).
4) To do this you should **identify** why the **adaptations** (**characteristics**) are useful and **explain** how they've become more common due to **natural selection**.
5) Here are some examples of the kinds of characteristics that can help organisms to survive:

Adaptations that could increase chance of survival	How the adaptations could increase survival
Streamlined body, camouflage, larger paws for running quicker etc.	They help to escape from predators.
Streamlined body, camouflage, larger paws for running quicker, larger claws, longer neck etc.	They help to catch prey/get food.
Shorter/longer hairs, large ears, increased water storage capacity etc.	They make the animal more suited to the climate.

Dave wasn't convinced his camouflage was working.

Antibiotic Resistance

Antibiotic Resistance Makes it Difficult to Treat Some Diseases

Diseases caused by bacteria are treated using **antibiotics**. Because bacteria are becoming resistant to different antibiotics through **natural selection** it's becoming more and more **difficult** to treat some bacterial infections, such as tuberculosis (**TB**) and methicillin-resistant *Staphylococcus aureus* (**MRSA**).

Tuberculosis

1) **TB** is a **lung disease** caused by bacteria.
2) TB was once a **major killer** in the UK, but the number of people dying from TB **decreased** with the development of **specific antibiotics** that killed the bacterium. Also the number of people catching TB **dropped** due to a vaccine (see p. 8)
3) More recently, some populations of TB bacteria have **evolved** resistance to the **most effective** antibiotics. **Natural selection** has led to populations that are resistant to a **range** of different antibiotics — the populations (strains) are **multidrug-resistant**.
4) To try to combat the **emergence** of resistance, TB treatment now involves taking a **combination** of different antibiotics for about **6 months**.
5) TB is becoming harder to treat as multidrug-resistant strains are **evolving quicker** than **drug companies** can develop new antibiotics.

There's more about TB on page 36.

MRSA

1) **Methicillin-resistant *Staphylococcus aureus*** (MRSA) is a strain of the *Staphylococcus aureus* bacterium that has evolved to be resistant to a number of commonly used antibiotics, including **methicillin**.
2) *Staphylococcus aureus* causes a **range** of illnesses from **minor skin infections** to **life-threatening diseases** such as **meningitis** and **septicaemia**.
3) The major problem with MRSA is that some strains are resistant to **nearly all** the antibiotics that are available.
4) Also, it can take a long time for **clinicians** to determine which antibiotics, if any, will **kill** the strain each individual is infected with. During this time the **patient** may become **very ill** and even **die**.
5) **Drug companies** are trying to **develop alternative ways** of treating MRSA to try to combat the emergence of resistance.

Lab tests are carried out to see if any antibiotics can kill a strain of MRSA.

Practice Questions

Q1 Briefly describe the process of natural selection.

Q2 Why is it becoming more difficult to treat TB infections?

Exam Questions

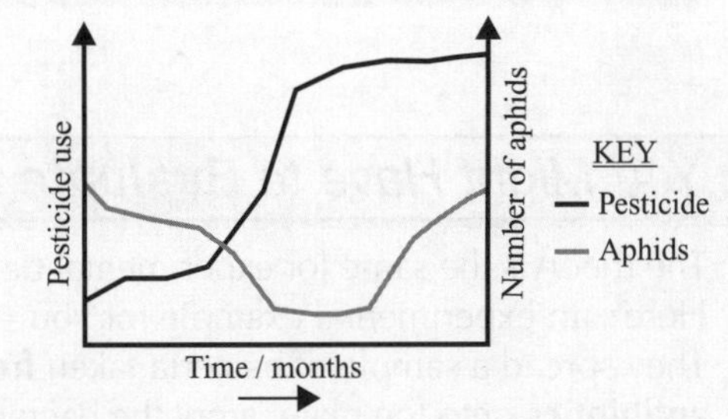

Q1 The graph shows the use of an anti-aphid pesticide on a farm and the number of aphids found on the farm over a period of time.

Describe and explain the change in aphid numbers shown in the graph. [6 marks]

Q2 The bat *Anoura fistulata* has a very long tongue (up to one and a half times the length of its body). The tongue enables the bat to feed on the nectar inside a deep tubular flower found in the forests of Ecuador.

Describe how natural selection can explain the evolution of such a long tongue. [3 marks]

Why do giraffes have long necks?...*

Adaptation and selection aren't that bad really... just remember that any characteristic that increases the chances of an organism getting more dinner, getting laid or avoiding being gobbled up by another creature will increase in the population (due to the process of natural selection). Now I know why mullets have disappeared... so unattractive...

*So they can reach food found high up. This means they're more likely to survive, reproduce and pass on their alleles (genes). So no, it's not because they've got smelly feet.

Evaluating Resistance Data

The number of infections caused by resistant bacteria is rising, so it's important to keep an eye on them and any new ones that pop up. Like they say, you need to know your enemies if you want to beat them (OK, so I don't know who says it, but someone does. I think I heard it on Catchphrase once...).

You Need to be Able to **Evaluate Data** About **Antibiotic Resistance**

It's very possible that you could get some data from a **study** into antibiotic resistance in the exam. You need to be able to **evaluate** the **methodology**, **data** and any **conclusions** drawn. Here's an example:

This study investigated the **number** of **death certificates mentioning** *Staphylococcus aureus* (***S. aureus***) and methicillin-resistant *S. aureus* (**MRSA**) in the UK between 1993 and 2002. The data was collected from **UK death certificates** issued between 1993 and 2002. The **results** are shown in the graph opposite.

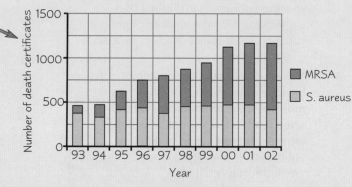

Graph to show the number of death certificates mentioning Staphylococcus aureus and MRSA between 1993 and 2002

Here are some of the things you might be asked to do:

1) **Describe the data** — This study shows that the number of death certificates mentioning **all forms** of *S. aureus* **increased** between 1993 and 2002. The number mentioning **MRSA increased** while the number mentioning ***S. aureus*** **stayed relatively level**.
2) **Check the evidence backs up any conclusions** — Dr Bottril said, 'This data shows that the number of deaths **caused** by MRSA is rising'. Does the data support this conclusion? No. The study only looked at death certificates **mentioning** MRSA, not deaths **caused** by MRSA.
3) **Other points to consider**
   - This study looked at the number of **death certificates** mentioning MRSA. Studying the number of **reported** MRSA **infections** each year may have been a **better way** of investigating the occurrence of bacterial resistance.
   - Some death certificates may not have mentioned MRSA because it wasn't the **cause of death**, but the people may have been **infected** at the time. This means the data **doesn't** reflect the number of infections.
   - Increased **awareness** of MRSA may have influenced the decision to include MRSA on the death certificate, **biasing** the data.

You Might Have to **Evaluate Experimental Data**

The theory's the same for experimental data — **evaluate** the **methodology**, **data** and **conclusions**. Here's an experimental example for you — A clinician needs to find out which antibiotics will treat a **patient's infection**. They spread a sample of bacteria taken **from the patient** onto an agar plate. Then they place paper discs **soaked** with **antibiotics** onto the plate, grow the bacteria and **measure the areas of growth inhibition** after a set period of time:

1) **Draw conclusions** — Be **precise** about what the data shows. A 250 mg dose of **streptomycin inhibited** growth the most. A 250 mg dose of **tetracycline inhibited** growth a small amount. The bacteria appear to be **resistant** to **methicillin** up to **250 mg**.
2) **Evaluate the methodology** — the experiment included a **negative control**, which is good. The negative control is a paper disc soaked in sterile water. The bacteria grew around this disc, which shows the paper disc **alone** doesn't kill the bacteria.

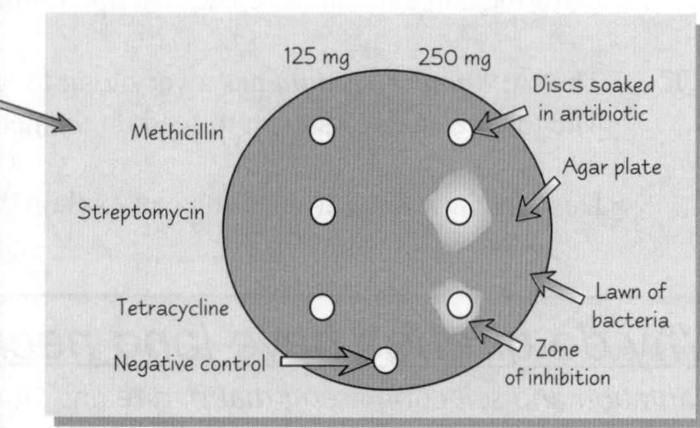

Evaluating Resistance Data

Decisions are *Made Using Scientific Knowledge*

Bacteria **will develop** antibiotic resistance by natural selection — it's nature. But scientific research has shown that certain things can be done to **slow down** the natural process. People working in the **public health sector**, together with patients, have to be made aware of recent **scientific findings** so that they can **act** upon them.
Here are two examples:

Scientific Knowledge: Using an **antiseptic gel** to wash hands can help to **reduce the spread** of infectious diseases by **person-to-person contact.**

Decision: Health workers should **reduce spread** by washing their hands with **antiseptic gel** (placed at all hand basins) before and after **visiting each patient** on a ward.

Scientific knowledge: Bacteria become resistant to antibiotics **more quickly** when antibiotics are **misused** and patients **don't finish the course.**

Decision: Doctors should only prescribe antibiotics when **absolutely necessary**. Patients have to be told the **importance** of finishing **all the antibiotics** even if they start to feel better.

*There are **Ethical Issues** Surrounding the Use of **Antibiotics***

People are very concerned about the spread of antibiotic-resistant bacteria. **Limiting the use** of antibiotics is one way of helping to slow down the emergence of resistance, but this raises some **ethical issues**.

1) Some people believe that antibiotics should only be used in **life-threatening situations** to reduce the increase of resistance. Others argue against this because people would take **more time off work** for illness, it could **reduce** people's **standard of living**, it could increase the **incidence of disease** and it could cause **unnecessary suffering**.
2) A few people believe doctors shouldn't prescribe antibiotics to those suffering **dementia**. They argue that they may forget to take them, increasing the chance of **resistance** developing. However, some people argue that **all patients** have the **right to medication**.
3) Some also argue that **terminally ill** patients shouldn't receive antibiotics because they're going to die. But **withholding** antibiotics from these patients could reduce their **length of survival** and **quality of life**.
4) Some people believe **animals shouldn't** be given antibiotics (as this may increase antibiotic resistance). Other people argue that this could cause **unnecessary suffering** to the animals.

Practice Questions

Q1 Briefly describe a method for testing the antibiotic resistance of bacteria.

Q2 Briefly describe one situation where scientific knowledge has affected the decision-making surrounding antibiotic resistance.

Exam Questions

Q1 Put forward arguments for and against giving antibiotics to someone suffering from dementia. [2 marks]

Q2 A mother takes her son, who is suffering from a mild chest infection, to the doctor to get some antibiotics.

a) Why might the doctor be reluctant to prescribe the child antibiotics? [2 marks]
b) Why might the mother disagree if the doctor refuses to prescribe antibiotics? [2 marks]

Q3 A study was carried out to determine if the increase in the national rate of bacterial resistance to antibiotic X is linked to people not finishing their course of the antibiotic. Part of the study involved sending out questionnaires to 300 patients from one GP surgery in East Anglia.

Evaluate the methodology of this study. [3 marks]

R-E-S-I-S-T-A-N-T — find out what it means to me...

You're probably a bit bored of me ramming it down your throat now but you need to be able to evaluate any data or study you're presented with. Remember to look at the methodology, evidence and conclusions, and look out for anywhere there may be problems. And if you're ever asked to consider any ethical issues, think of the arguments for and against.

Human Impacts on Diversity

Human activity has an impact on species diversity — and I'm not just talking about stepping on bugs. There are plenty of examples, but you need to know how agriculture and deforestation affect diversity...

Species Diversity is the Number of Species Present in a Community

1) Species diversity is the number of **different** species and the **abundance** of each species within a **community**.
2) The **higher** the species diversity of **plants and trees** in an area, the **higher** the species diversity of **insects, animals** and **birds**. This is because there are **more habitats** (places to live) and a larger and more varied **food source**.
3) Diversity can be **measured** to help us monitor ecosystems and identify areas where it has been **dramatically reduced**.

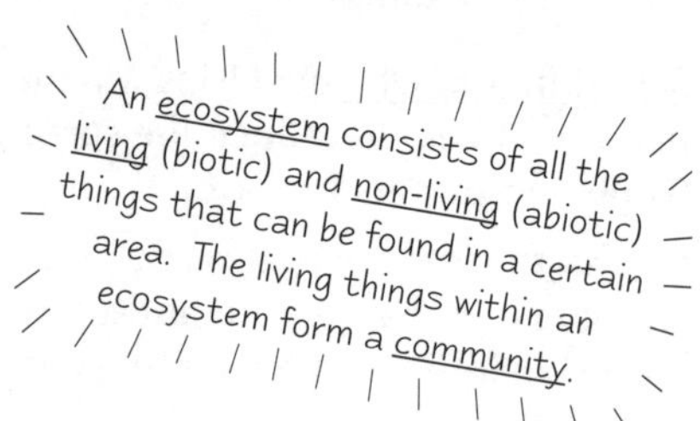

An ecosystem consists of all the living (biotic) and non-living (abiotic) things that can be found in a certain area. The living things within an ecosystem form a community.

Species Diversity is Measured using the Index of Diversity

1) The simplest way to measure diversity is just to **count up** the number of **different species**.
2) But that **doesn't** take into account the **population size** of each species.
3) Species that are in a community in very **small** numbers shouldn't be treated the same as those with **bigger** populations.
4) The **index of diversity** is calculated using an equation that takes different population sizes into account. You calculate the index of diversity (**d**) of a community using this formula:

$$d = \frac{N(N-1)}{\sum n(n-1)}$$

Where...
N = **Total number** of organisms of **all** species
n = **Total number** of **one** species
Σ = '**Sum of**' (i.e. added together)

The **higher** the number the **more diverse** the area is. If all the individuals are of the same species (i.e. no diversity) the diversity index is 1. Here's an example:

There are 3 different species of flower in this field — a red species, a white and a blue.
There are 11 organisms altogether, so N = 11.
There are 3 of the red species, 5 of the white and 3 of the blue.
So the species diversity index of this field is:

$$d = \frac{11\,(11-1)}{3\,(3-1) + 5\,(5-1) + 3\,(3-1)} = \frac{110}{6 + 20 + 6} = 3.44$$

When calculating the bottom half of the equation you need to work out the n(n–1) bit for each different species then add them all together.

Deforestation Decreases Species Diversity...

We cut down forests to get **wood** and **create land** for **farming** and **settlements**.
Here are some reasons why this affects diversity:

1) Deforestation **directly** reduces the **number** of **trees** and sometimes the **number** of **different tree species**.
2) Deforestation also **destroys habitats**, so some species could lose their **shelter** and **food source**. This means that these species will **die** or be forced to **migrate** to another suitable area, further **reducing** diversity.
3) The migration of organisms into increasingly smaller areas of remaining forest may **temporarily increase species diversity** in those areas.

Human Impacts on Diversity

...as Does Agriculture

Farmers try to **maximise** the **amount of food** that they can produce from a given area of land. Many of the methods they use reduce diversity.

1) **Woodland clearance** — this is done to **increase** the **area** of farmland. This **reduces** species diversity for the same reasons as **deforestation** (see previous page).
2) **Hedgerow removal** — this is also done to **increase** the **area** of farmland by turning **lots of small fields** into **fewer large fields**. This **reduces** species diversity for the same reasons as **woodland clearance** and **deforestation**.
3) **Monoculture** — this is when farmers grow fields containing only **one type of plant**. A **single type** of plant will support **fewer species**, so diversity is **reduced**.
4) **Pesticides** — these are chemicals that **kill** organisms (**pests**) that feed on **crops**. This **reduces** diversity by **directly killing** the pests. Also, any species that feed on the pests will **lose** a food source, so their numbers could **decrease** too.
5) **Herbicides** — these are chemicals that kill **unwanted plants (weeds)**. This **reduces** plant diversity and could **reduce** the number of organisms that feed on the weeds.

Pete wasn't sure that the company's new increased diversity policy would be good for productivity.

Practice Questions

Q1 Why does greater plant and tree diversity increase insect, animal and bird diversity?

Q2 How is species diversity calculated?

Q3 Why does deforestation lead to reduced species diversity?

Q4 What is monoculture and how does it reduce species diversity?

Q5 Why do pesticides and herbicides reduce species diversity?

Exam Question

Q1 A study was conducted to investigate the impact of introducing enhanced field margins on the diversity of bumblebees. Enhanced field margins are thick bands of land around the edges of fields that are not farmed, but instead are planted with plants that are good for wildlife. Scientists studied two wheat fields, one where the farmer sowed crops right to the edge of the field and another where the farmer created enhanced field margins.

a) What is species diversity a measure of? [2 marks]

b) Use the data below to calculate the index of diversity for each site.

Site 1 — No Field Margins		Site 2 — Enhanced Field Margins	
Bombus lucorum	15	*Bombus lucorum*	35
Bombus lapidarius	12	*Bombus lapidarius*	25
Bombus pascuorum	24	*Bombus pascuorum*	34
		Bombus ruderatus	12
		Bombus terrestris	26

[4 marks]

c) What conclusions can be drawn from the findings of this study? [2 marks]

TIM-BER! — deforestation brings species diversity crashing down...

There's nothing special about species diversity — just remember it's the number of species in a community and the abundance of each. Simple. Now for the maths bit... practise the index of diversity equation until you can't imagine life without it, then when it comes up in the exam it'll be a breeze — a nice warm one, not a spine-chilling one.

Interpreting Diversity Data

I know it seems like there are lots of risks associated with human activity, but there are plenty of benefits too — we don't just cut trees down for the fun of it you know.

Human Activity *in an Area Has* ***Benefits*** *and* ***Risks***

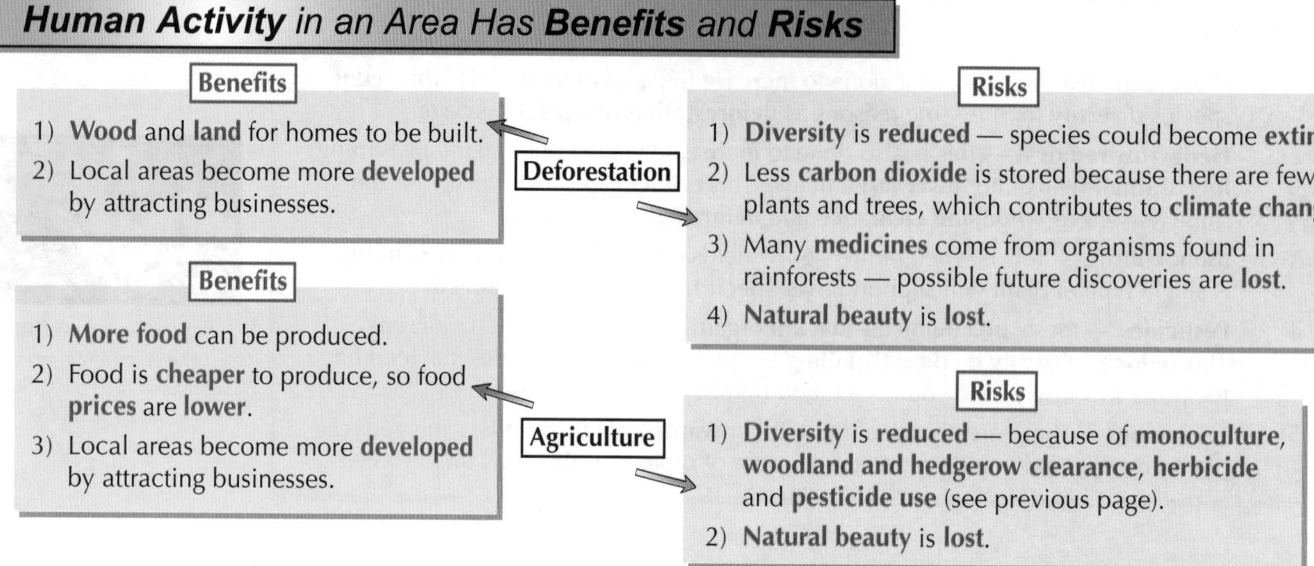

Benefits
1) Wood and land for homes to be built.
2) Local areas become more developed by attracting businesses.
Deforestation
Risks
1) Diversity is reduced — species could become extinct.
2) Less carbon dioxide is stored because there are fewer plants and trees, which contributes to climate change.
3) Many medicines come from organisms found in rainforests — possible future discoveries are lost.
4) Natural beauty is lost.
Benefits
1) More food can be produced.
2) Food is cheaper to produce, so food prices are lower.
3) Local areas become more developed by attracting businesses.
Agriculture
Risks
1) Diversity is reduced — because of monoculture, woodland and hedgerow clearance, herbicide and pesticide use (see previous page).
2) Natural beauty is lost.

You Might Have to ***Interpret Data*** *About How* ***Human Activity*** *Affects* ***Diversity***

You might have to interpret some data in the exam. Here are three examples of the kind of thing you might get:

Example 1 — Herbicides

Herbicides kill **unwanted plants** (weeds) whilst leaving **crops unharmed**.
The crops can then **grow better** because they're not **competing for resources** with weeds.
The graph below shows plant diversity in an **untreated** field and a field **treated annually** with **herbicide**.

Describe the data:

- Plant diversity in the **untreated field** showed a **slight increase** in the seven years. Plant diversity **decreased** a lot in the **treated field** when the **herbicide** was **first applied**. The diversity then **recovered** throughout each year, but was **reduced again** by **each** annual application of herbicide.

Explain the data:

- When the herbicide was applied, the weeds were **killed**, **reducing species diversity**. **In between** applications, diversity **increased** as **new weeds grew**. These were then killed again by each annual application of herbicide.

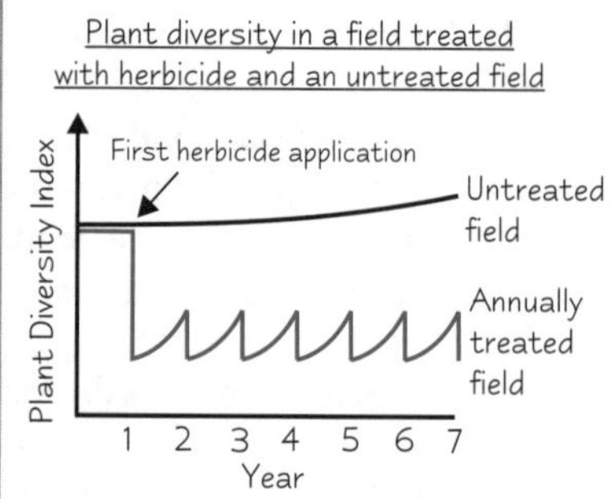

Example 2 — Skylark Population

Since the 1970s, farmers have been planting many crops in the **winter** instead of the **spring** to **maximise production**. This means there are **fewer** fields left as **stubble** over the winter. This is a problem because **skylarks** like to **nest** in fields with **stubble**. The graph below shows how the skylark population has **changed** in the UK since 1970.

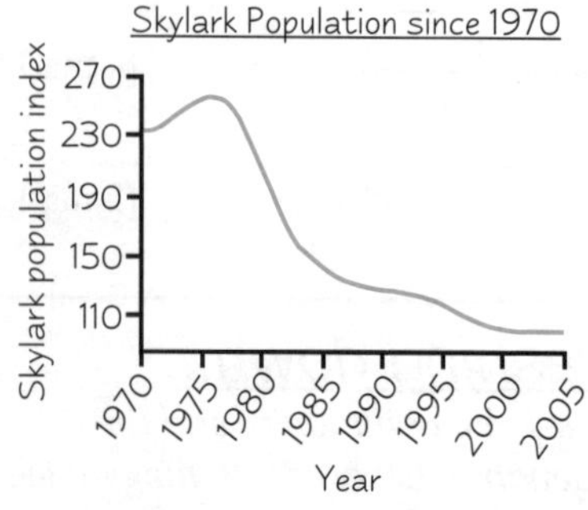

Describe the data:

- Skylark diversity showed a **small increase** from **1970** to the **late 1970s**. Since the late 1970s skylark diversity has **decreased** a lot. The skylark diversity has remained roughly **constant** from **2000** to **2005**.

Explain the data:

- With **fewer nesting sites** available, **fewer offspring** could be successfully raised, leading to a reduction in the number of skylarks.

Interpreting Diversity Data

Example 3 — Loss of Rainforest

The graph below shows the results of a study that **compared** the **diversity** in a **rainforest** with the diversity in a **deforested area** that had been **cleared** for agricultural use.

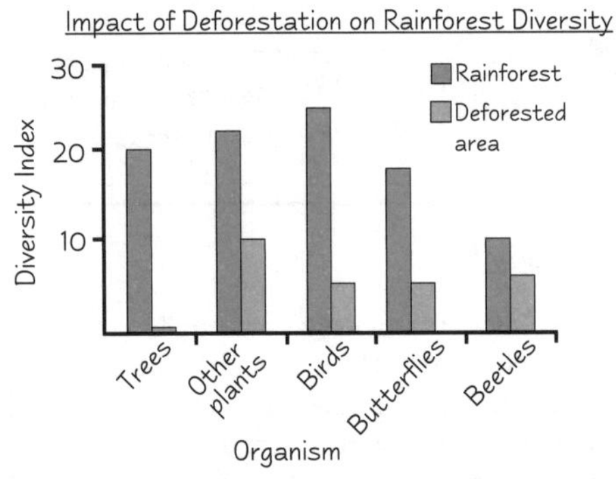

Describe the data:

- For **all types** of organism studied, **species diversity** is **higher** in the **rainforest** than in the **deforested area**.
- **Deforestation** has **reduced** the species diversity of **trees** the most.

Explain the data:

- Many organisms **can't adapt** to the **change in habitat** and must **migrate** or **die** — **reducing** diversity in the area.
- Reduced **tree diversity** leads to a reduction in the diversity of **all other organisms** (see p. 86).

Society Uses Diversity Data to Make Decisions

Diversity data can be used to see which **species or areas** are being **affected** by **human activity**.
This information can then be used by **society** to **make decisions** about human activities. For example:

Scientific Finding	Decision Made
Fewer hedgerows reduces diversity.	The UK government offers farmers money to encourage them to plant hedgerows, and to cover the cost of not growing crops on these areas.
Deforestation reduces diversity.	Some governments encourage sustainable logging (a few trees are taken from lots of different areas and young trees are planted to replace them).
Human development reduces diversity.	Many governments are setting up protected areas (e.g. national parks) where human development is restricted to help conserve diversity.
Some species are facing extinction.	Breeding programmes in zoos help to increase the numbers of endangered species in a safe environment before reintroducing them to the wild.

Practice Questions

Q1 Describe one risk associated with deforestation.

Q2 Describe one benefit of agricultural activities.

Exam Question

Q1 The graph on the right shows the results from a study conducted into wild bird populations in the UK between 1970 and 2006. It shows the pattern of change for woodland and farmland species.

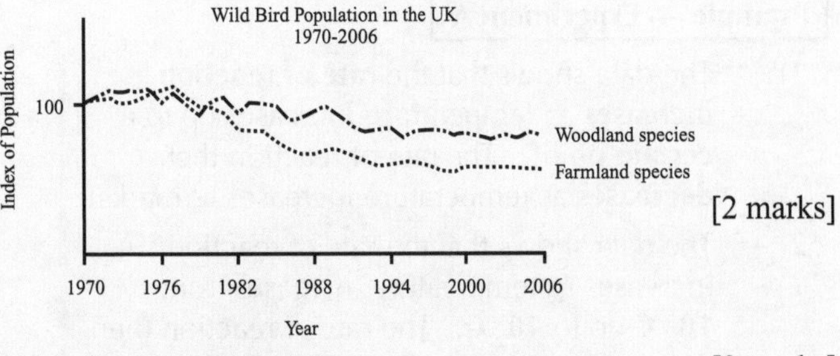

a) Describe the data. [2 marks]

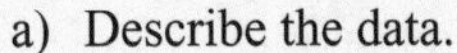

b) Human activity has significantly affected wild bird populations. Suggest reasons why the woodland and farmland species have changed in the way shown on the graph. [3 marks]

c) Discuss the potential benefits of agriculture and deforestation and the associated risks to diversity. [8 marks]

Diver-city — it's a wonderful place where you get to jump off stuff...

Interpreting data is just about understanding what the graphs or tables are telling us. Describing it is simple — just say what you see. Then you've usually got to explain it — say what the reasons might be for what you've described. After that they might ask you to do a little song, or a little dance, or maybe jump through a couple of hoops...

How to Interpret Experiment and Study Data

Science is all about getting good evidence to test your theories... so scientists need to be able to spot a badly designed experiment or study a mile off, and be able to interpret the results of an experiment or study properly. Being the cheeky little monkeys they are, your exam board will want to make sure you can do it too. Here's a quick reference section to show you how to go about interpreting data-style questions.

Here Are Some **Things** You Might be **Asked** to do...

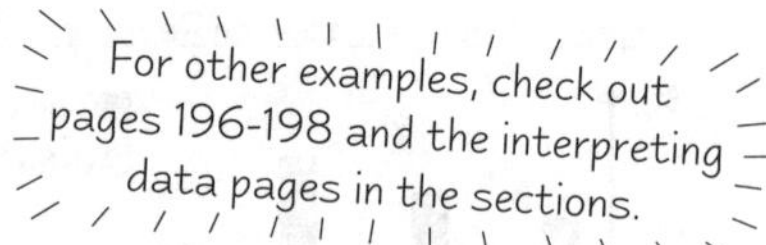

For other examples, check out
pages 196-198 and the interpreting
data pages in the sections.

Here are two examples of the kind of data you could expect to get:

Experiment A

Experiment A examined the effect of temperature on the rate of an enzyme-controlled reaction. The rate of reaction for enzyme X was measured at six different temperatures (from 10 to 60 °C). All other variables were kept constant. A negative control containing all solutions except the enzyme was included. The rate of reaction for the negative control was zero at each temperature used. The results are shown in the graph below.

The effect of temperature on the rate of an enzyme-controlled reaction

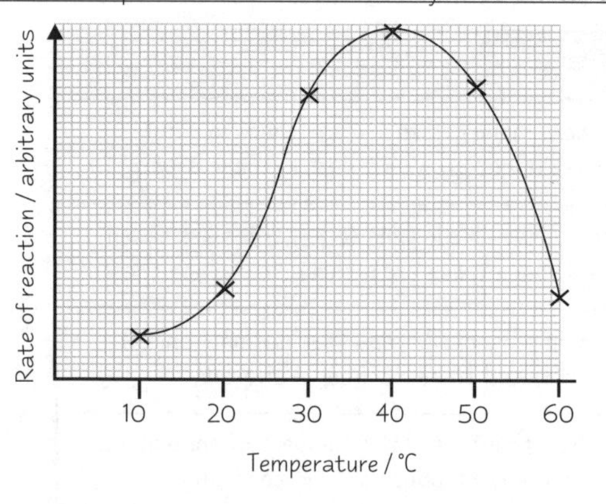

Study B

Study B examined the effect of farm hedgerow length on the number of species in a given area. The number of species present during a single week on 12 farms was counted by placing ground-level traps. All the farms were a similar area. The traps were left out every day, at 6 am for two hours and once again at 6 pm for two hours. The data was plotted against hedgerow length. The results are shown in the scattergram below.

The effect of hedgerow length on number of species

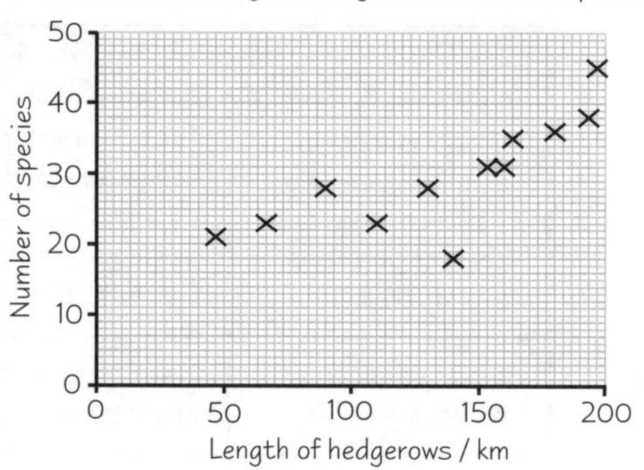

1) **Describe** the **Data**

You need to be able to **describe** any data you're given. The level of **detail** in your answer should be appropriate for the **number of marks** given. Loads of marks = more detail, few marks = less detail.
For the two examples above:

Example — Experiment A

1) The data shows that the **rate of reaction increases** as **temperature increases** up to a **certain point**. The rate of reaction then **decreases** as temperature increases (2 marks).
2) The data shows that the rate of reaction **increases** as temperature increases from **10 °C** up to **40 °C**. The rate of reaction then **decreases** as temperature increases from **40 °C to 60 °C** (4 marks).

Example — Study B

The data shows a **positive correlation** between the length of hedgerows and the number of species in the area (1 mark).

Correlation describes the **relationship** between two variables — the one that's been changed and the one that's been measured. Data can show **three** types of correlation:

1) **Positive** — as one variable **increases** the other **increases**.
2) **Negative** — as one variable **increases** the other **decreases**.
3) **None** — there is **no relationship** between the two variables.

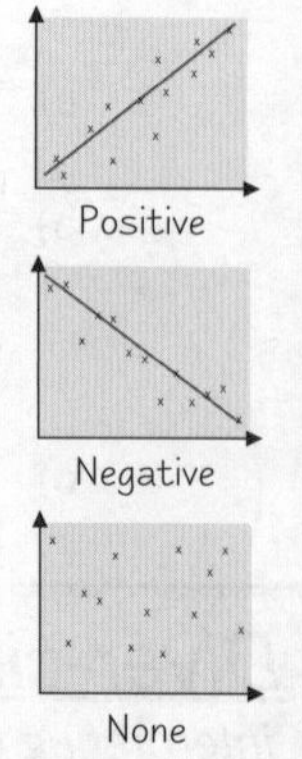

How to Interpret Experiment and Study Data

2) *Draw* or *Check* the *Conclusions*

1) Ideally, only **two** quantities would ever change in any experiment or study — everything else would be **constant**.
2) If you can keep everything else constant and the results show a correlation then you **can** conclude that the change in one variable **does cause** the change in the other. ⟹
3) But usually all the variables **can't** be controlled, so other **factors** (that you **couldn't** keep constant) could be having an **effect**.
4) Because of this, scientists have to be very careful when **drawing conclusions**. Most results show a **link** (correlation) between the variables, but that **doesn't prove that a change in one causes the change in the other**. ⟹
5) The **data** should always **support** the conclusion. This may sound obvious but it's easy to **jump** to conclusions. Conclusions have to be **precise** — not make sweeping generalisations. ⟹

Example — Experiment A

All other variables were **kept constant**. E.g. pH, enzyme concentration and substrate concentration **stayed the same** each time, so these **couldn't** have influenced the change in the rate of reaction. So you **can say** that an increase in temperature **causes** an increase in the rate of reaction up to a certain point.

Example — Study B

The length of hedgerows shows a **positive correlation** with the number of species in that area. But you **can't** conclude that fewer hedgerows **causes** fewer species. **Other factors** may have been involved, e.g. the number of **predators** of the species studied may have increased in some areas, the farmers may have used **more pesticide** in one area, or something else you hadn't thought of could have caused the pattern...

Example — Experiment A

A science magazine **concluded** from this data that enzyme X works best at **40 °C**. The data **doesn't** support this. The enzyme **could** work best at 42 °C, or 47 °C but you can't tell from the data because **increases** of **10 °C** at a time were used. The rates of reaction at in-between temperatures **weren't** measured.

3) *Comment on the* ***Reliability*** *of the* ***Results***

Reliable means the results can be **consistently reproduced** in independent experiments. And if the results are reproducible they're more likely to be **true**. If the data isn't reliable for whatever reason you **can't draw** a valid **conclusion**. Here are some of the things that affect the reliability of data:

Davina wasn't sure she'd got a large enough sample size.

1) Size of the data set — For experiments, the **more repeats** you do, the **more reliable** the data. If you get the **same result** twice, it could be the correct answer. But if you get the same result **20 times**, it's much more reliable. The general rule for **studies** is the **larger** the sample size, the more **reliable** the **data** is.

   E.g. Study B is quite **small** — they only used 12 farms. The **trend** shown by the data may not appear if you studied **50 or 100 farms**, or studied them for a longer period of time.

2) Variables — The **more variables** you **control**, the **more reliable** your data is. In an experiment you would control all the variables, but when doing a study this isn't always possible. You try to control **as many as possible** or use **matched groups** (see page 2).

   E.g. ideally, all the farms in Study B would have a similar **type** of land, similar **weather**, have the same **crops** growing, etc. Then you could be more sure that the one factor being **investigated** (hedgerows) is having an **effect** on the thing being **measured** (number of species). In Experiment A, **all** other variables were controlled, e.g. pH, concentrations, volumes, so you can be sure the temperature is causing the **change** in the **reaction rate**.

3) Data collection — think about all the **problems** with the **method** and see if **bias** has slipped in. For example, members of the public sometimes tell **little porkies**, so it's easy for studies involving **questionnaires** to be **biased**. E.g. people often underestimate how much alcohol they drink or how many cigarettes they smoke.

   E.g. in Study B, the traps were placed on the **ground**, so species like birds weren't included. The traps weren't left overnight, so **nocturnal** animals wouldn't get counted, etc. This could have affected the results.

Jane rarely ate chocolate, honestly.

How to Interpret Experiment and Study Data

4) **Controls** — without controls, it's very difficult to **draw valid conclusions**. **Negative controls** are used to make sure that nothing you're doing in the experiment has an effect, **other than** what you're testing. But it's not always possible to have controls in studies (study controls usually involve a group where **nothing changes**, e.g. a group of patients aren't given a new long-term treatment to make sure any effects detected in the patients having the treatment aren't due to the fact that they've had two months to recover).

E.g. in Experiment A, the **negative control** contained everything from the experiment **except** the enzyme. This was used to show that the change in reaction rate was caused by the effect of **temperature** on the **enzyme**, and nothing else. If something else in the experiment (e.g. the water, or something in the test tube) was causing the change, you would get the **same results** in the negative control (and you'd know something was up).

5) **Repetition by other scientists** — for theories to become accepted as 'fact' other scientists need to **repeat** the work (see page 1). If **multiple studies** or **experiments** come to the same conclusion, then that conclusion is **more reliable**.

E.g. if a second group of scientists carried out the same experiment for enzyme X and got the same results, the results would be **more reliable**.

4) Analyse the Data

Sometimes it's easier to **compare data** by making a few calculations first, e.g. converting raw data into **ratios** or **percentages**.

Example **Three** UK hospitals have been trying out three **different methods** to **control the spread** of chest infections. A study investigated the number of people suffering from chest infections in those hospitals over a **three month period**. The table opposite shows the results. If you just look at the **number of cases** in the **last month** (March) then the method of hospital 3 appears to have worked **least well**, as they have the **highest number** of infections. But if you look at the **percentage increase** in infections you get a different picture: hospital 1 = 30%, hospital 2 = 293%, and hospital 3 = 18%. So hospital 3 has the lowest percentage increase, suggesting their method of control is **working the best**.

	Number of cases per 6000 patients		
Hospital	Jan	Feb	March
1	60	65	78
2	14	24	55
3	93	96	110

Calculating percentage increase, hospital 1:

$$\frac{(78 - 60)}{60} \times 100 = \frac{18}{60} \times 100 = 30\%$$

There Are a Few Technical Terms You Need to Understand

I'm sure you probably know these all off by heart, but it's easy to get mixed up sometimes. So here's a quick recap of some words **commonly used** when assessing and analysing experiments and studies:

1) **Variable** — A variable is a **quantity** that has the **potential to change**, e.g. weight. There are two types of variable commonly referred to in experiments:
- **Independent variable** — the thing that's **changed** in an experiment.
- **Dependent variable** — the thing that you **measure** in an experiment.

When drawing graphs, the dependent variable should go on the **y-axis** (the vertical axis) and the independent on the **x-axis** (the horizontal axis).

2) **Accurate** — Accurate results are those that are **really close** to the **true** answer.

3) **Precise results** — These are results taken using **sensitive instruments** that measure in **small increments**, e.g. pH measured with a meter (pH 7.692) will be **more precise** than pH measured with paper (pH 8).

It's possible for results to be precise **but not** accurate, e.g. a balance that weighs to 1/1000 th of a gram will give precise results, but if it's not **calibrated** properly the results won't be accurate.

4) **Qualitative** — A **qualitative** test tells you **what's** present, e.g. an acid or an alkali.

5) **Quantitative** — A **quantitative** test tells you **how much** is present, e.g. an acid that's pH 2.46.

Controls — I think I prefer the remote kind...

*These pages should give you a fair idea of the points to think about when interpreting data. Just use your head and remember the three main points in the checklist — **d**escribe the **d**ata, **c**heck the **c**onclusions and make sure the **r**esults are **r**eliable.*

A2-Level Biology

Exam Board: AQA

Populations and Ecosystems

Now, I need something momentous to say at the start of the A2 part of the book — these pages are one small step for an A-level biology student, but one giant leap for the subject of biology... oh, someone's said that before. Nuts.

*You Need to **Learn Some Definitions** to get you **Started***

Habitat	—	The **place** where an organism **lives**, e.g. a rocky shore or a field.
Population	—	**All** the organisms of **one species** in a **habitat**.
Community	—	Populations of **different species** in a habitat make up a **community**.
Ecosystem	—	**All** the **organisms** living in a **particular area** and all the **non-living** (abiotic) conditions, e.g. a freshwater ecosystem such as a lake.
Abiotic conditions	—	The **non-living** features of the ecosystem, e.g. **temperature** and **availability of water**.
Biotic conditions	—	The **living** features of the ecosystem, e.g. the presence of **predators** or **food**.
Niche	—	The **role** of a species within its habitat, e.g. what it eats, where and when it feeds.
Adaptation	—	A **feature** that members of a species have that **increases** their chance of **survival** and **reproduction**, e.g. **giraffes** have **long necks** to help them reach vegetation that's high up. This increases their chances of survival when food is **scarce**.

Being a member of the undead made it hard for Mumra to know whether he was a living or a non-living feature of the ecosystem.

Every Species** Occupies a **Different Niche

1) The **niche** a species occupies within its habitat includes:
   - Its **biotic** interactions — e.g. the organisms it **eats**, and those it's **eaten by**.
   - Its **abiotic** interactions — e.g. the **oxygen** an organism breathes in, and the **carbon dioxide** it breathes out.

2) Every species has its own **unique niche** — a niche can only be occupied by **one species**.

Don't get confused between habitat (where a species lives) and niche (what it does in its habitat).

3) It may **look** like **two species** are filling the **same niche** (e.g. they're both eaten by the same species), but there'll be **slight differences** (e.g. variations in what they eat).
4) If two species **try** to occupy the **same niche**, they will **compete** with each other. One species will be **more successful** than the other, until **only one** of the species is **left**.
5) Here are a couple of examples of niches:

<u>Common pipistrelle bat</u>

This bat lives throughout Britain on **farmland**, **open woodland**, **hedgerows** and **urban areas**. It feeds by **flying** and catching **insects** using **echolocation** (**high-pitched sounds**) at a **frequency** of around **45 kHz**.

<u>Soprano pipistrelle bat</u>

This bat lives in Britain in **woodland** areas, close to **lakes** or **rivers**. It feeds by **flying** and catching **insects** using **echolocation**, at a **frequency** of **55 kHz**.

It may **look like** both species are filling the **same niche** (e.g. they both eat insects), but there are **slight differences** (e.g. they use **different frequencies** for their echolocation).

Populations and Ecosystems

Organisms are Adapted to Biotic and Abiotic Conditions

1) As you know, **adaptations** are features that **increase** an organism's chance of **survival** and **reproduction**.
2) They can be **physiological** (processes **inside** their body), **behavioural** (the way an organism **acts**) or **anatomical** (**structural features** of their body).
3) Organisms with better adaptations are **more likely** to **survive**, **reproduce** and **pass on** the alleles for their adaptations, so the adaptations become **more common** in the population. This is called **natural selection**.
4) Every species is adapted to **use** an **ecosystem** in a way that **no other** species can. For example, only giant anteaters can **break into** ant nests and **reach** the ants. They have **claws** to rip open the nest, and a **long, sticky tongue** which can move **rapidly** in and out of its mouth to **pick up** the ants.
5) Organisms are **adapted** to both the **abiotic conditions** (e.g. how much **water** is available) and the **biotic conditions** (e.g. what **predators** there are) in their ecosystem.

Here are a few ways that **different organisms** are **adapted** to the **abiotic** or the **biotic** conditions in their ecosystems:

Adaptations to abiotic conditions

- **Otters** have **webbed paws** — this means they can both **walk** on land and **swim** effectively. This increases their chance of survival because they can **live** and **hunt** both on land and in water.
- **Whales** have a **thick layer** of **blubber** (fat) — this helps to keep them **warm** in the **coldest seas**. This increases their chance of survival because they can **live** in places where food is plentiful.
- **Brown bears hibernate** — they **lower their metabolism** (all the chemical reactions taking place in their body) over **winter**. This increases their chance of survival because they can **conserve energy** during the **coldest** months.

Adaptations to biotic conditions

- **Sea otters** use **rocks** to **smash open** shellfish and clams. This increases their chance of survival because it gives them **access** to **another source** of food.
- **Scorpions dance** before **mating** — this makes sure they **attract** a **mate** of the **same species**. This increases their chance of reproduction by making **successful mating** more likely.
- Some **bacteria** produce **antibiotics** — these **kill other species** of bacteria in the **same area**. This increases their chance of survival because there's **less competition** for **resources**.

Take your partner 1, 2, 3, swing them round a sycamore tree.

Practice Questions

Q1 What is the name given to all the organisms of one species in a habitat?

Q2 Define a community.

Q3 Give the term for the non-living features of an ecosystem.

Q4 What happens when two species try to occupy the same niche in an ecosystem?

Exam Question

Q1 Common pipistrelle bats have light, flexible wings, which means they can fly fast and are manoeuvrable. They hunt insects at night using echolocation and live on farmland, in open woodland, hedgerows and urban areas. They make unique mating calls to find mates, hibernate through the winter, and roost in cracks in trees and buildings during the day.

a) Describe the habitat of the common pipistrelle bat. [2 marks]

b) Explain how the common pipistrelle bat is adapted to the biotic conditions in its ecosystem. [3 marks]

Unique quiche niche — say it ten times really fast...

All this population and ecosystem stuff is pretty wordy I'm afraid, but I'll tell you what, you'll be missing it when you get onto the really sciencey stuff later. You just need to learn and re learn all the key words here, then when they ask you to interpret some bat-related babble in the exam, you'll know exactly what they're talking about. Niche work.

Investigating Populations

Examiners aren't happy unless you're freezing to death in the rain in a field somewhere in the middle of nowhere. Still, it's better than being stuck in the classroom being bored to death learning about fieldwork techniques...

You need to be able to **Investigate Populations** of **Organisms**

Investigating **populations** of organisms involves looking at the **abundance** and **distribution** of **species** in a particular **area**.

1) **Abundance** — the **number of individuals** of **one species** in a **particular area**.
   The abundance of **mobile organisms** and **plants** can be estimated by simply counting the **number** of individuals in samples taken. There are other measures of abundance that can be used too:
   - **Frequency** — the **number of samples** a species is **recorded in**, e.g. 70% of samples.
   - **Percentage cover** (for plants only) — **how much** of the area you're investigating is **covered** by a species.
2) **Distribution** — this is **where** a particular species is within the **area you're investigating**.

You need to take a **Random Sample** from the **Area You're Investigating**

Most of the time it would be too **time-consuming** to measure the **number of individuals** and the **distribution** of every species in the **entire area** you're investigating, so instead you take **samples**:

1) **Choose** an **area to sample** — a **small** area **within** the area being investigated.
2) Samples should be **random** to **avoid bias**, e.g. if you were investigating a field you could pick random sample sites by dividing the field into a **grid** and using a **random number generator** to select **coordinates**.
3) Use an **appropriate technique** to take a sample of the population (see below and the next page).
4) **Repeat** the process, taking as many samples as possible. This gives a more **reliable** estimate for the **whole area**.
5) The **number of individuals** for the **whole area** can then be **estimated** by taking an **average** of the data collected in each sample and **multiplying** it by the size of the whole area. The **percentage cover** for the whole area can be estimated by taking the average of all the samples.

Different Methods are Used to **Investigate Different Organisms**

1 **Pitfall Traps** and **Pooters** are used to **Investigate Ground Insects**

Pitfall traps

1) **Pitfall traps** are **steep-sided containers** that are sunk in a **hole** in the ground. The top is **partially open**.
2) Insects **fall** into the container and **can't get out** again — they're **protected** from **rain** and **some predators** by a **raised lid**.
3) The sample can be affected by **predators small enough** to fall into the pitfall trap though — they may **eat** other insects, **affecting** the **results**.

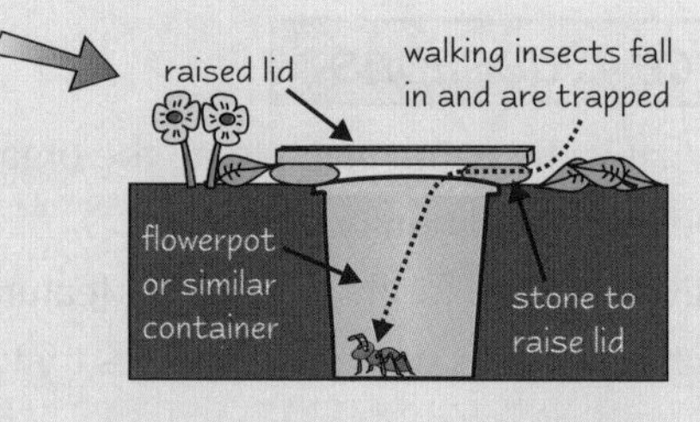

Pooters

1) **Pooters** are **jars** that have **rubber bungs** sealing the top, and **two tubes** stuck through the bung.
2) The **shorter tube** has **mesh** over the end that's in the jar. The **longer tube** is **open** at both ends.
3) When you **inhale** through the shorter tube, **air is drawn** through the longer tube. If you **place** the end of the **longer tube** over an insect it'll be **sucked** into the jar.
4) It can take a **long time** (or **lots of people**) to get a **large sample** using pooters. Some species may be **missed** if the sample **isn't large enough**.

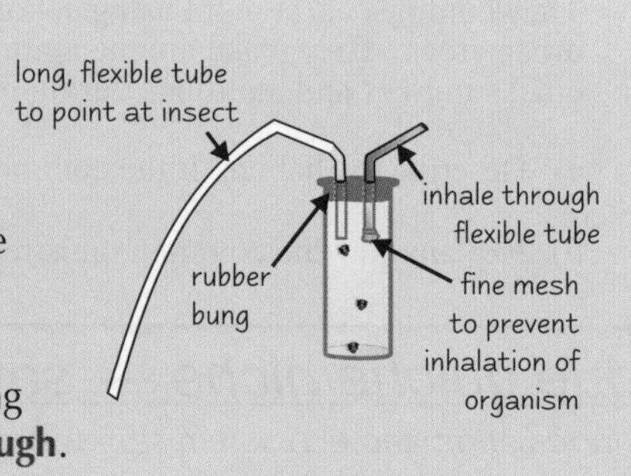

Investigating Populations

2 Quadrats and Transects are used to Investigate Plant Populations

Quadrats

1) A **quadrat** is a **square** frame divided into a **grid** of 100 **smaller squares** by strings attached across the frame.
2) Quadrats are **placed on the ground** at **different points** within the area you're investigating.
3) The **species frequency** or the **number of individuals** of each species is recorded in **each quadrat**.
4) The **percentage cover** of a species can also be measured by counting how much of the quadrat is **covered** by the species — you count a square if it's **more than half-covered**. Percentage cover is a **quick** way to investigate populations and you **don't** have to **count** all the **individual** plants.
5) Quadrats are useful for **quickly** investigating areas with plant species that **fit** within a **small quadrat** — areas with **larger plants** and **trees** need **very large** quadrats.

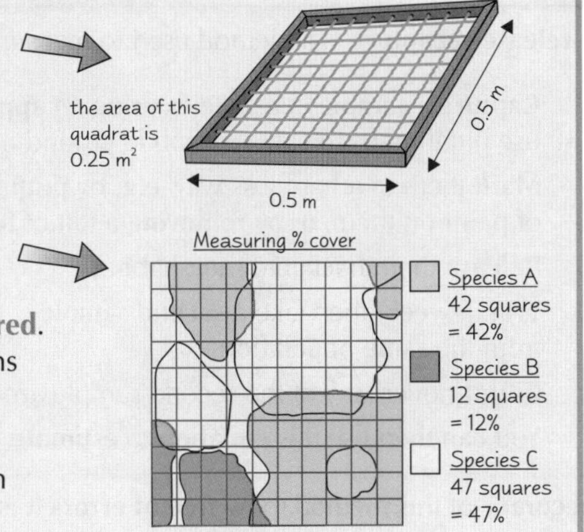

Transects

You can use **lines** called **transects** to help find out how plants are **distributed across** an area, e.g. how species change from a hedge towards the middle of a field. There are three types:

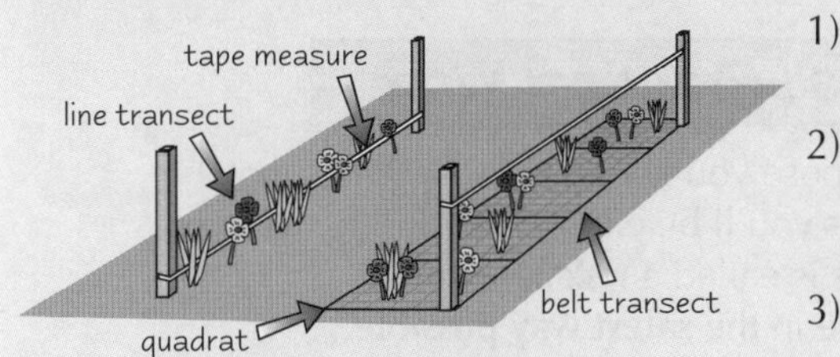

1) **Line transects** — a **tape measure** is placed **along** the transect and the species that **touch** the tape measure are **recorded**.
2) **Belt transects** — **quadrats** are placed next to each other **along** the transect to work out **species frequency** and **percentage cover** along the transect.
3) **Interrupted transects** — instead of investigating the whole transect of either a line or a belt, you can take **measurements** at **intervals**.

3 Beating Trays are used to Investigate Insects Found in Vegetation

Beating trays are usually white so you can see the insects.

1) A **beating tray** is a **tray** or **sheet** held **under** a plant or tree.
2) The plant or tree is **shaken** and a sample of insects **falls onto** the beating tray.
3) You can take **large samples** using beating trays, giving **good estimates** of the **abundance** of each species.
4) However, the **sample** may **not be random** because most of it will be made up of insects that **fall easily** when the vegetation is **shaken**.

Practice Questions

Q1 Explain why samples of a population are taken.

Q2 Give one drawback of using beating trays to investigate insect populations.

Exam Question

Q1 A student wants to sample a population of daffodils in a field.

a) How could she avoid bias in her investigation? [1 mark]

b) Describe how she could investigate the percentage cover of daffodils in the field. [4 marks]

Beating trays — as used by wrestlers on TV...

There are plenty of pitfalls to avoid when investigating populations, but if you give these pages a good read you'll leap straight over them. Just be aware that every technique for collecting data on populations has its drawbacks.

Investigating Populations

More practical fun... no wait, there's some data interpretation on these pages too. Sadly, dealing with data is pretty important — I'd bet my phone, car and beloved bike that there'll be at least one data interpretation question in the exam.

Mark-Release-Recapture is Used to Investigate More Mobile Species

Mark-release-recapture is a method used to measure the **abundance** of more **mobile** species. Here's how it's done:

1) **Capture** a sample of a species using an **appropriate technique**, e.g. you could use pitfall traps to capture mobile ground insects (see p. 96), and **count** them.
2) **Mark** them in a harmless way, e.g. by putting a spot of **paint** on them, or by **removing** a tuft of **fur**.
3) **Release** them back into their habitat.
4) Wait a week, then take a **second sample** from the **same population**.
5) **Count** how many of the second sample **are marked**.
6) You can then use this **equation** to **estimate** the **total** population size.

$$\text{Total population size} = \frac{\text{Number caught in 1st sample} \times \text{Number caught in 2nd sample}}{\text{Number marked in 2nd sample}}$$

The **accuracy** of this method (how **free of errors** it is) depends on a few **assumptions**:

1) The marked sample has had enough **time** and **opportunity** to **mix** back in with the population.
2) The marking hasn't affected the individuals' **chances of survival**, and is **still visible**.
3) **Changes** in **population size** due to **births**, **deaths** and **migration** are **small** during the period of the study.

You Need to Carry Out a Risk Assessment for all Practical Work

When you're carrying out fieldwork to investigate populations you expose yourself to **risks** — things that could **potentially** cause you **harm**. You need to think about **what risks** you'll be exposed to during fieldwork, so you can **plan** ways to **reduce** the **chance** of them happening — this is called a **risk assessment**. Risk assessments are always carried out to ensure that fieldwork's done in the **safest way possible**.

Here are some **examples** of the fieldwork risks when investigating populations and the **ways** to **reduce** the risks:

Falls and slips	Wear suitable footwear for the terrain, e.g. wellies on wet or boggy ground, and take care on rough terrain. Make sure the study area isn't near any cliffs or on steep ground.
Bad weather	Check the weather forecast beforehand and take precautions, e.g. wear warm or waterproof clothing on cold or wet days. If the weather is too bad, do the fieldwork another day.
Stings and bites	Wear insect repellent or, if you have an allergy, take medication with you.

OK chaps, just get out your lightning-proof suits and let's crack on with the fieldwork.

There are Ethical Issues to Consider When Doing Fieldwork

All fieldwork **affects** the **environment** where it's carried out, e.g. lots of people **walking around** may cause **soil erosion**. Some people don't think it's **right** to **damage** the **environment** when doing fieldwork, so investigations should be planned to have the **smallest impact possible**, e.g. people should restrict **where they walk** to the area being studied.

Some fieldwork **affects** the **organisms** being studied, e.g. **capturing** an organism for study may cause it **stress**. Some people don't think it's **right** to **distress** organisms **at all** when doing fieldwork, so investigations should be planned so that organisms are treated with **great care**, and are **kept** and **handled** as **little** as possible. They should also be **released** as soon as possible after they have been captured.

Investigating Populations

You need to *Analyse* and *Interpret Data* on the *Distribution of Organisms*

Here's an **example** of the kind of thing you might get in the **exam**:

A group of students investigated how the **distribution** of plant species changed with **distance** from a path. They used a **belt transect** (see p. 97) and measured **percentage cover** of plant species in each quadrat. The students also carried out a **survey** at the **same** location to record how many people **strayed away** from the path, and **how far** they strayed. Here's a **table** and a **graph** showing their results.

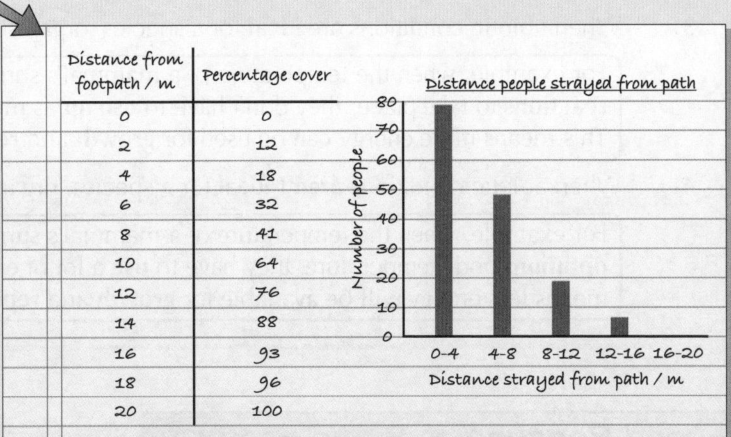

Distance from footpath / m	Percentage cover
0	0
2	12
4	18
6	32
8	41
10	64
12	76
14	88
16	93
18	96
20	100

You might have to:

1) **Describe the data:**
   - The table shows **low** percentage cover of plants **near** the path, e.g. **2 m** from the path it was **12%**, but **higher** percentage cover **away** from the path, e.g. **20 m** from the path it was **100%**.
   - The graph shows **lots** of walkers **near** the path, e.g. **0-4 m** from the path there were **79** walkers, but **fewer** walkers away from the path, e.g. **16-20 m** from the path there **were none**.

2) **Draw conclusions:**
   - There's a **positive correlation** between **distance** from the path and **percentage cover** of plants — as distance from the path **increases**, the percentage cover of plants **increases**.
   - There's a **negative correlation** between **distance** from the path and the **number** of people that walk there — as distance from the path **increases**, the number of people that walk there **decreases**.
   - There's a **negative correlation** between the **number of walkers** and the **percentage cover** of plants — the **higher** the number of people that walk over an area, the **lower** the percentage cover of plants.

   You **can't conclude** that the **lower percentage cover** of plants **near** the path is **caused** by the **higher number** of **people** walking there. There could be **other factors** involved that affect the percentage cover of plants, e.g. the path may be covered by **stones** or **gravel**, so plants won't grow **on** or **near** the path regardless of how many people walk on it.

See pages 90-92 and 196-198 for more on interpreting data.

3) **Suggest explanations for your conclusions:**

   As you move **away** from the path the **number** of people that trample the ground **decreases** because people tend to **follow the path**. As you move **away** from the path the **percentage cover** of plants **increases** because plants **grow** and **survive better** where they're trodden on **less**.

Practice Questions

Q1 Name one fieldwork risk when investigating populations.

Q2 Give an example of an ethical issue associated with fieldwork into populations.

Exam Question

Number of snails caught in first sample	Number of snails caught in second sample	Number of marked snails caught in second sample
52	38	14

Q1 The size of a snail population was investigated using the mark-release-recapture method. The table shows the results.

a) Describe the method that could have been used to collect the data. [5 marks]

b) Calculate the total population size. [2 marks]

Risks associated with this book — laughter, increased intelligence...

Mark-release-recapture isn't too bad — it's exactly what it sounds like. Risk assessments aren't too bad either, they usually just involve a bit of common-sense thinking to work out what might be dangerous. As always, interpreting data can be a pain — just make sure you're clear that even if two things correlate, it doesn't mean that one is caused by the other.

Variation in Population Size

Uh-oh, anyone who loves cute little bunny-wunnys look away now — these pages are about how the population sizes of organisms fluctuate and the reasons why. One of the reasons, I'm sad to say, is because the little rabbits get eaten.

Population Size Varies Because of Abiotic Factors...

Remember — abiotic factors are the non-living features of the ecosystem.

1) **Population size** is the **total number** of organisms of **one species** in a **habitat**.
2) The **population size** of any species **varies** because of **abiotic** factors, e.g. the amount of **light**, **water** or **space** available, the **temperature** of their surroundings or the **chemical composition** of their surroundings.
3) When abiotic conditions are **ideal** for a species, organisms can **grow fast** and **reproduce successfully**.

   For example, when the temperature of a mammal's surroundings is the ideal temperature for **metabolic reactions** to take place, they don't have to **use up** as much energy **maintaining** their **body temperature**. This means more energy can be used for **growth** and **reproduction**, so their population size will **increase**.

4) When abiotic conditions **aren't ideal** for a species, organisms **can't** grow as **fast** or reproduce as **successfully**.

   For example, when the temperature of a mammal's surroundings is significantly **lower** or **higher** than their **optimum** body temperature, they have to **use** a lot of **energy** to maintain the right **body temperature**. This means less energy will be available for **growth** and **reproduction**, so their population size will **decrease**.

...and Because of Biotic Factors

Remember — biotic factors are the living features of the ecosystem.

1 *Interspecific Competition — Competition Between Different Species*

1) Interspecific competition is when organisms of **different species compete** with each other for the **same resources**, e.g. **red** and **grey** squirrels compete for the same **food sources** and **habitats** in the **UK**.
2) Interspecific competition between two species can mean that the **resources available** to **both** populations are **reduced**, e.g. if they share the **same** source of food, there will be **less** available to both of them. This means both populations will be **limited** by a lower amount of food. They'll have less **energy** for **growth** and **reproduction**, so the population sizes will be **lower** for both species. E.g. in areas where both **red** and **grey** squirrels live, both populations are **smaller** than they would be if there was **only one** species there.
3) If **two** species are competing but one is **better adapted** to its surroundings than the other, the less well adapted species is likely to be **out-competed** — it **won't** be able to **exist** alongside the better adapted species. E.g. since the introduction of the **grey squirrel** to the UK, the native **red squirrel** has **disappeared** from large areas. The grey squirrel has a better chance of **survival** because it's **larger** and can store **more fat** over winter. It can also eat a **wider range** of **food** than the red squirrel.

Never mind what the doctors said, Nutkin knew his weight problem would increase his chance of survival.

2 *Intraspecific Competition — Competition Within a Species*

Intraspecific competition is when organisms of the **same species compete** with each other for the **same resources**.

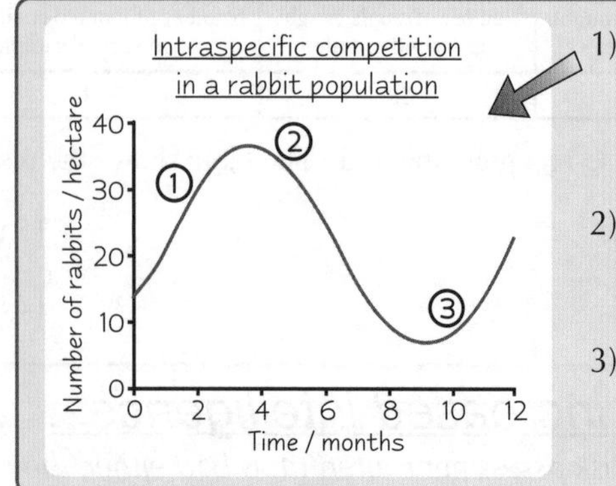

1) The **population** of a species **increases** when resources are **plentiful**. As the population increases, there'll be **more** organisms competing for the **same amount** of **space** and **food**.
2) Eventually, resources such as food and space become **limiting** — there **isn't enough** for all the organisms. The population then begins to **decline**.
3) A **smaller** population then means that there's **less competition** for space and food, which is **better** for **growth** and **reproduction** — so the population starts to **grow** again.

Variation in Population Size

3) Predation — Predator and Prey Population Sizes are Linked

Predation is where an organism (the predator) kills and eats another organism (the prey), e.g. lions kill and eat (**predate** on) buffalo. The **population sizes** of predators and prey are **interlinked** — as the population of one **changes**, it **causes** the other population to **change**:

1) As the **prey** population **increases**, there's **more food** for predators, so the **predator** population **grows**. E.g. in the graph on the right the **lynx** population **grows** after the **snowshoe hare** population has **increased** because there's **more food** available.
2) As the **predator** population **increases**, **more prey** is **eaten** so the **prey** population then begins to **fall**. E.g. **greater numbers** of lynx eat lots of snowshoe hares, so their population **falls**.
3) This means there's **less food** for the **predators**, so their population **decreases**, and so on. E.g. **reduced** snowshoe hare numbers means there's **less food** for the lynx, so their population **falls**.

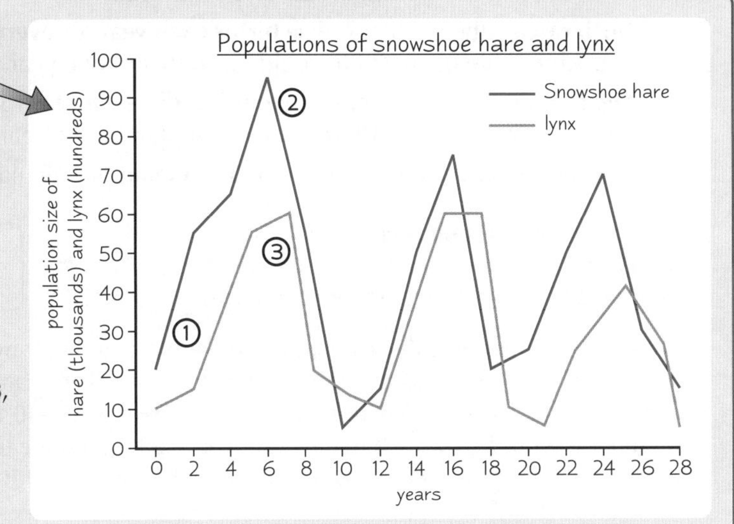

Predator-prey relationships are usually more **complicated** than this though because there are **other factors** involved, like availability of **food** for the **prey**. E.g. it's thought that the population of snowshoe hare initially begins to **decline** because there's **too many** of them for the amount of **food available**. This is then **accelerated** by **predation** from the lynx.

Practice Questions

Q1 Give one example of how an abiotic factor can affect population size.

Q2 What is interspecific competition?

Q3 What will be the effect of interspecific competition on the population size of a species?

Q4 What does it mean when a species is out-competed?

Q5 Give one example of interspecific competition.

Q6 Define intraspecific competition.

Exam Question

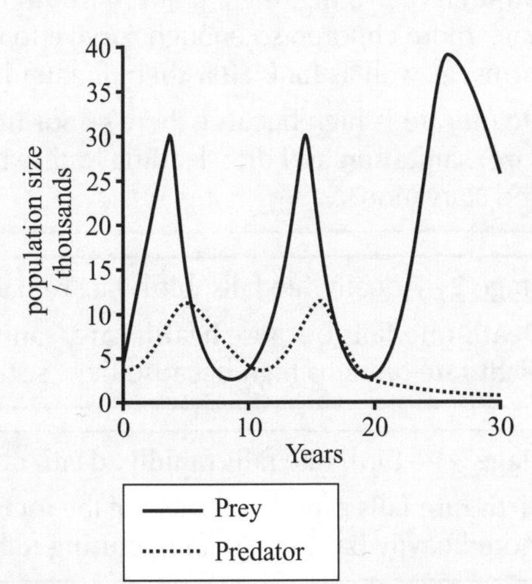

Q1 The graph on the right shows the population size of a predator species and a prey species over a period of 30 years.

a) Using the graph, describe and explain how the population sizes of the predator and prey species vary over the first 20 years. [7 marks]

b) The numbers of species B declined after year 20 because of a disease. Describe and explain what happened to the population of species A. [4 marks]

Predator-prey relationships — they don't usually last very long...

You'd think they could have come up with names a little more different than inter- and intraspecific competition. I always remember it as int-er means diff-er-ent species. The factors that affect population size are divided up nicely for you here — abiotic factors, competition and predation — just like predators like to nicely divide up their prey into bitesize chunks.

Human Populations

These pages are about how human populations change, so they're about joyful births... and not so joyful deaths.

Human Population Growth is Calculated using Birth and Death Rates

Human population sizes constantly **change**. Whether they're **growing** or **shrinking** (and by **how much**) depends on the population's **birth rate** and **death rate**.

1) **Birth rate** — the number of **live births each year** for **every 1000** people in the population, e.g. a birth rate of **10/1000** would mean that in one year there were **10 live births** for every **1000 people**.
2) **Death rate** — the number of people that **die each year** for every **1000** people in the population, e.g. a death rate of **10/1000** would mean that in one year there were **10 deaths** for every **1000 people**.

You can work out **how fast** the population's **changing** by calculating the **population growth rate**:

Population growth rate is how much the **population** size **increases** or **decreases** in a **year**. You can work it out using the **birth** and **death rate**:

population growth rate (per 1000 people per year) = birth rate – death rate

This gives you the **overall (net) number of people** that the population **grows** or **shrinks by** in a **year** for every **1000 people**. For example, if the birth rate was **13/1000** and the death rate was **10/1000** the population would grow by **3 people** for every **1000 people each year** (or **3/1000** people per year). It's normally given as a **percentage**, so a growth rate of **3/1000** people per year would be **0.3%** (3/1000 × 100%).

3/1000 = 13/1000 – 10/1000
3/1000 × 100% = 0.3%

The Demographic Transition Model shows Trends in Human Populations

The **Demographic Transition Model** (**DTM**) is a graph that shows changes in **birth rate**, **death rate** and **total population size** for a **human population** over a **long period** of time. It's divided into **five** stages:

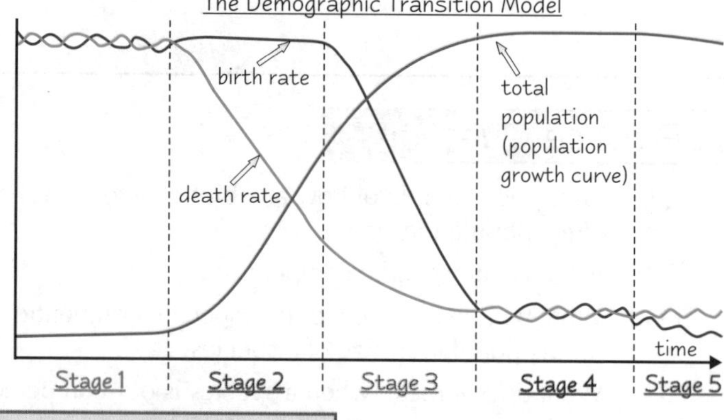

Stage 1 — birth rate and death rate fluctuate at a **high level**. The population stays **low**.
Birth rate is high because there's **no birth control** or **family planning** and **education** is **poor**. Lots of children **die young** (high **infant mortality**), so parents have more children so enough **survive** to **work** on farms, as well as **look after** them in later life.
Death rate is high because there's **poor health care, sanitation** and **diet**, leading to disease and starvation.

Stage 2 — death rate **falls**, birth rate **remains high**. The population **increases rapidly**.
Death rate falls because **health care**, **sanitation** and **diet improve**.
Birth rate remains **high** because there's still little **birth control** or **family planning**.

'Demographic' means it's to do with human populations.

Stage 3 — birth rate **falls rapidly**, death rate **falls more slowly**. The population **increases** at a **slower rate**.
Birth rate falls rapidly because of the **increased** use of **birth control** and **family planning**. Also, the economy becomes more heavily based on **manufacturing** rather than agriculture, so **fewer** children are needed to work on **farms**.

Stage 4 — birth rate and death rate fluctuate at a **low level**. The population remains **stable** but **high**.
Birth rate **stays low** because there's an **increased demand** for **luxuries** and **material possessions**, so **less money** is **available** to raise children. They're not needed to work to **provide income**, so parents have **fewer children**.

Stage 5 — birth rate begins to **fall**, death rate **remains stable**. The population begins to **decrease**.
Birth rate falls because children are **expensive** to raise and people often have **dependent elderly relatives**.
Death rate remains **steady** despite continued health care advances as **larger generations** of **elderly people die**.

Human Populations

Human Population Data can be Plotted in Different Ways

1 Population Growth Curves show Change in Population Size

Population change can be shown by a **population growth curve** (the **DTM** has one, see previous page). They're made by plotting data for **population size** against **time**.

1) **Growth curves** show whether the population was **increasing** or **decreasing** by the direction of the curve (**up** or **down**).

2) The **steepness** of the curve shows **how fast** the population was **changing** (the **steeper** the curve, the **faster** it was changing). You can use the curve to **calculate** the **rate of change**. For example, between **1850** and **1900** this population **increased** from **5** to **45** million. An increase of **40** million in **50** years meant the population **increased** at a rate of **800 000 people per year** (40 000 000 ÷ 50 = 800 000). Between **1950** and **2000**, the population **decreased** by **5** million in **50** years, so it **decreased** at a rate of **100 000 people per year** (5 000 000 ÷ 50 = 100 000).

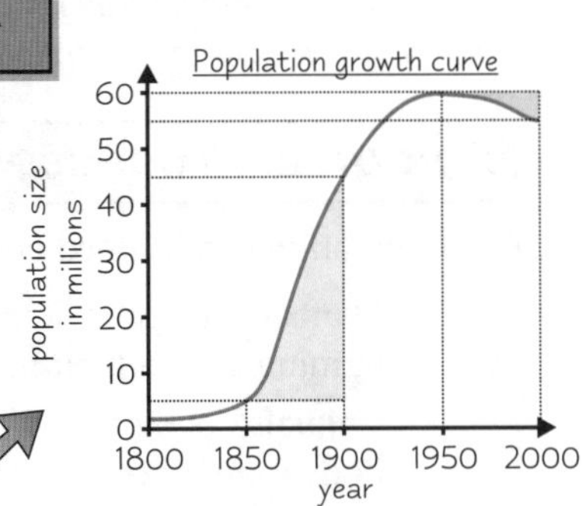

2 Survival Curves show Survival Rates

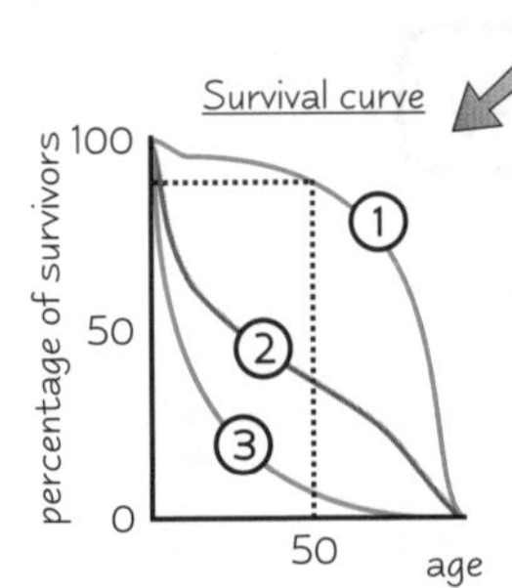

Survival curves show the **percentage** of all the individuals that were **born** in a population that are still **alive** at any **given age**. This gives a **survival rate** for any given age, e.g. population 1 has a survival rate of around **90%** for individuals at the age of **50** — 90% of people survive to be 50.

> **Population 1** — **few** people die at a young age, **lots** of people **survive** to an old age.
> **Population 2** — **many** people die at a young age, but **some survive** to an old age.
> **Population 3** — **most people die** at a young age, very **few survive** to an old age.

Life expectancy is the **age** that a person born into a population is expected to **live to** — it's worked out by calculating the **average age** that people **die**.

3 Age-population Pyramids show Population Structure

Population structure can be shown using **age-population pyramids**. These show how many **males** and **females** there are in different **age groups** within a population.

1) This population has **a lot** of **young people** with **very few** surviving to **old age** — there's a **low** life expectancy (**DTM stage 1**).

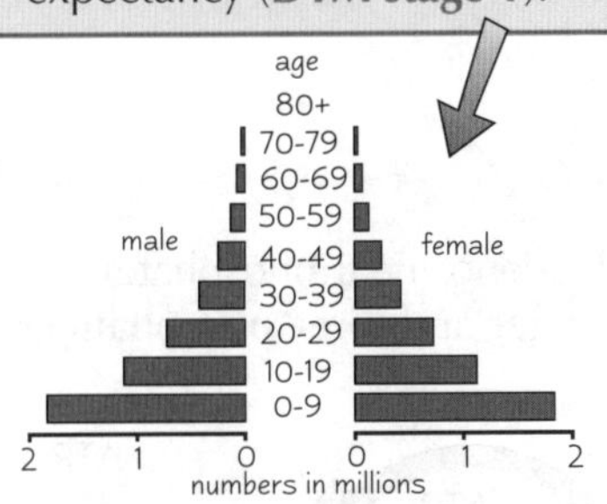

2) This population has **a lot** of **young people** with **more surviving** to **old age** — life expectancy is **higher** (**DTM stage 2**).

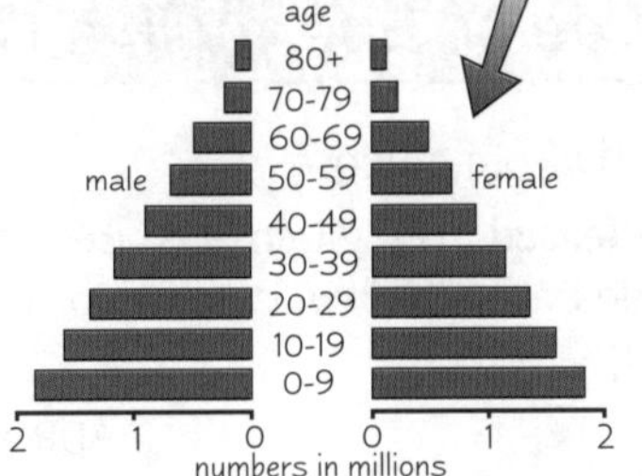

3) This population has **fewer young people** with **a lot of older people** — life expectancy is **high** (**DTM stage 5**).

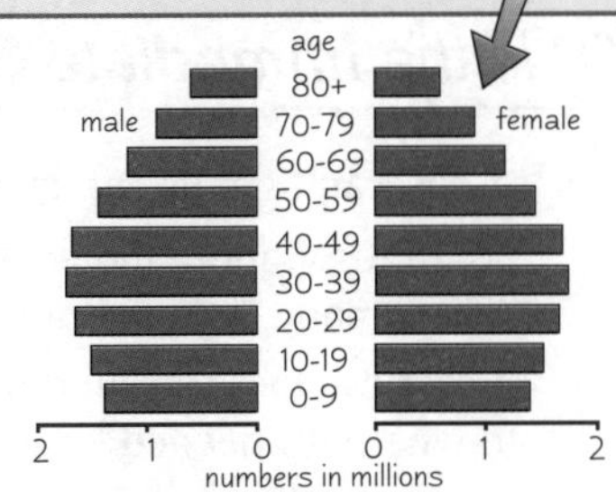

Practice Questions

Q1 How do you calculate population growth rate from birth and death rate?

Q2 What is shown by an age-population pyramid?

Exam Question

Q1 Describe the differences in population size and structure at stage 1 compared to stage 5 of the DTM. [4 marks]

Population's growth rate — almost 20 cm a year now he's a teenager...

Boy, when it comes to human populations these biologists love their graphs. Even if you feel like your brain's turning to custard, you need to understand what the graphs are showing — you might have to interpret them in the exam.

Photosynthesis, Respiration and ATP

All organisms need energy for life processes (and you'll need some for revising A2 Biology), so it's pretty important stuff. Annoyingly, it's pretty complicated stuff too, but 'cos I'm feeling nice today we'll take it slowly, one bit at a time...

Biological Processes Need Energy

Plant and animal cells **need energy** for biological processes to occur:

- **Plants** need energy for things like **photosynthesis**, **active transport** (e.g. to take in minerals via their roots), **DNA replication**, **cell division** and **protein synthesis**.
- **Animals** need energy for things like **muscle contraction**, maintenance of **body temperature**, **active transport**, **DNA replication**, **cell division** and **protein synthesis**.

Without energy, these biological processes would stop and the plant or animal would die.

Photosynthesis Stores Energy in Glucose

1) **Photosynthesis** is the process where **energy** from **light** is used to **make glucose** from H_2O and CO_2 (the light energy is **converted** to **chemical energy** in the form of glucose).
2) Photosynthesis occurs in a **series** of **reactions**, but the overall equation is:

$$6CO_2 + 6H_2O + \text{Energy} \longrightarrow C_6H_{12}O_6 \text{ (glucose)} + 6O_2$$

3) So, energy is **stored** in the **glucose** until the plants **release** it by **respiration**.
4) Animals obtain glucose by **eating plants** (or **other animals**), then respire the glucose to release energy.

Cells Release Energy from Glucose by Respiration

1) **Plant** and **animal** cells **release energy** from **glucose** — this process is called **respiration**.
2) This energy is used to power all the **biological processes** in a cell.
3) There are two types of respiration:
   - **Aerobic respiration** — respiration **using oxygen**.
   - **Anaerobic respiration** — respiration **without oxygen**.
4) Aerobic respiration produces **carbon dioxide** and **water**, and releases **energy**. The overall equation is:

$$C_6H_{12}O_6 \text{ (glucose)} + 6O_2 \longrightarrow 6CO_2 + 6H_2O + \text{Energy}$$

ATP is the Immediate Source of Energy in a Cell

1) A cell **can't** get its energy **directly** from glucose.
2) So, in respiration, the **energy released** from glucose is used to **make ATP** (adenosine triphosphate). ATP is made from the nucleotide base **adenine**, combined with a **ribose sugar** and **three phosphate groups**.
3) It **carries energy** around the cell to where it's **needed**.
4) **ATP is synthesised** from **ADP** and **inorganic phosphate** (P_i) using energy from an **energy-releasing** reaction, e.g. the **breakdown** of **glucose** in **respiration**. The energy is stored as **chemical energy** in the **phosphate bond**. The enzyme **ATP synthase** catalyses this reaction.
5) ATP **diffuses** to the part of the cell that **needs** energy.
6) Here, it's **broken down** back into **ADP** and **inorganic phosphate** (P_i). Chemical **energy** is **released** from the phosphate bond and used by the cell. **ATPase** catalyses this reaction.
7) The ADP and inorganic phosphate are **recycled** and the process starts again.

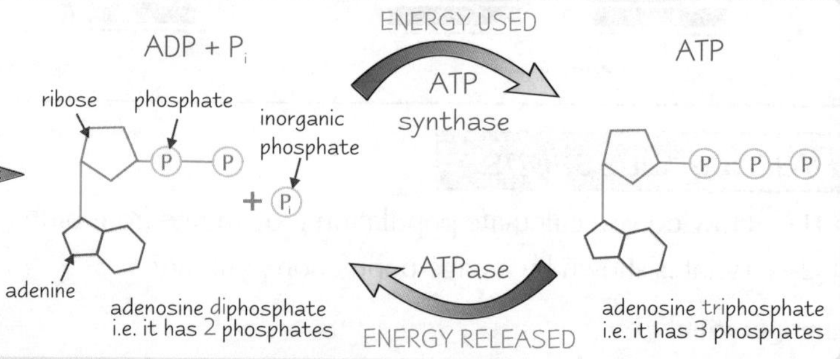

Inorganic phosphate (P_i) is just the fancy name for a single phosphate.

Photosynthesis, Respiration and ATP

ATP has Specific Properties that Make it a Good Energy Source

1) ATP stores or releases only a **small**, **managable amount** of energy at a time, so **no** energy is **wasted**.
2) It's a **small**, **soluble** molecule so it can be **easily transported** around the cell.
3) It's **easily broken down**, so energy can be **easily released**.
4) It can **transfer energy** to another molecule by transferring one of its **phosphate groups**.
5) ATP **can't pass out** of the **cell**, so the cell **always** has an immediate supply of energy.

Karen needed a lot of energy just to keep her headdress on...

You Need to Know Some Basics Before You Start

There are some pretty confusing technical terms in this section that you need to get your head around:

- **Metabolic pathway** — a **series** of **small reactions** controlled by **enzymes**, e.g. **respiration** and **photosynthesis**.
- **Phosphorylation** — **adding phosphate** to a molecule, e.g. **ADP** is phosphorylated to **ATP** (see previous page).
- **Photophosphorylation** — **adding phosphate** to a molecule using **light**.
- **Photolysis** — the **splitting** (lysis) of a molecule using **light** (photo) energy.
- **Hydrolysis** — the **splitting** (lysis) of a molecule using **water** (hydro).
- **Decarboxylation** — the **removal** of **carbon dioxide** from a molecule.
- **Dehydrogenation** — the **removal** of **hydrogen** from a molecule.
- **Redox reactions** — reactions that involve **oxidation** and **reduction**.

Remember redox reactions:

1) If something is **reduced** it has **gained electrons** (e^-), and may have **gained hydrogen** or lost oxygen.
2) If something is **oxidised** it has **lost electrons**, and may have **lost hydrogen** or gained oxygen.
3) Oxidation of one molecule **always** involves reduction of another molecule.

One way to remember electron and hydrogen movement is OILRIG. Oxidation Is Loss, Reduction Is Gain.

Photosynthesis and Respiration Involve Coenzymes

1) A **coenzyme** is a molecule that **aids** the **function** of an **enzyme**.
2) They work by **transferring** a **chemical group** from one molecule to another.
3) A coenzyme used in **photosynthesis** is **NADP**. NADP transfers **hydrogen** from one molecule to another — this means it can **reduce** (give hydrogen to) or **oxidise** (take hydrogen from) a molecule.
4) Examples of coenzymes used in **respiration** are: **NAD**, **coenzyme A** and **FAD**.
   - NAD and FAD transfer **hydrogen** from one molecule to another — this means they can **reduce** (give hydrogen to) or **oxidise** (take hydrogen from) a molecule.
   - **Coenzyme A** transfers **acetate** between molecules (see pages 113-114).

When hydrogen is transferred between molecules, electrons are transferred too.

Practice Questions

Q1 Write down three biological processes in animals that need energy.

Q2 What is photosynthesis?

Q3 What is the overall equation for aerobic respiration?

Q4 How many phosphate groups does ATP have?

Q5 Give the name of a coenzyme involved in photosynthesis.

Exam Question

Q1 ATP is the immediate source of energy inside a cell.
Describe how the synthesis and breakdown of ATP meets the energy needs of a cell. [6 marks]

Oh dear, I've used up all my ATP on these two pages...

Well, I won't beat about the bush, this stuff is pretty tricky... nearly as hard as a cross between Mr T, Hulk Hogan and Arnie. But, with a little patience and perseverance (and plenty of [chocolate] [coffee] [marshmallows] — delete as you wish), you'll get there. Once you've got these pages straight in your head, the next ones will be easier to understand.

Photosynthesis

Right, pen at the ready. Check. Brain switched on. Check. Cuppa piping hot. Check. Sweets on standby. Check. Okay, I think you're all sorted to start photosynthesis. Finally, take a deep breath and here we go...

Photosynthesis Takes Place in the Chloroplasts of Plant Cells

1) **Chloroplasts** are **small, flattened organelles** found in **plant cells**.
2) They have a **double membrane** called the **chloroplast envelope**.
3) **Thylakoids** (fluid-filled sacs) are **stacked up** in the chloroplast into structures called **grana** (singular = **granum**). The grana are **linked** together by bits of thylakoid membrane called **lamellae** (singular = **lamella**).
4) Chloroplasts contain **photosynthetic pigments** (e.g. **chlorophyll a, chlorophyll b** and **carotene**). These are **coloured substances** that **absorb** the **light energy** needed for photosynthesis. The pigments are found in the **thylakoid membranes** — they're attached to **proteins**. The protein and pigment is called a **photosystem**.
5) There are **two** photosystems used by plants to capture light energy. **Photosystem I** (or PSI) absorbs light best at a wavelength of **700 nm** and **photosystem II** (PSII) absorbs light best at **680 nm**.
6) Contained within the inner membrane of the chloroplast and **surrounding** the thylakoids is a gel-like substance called the **stroma**. It contains **enzymes, sugars** and **organic acids**.
7) Carbohydrates produced by photosynthesis and not used straight away are stored as **starch grains** in the **stroma**.

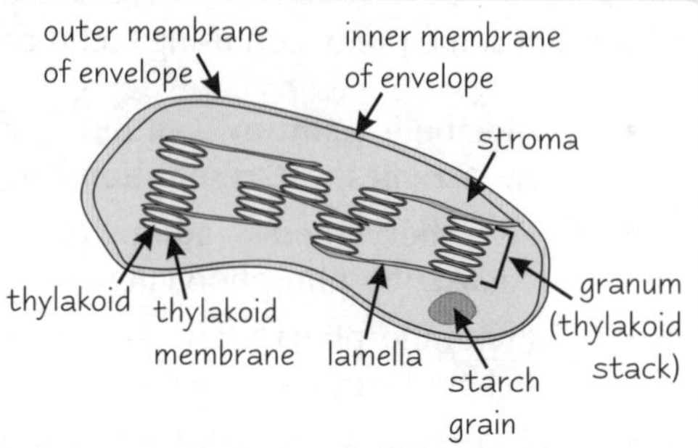

Photosynthesis can be Split into Two Stages

There are actually **two stages** that make up **photosynthesis**:

See p. 108 for loads more information on the Calvin cycle.

1 The Light-Dependent Reaction

1) As the name suggests, this reaction **needs light energy**.
2) It takes place in the **thylakoid membranes** of the chloroplasts.
3) Here, light energy is absorbed by **photosynthetic pigments** in the **photosystems** and converted to **chemical energy**.
4) The light energy is used to add a phosphate group to ADP to form **ATP**, and to reduce NADP to form **reduced NADP**. **ATP transfers energy** and reduced **NADP transfers hydrogen** to the light-independent reaction.
5) During the process H_2O is **oxidised** to O_2.

2 The Light-Independent Reaction

1) This is also called the **Calvin cycle** and as the name suggests it **doesn't use light energy** directly. (But it does **rely** on the **products** of the light-dependent reaction.)
2) It takes place in the **stroma** of the chloroplast.
3) Here, the **ATP** and **reduced NADP** from the light-dependent reaction supply the **energy** and **hydrogen** to make **glucose** from CO_2.

This diagram shows how the two reactions link together in the chloroplast:

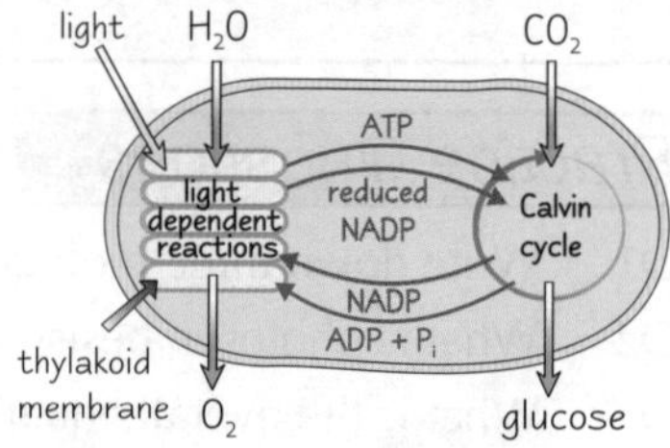

In the Light-Dependent Reaction ATP is Made by Photophosphorylation

In the light-dependent reaction, the **light energy** absorbed by the photosystems is used for **three** things:

1) Making **ATP** from **ADP** and **inorganic phosphate**. This reaction is called **photophosphorylation** (see p. 105).
2) Making **reduced NADP** from **NADP**.
3) Splitting **water** into **protons** (H^+ ions), **electrons** and **oxygen**. This is called **photolysis** (see p. 105).

The light-dependent reaction actually includes **two types** of **photophosphorylation** — **non-cyclic** and **cyclic**. Each of these processes has **different products**.

Photosynthesis

Non-cyclic Photophosphorylation Produces ATP, Reduced NADP and O_2

To understand the process you need to know that the photosystems (in the thylakoid membranes) are **linked** by **electron carriers**. Electron carriers are **proteins** that **transfer electrons**. The photosystems and electron carriers form an **electron transport chain** — a **chain** of **proteins** through which **excited electrons flow**. All the processes in the diagrams are happening together — I've just split them up to make it easier to understand.

1) Light energy excites electrons in chlorophyll

- **Light energy** is absorbed by **PSII**.
- The light energy **excites electrons** in **chlorophyll**.
- The electrons move to a **higher energy level** (i.e. they have more energy).
- These high-energy electrons **move along** the **electron transport chain** to **PSI**.

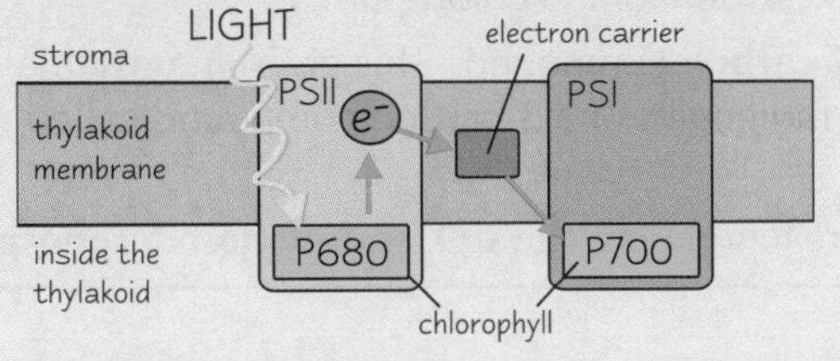

2) Photolysis of water produces protons (H^+ ions), electrons and O_2

- As the excited electrons **from chlorophyll leave PSII** to **move along** the electron transport chain, they must be **replaced**.
- **Light** energy splits **water** into **protons** (H^+ ions), **electrons** and **oxygen**. (So the O_2 in photosynthesis comes from water.)
- The reaction is: $H_2O \longrightarrow 2H^+ + \frac{1}{2}O_2$

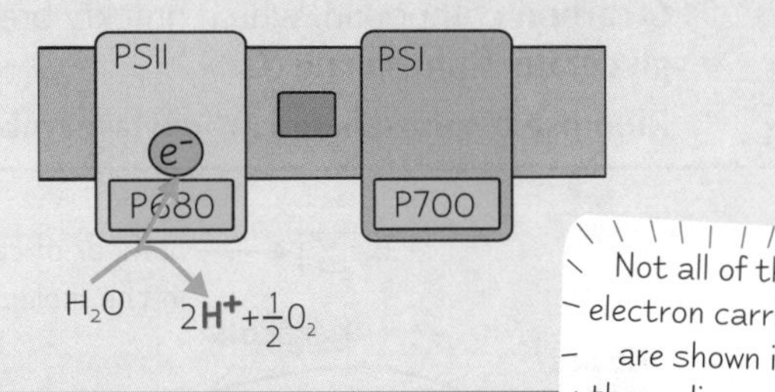

Not all of the electron carriers are shown in these diagrams.

3) Energy from the excited electrons makes ATP...

- The excited electrons **lose energy** as they **move along** the **electron transport chain**.
- This energy is used to **transport protons into** the **thylakoid** so that the thylakoid has a **higher concentration** of protons than the stroma. This forms a **proton gradient** across the membrane.
- Protons move **down** their concentration gradient, into the stroma, **via** an enzyme called **ATP synthase**. The energy from this movement combines **ADP** and **inorganic phosphate** (P_i) to form **ATP**.

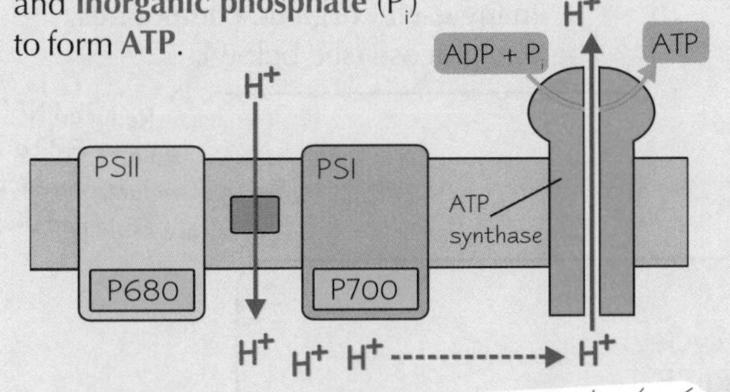

Chemiosmosis is the name of the process where the movement of H^+ ions across a membrane generates ATP. This process also occurs in respiration (see p. 115).

4) ...and generates reduced NADP.

- Light energy is **absorbed** by PSI, which excites the electrons again to an **even higher** energy level.
- Finally, the electrons are **transferred** to **NADP**, along with a **proton** (H^+ ion) from the **stroma**, to form **reduced NADP**.

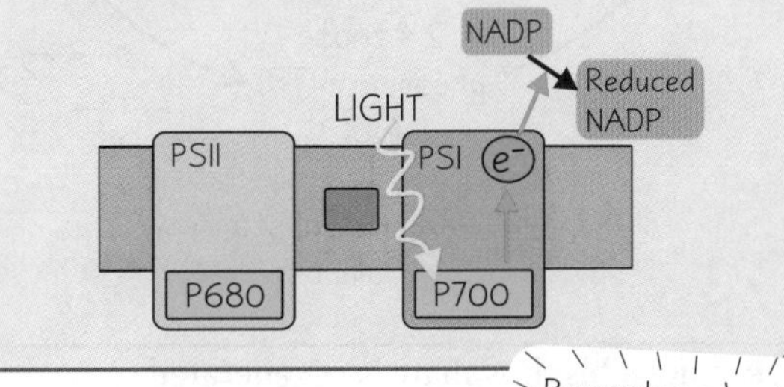

Remember a 'proton' is just another word for a hydrogen ion (H^+).

Cyclic Photophosphorylation Only Produces ATP

Cyclic photophosphorylation **only uses PSI**. It's called 'cyclic' because the electrons from the chlorophyll molecule **aren't** passed onto NADP, but are **passed back** to PSI via electron carriers. This means the electrons are **recycled** and can repeatedly flow through PSI. This process doesn't produce any reduced NADP or O_2 — it **only produces** small amounts of **ATP**.

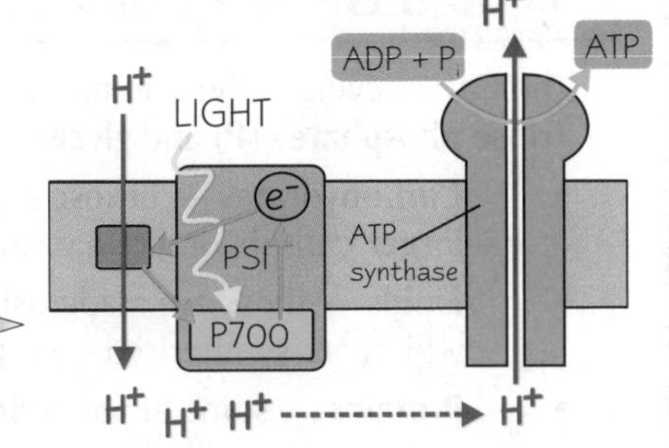

Photosynthesis

Don't worry, you're over the worst of photosynthesis now. Instead of electrons flying around, there's a nice cycle of reactions to learn. What more could you want from life? Money, fast cars and nice clothes have nothing on this...

The Light-Independent Reaction is also called the Calvin Cycle

1) The Calvin cycle takes place in the **stroma** of the chloroplasts.
2) It makes a molecule called **triose phosphate** from CO_2 and **ribulose bisphosphate** (a 5-carbon compound). Triose phosphate can be used to make **glucose** and other **useful organic substances** (see below).
3) There are a few steps in the cycle, and it needs **ATP** and **H^+ ions** to keep it going.
4) The reactions are linked in a **cycle**, which means the starting compound, **ribulose bisphosphate**, is **regenerated**.

The Calvin cycle is also called carbon fixation, because carbon from CO_2 is 'fixed' into an organic molecule.

Here's what happens at each stage in the cycle:

1) Carbon dioxide is combined with ribulose bisphosphate to form two molecules of glycerate 3-phosphate

- CO_2 enters the leaf through the **stomata** and diffuses into the **stroma** of the chloroplast.
- Here, it's combined with **ribulose bisphosphate (RuBP)**, a **5-carbon** compound. This gives an **unstable 6-carbon** compound, which quickly breaks down into **two** molecules of a **3-carbon** compound called **glycerate 3-phosphate (GP)**.
- **Ribulose bisphosphate carboxylase (rubisco)** catalyses the reaction between CO_2 and **ribulose bisphosphate**.

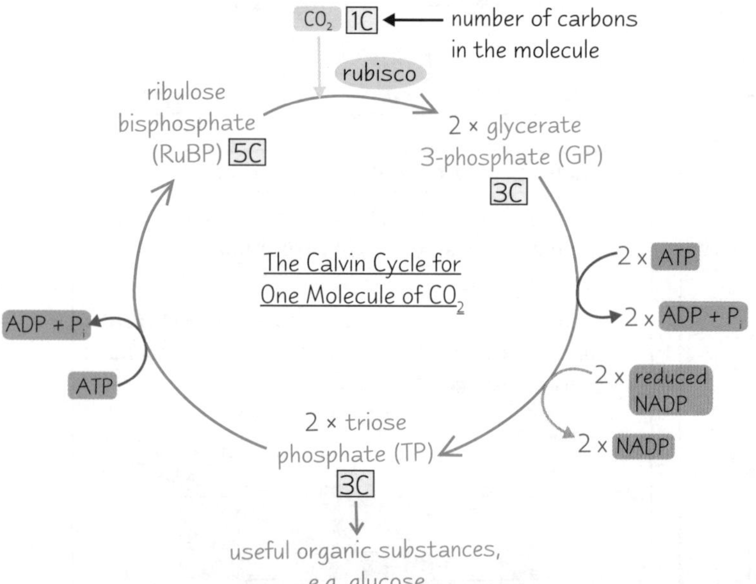

2) ATP and reduced NADP are required for the reduction of GP to triose phosphate

- Now **ATP**, from the **light-dependent** reaction, **provides energy** to turn the **3-carbon** compound, **GP**, into a **different** 3-carbon compound called **triose phosphate (TP)**.
- This reaction also requires **H^+ ions**, which come from **reduced NADP** (also from the **light-dependent reaction**). Reduced NADP is **recycled** to **NADP**.
- **Triose phosphate** is then converted into many **useful organic compounds**, e.g. glucose (see below).

Reduced NADP reduces GP to TP — reduction reactions are explained on p. 105.

3) Ribulose bisphosphate is regenerated

- **Five** out of every **six** molecules of **TP** produced in the cycle aren't used to make hexose sugars, but to **regenerate RuBP**.
- Regenerating RuBP uses the **rest** of the **ATP** produced by the **light-dependent reaction**.

TP and GP are Converted into Useful Organic Substances like Glucose

The Calvin cycle is the starting point for making **all** the organic substances a plant needs.
Triose phosphate (TP) and **glycerate 3-phosphate** (GP) molecules are used to make **carbohydrates**, **lipids** and **proteins**:

- **Carbohydrates** — **hexose sugars** (e.g. glucose) are made by joining **two triose phosphate molecules** together and **larger** carbohydrates (e.g. sucrose, starch, cellulose) are made by joining **hexose sugars** together in **different ways**.
- **Lipids** — these are made using **glycerol**, which is synthesised from **triose phosphate**, and **fatty acids**, which are synthesised from **glycerate 3-phosphate**.
- **Proteins** — some **amino acids** are made from **glycerate 3-phosphate**, which are joined together to make proteins.

Photosynthesis

The **Calvin Cycle** Needs to Turn **Six Times** to Make **One Hexose Sugar**

Here's the reason why:

1) **Three turns** of the cycle produces **six** molecules of **triose phosphate** (TP), because two molecules of TP are made for every one CO_2 molecule used.
2) **Five** out of **six** of these TP molecules are used to **regenerate ribulose bisphosphate** (RuBP).
3) This means that for **three turns** of the cycle only **one TP** is produced that's used to make a **hexose sugar**.
4) A hexose sugar has **six carbons** though, so **two TP** molecules are needed to form one hexose sugar.
5) This means the cycle must turn **six times** to produce **two molecules** of **TP** that can be used to make **one hexose sugar**.
6) Six turns of the cycle need **18 ATP** and **12 reduced NADP** from the light-dependent reaction.

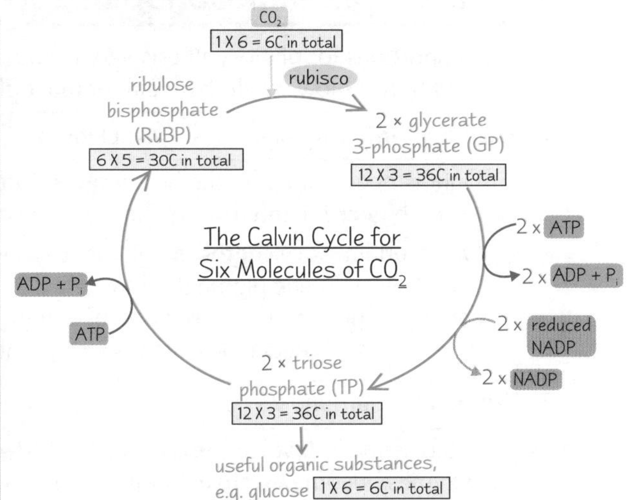

This might seem a bit inefficient, but it keeps the cycle going and makes sure there's always **enough RuBP** ready to combine with CO_2 taken in from the atmosphere.

Morag had to turn one million times to make a sock... two million for a scarf.

Practice Questions

Q1 Name two photosynthetic pigments in the chloroplasts of plants.

Q2 At what wavelength does photosystem I absorb light best?

Q3 What three substances does non-cyclic photophosphorylation produce?

Q4 Which photosystem is involved in cyclic photophosphorylation?

Q5 Where in the chloroplasts does the light-independent reaction occur?

Q6 How many carbon atoms are there in a molecule of TP?

Q7 Name two organic substances made from triose phosphate.

Q8 How many CO_2 molecules need to enter the Calvin cycle to make one hexose sugar?

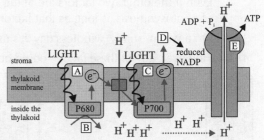

Exam Questions

Q1 The diagram above shows the light-dependent reaction of photosynthesis.
- a) Where precisely in a plant does the light-dependent reaction of photosynthesis occur? [1 mark]
- b) What is A? [1 mark]
- c) Describe process B and explain its purpose. [4 marks]
- d) What is reactant D? [1 mark]

Q2 Rubisco is an enzyme that catalyses the first reaction of the Calvin cycle. CA1P is an inhibitor of rubisco.
- a) Describe how triose phosphate is produced in the Calvin cycle. [6 marks]
- b) Briefly explain how ribulose bisphosphate (RuBP) is regenerated in the Calvin cycle. [2 marks]
- c) Explain the effect that CA1P would have on glucose production. [3 marks]

Calvin cycles — bikes made by people that normally make pants...

Next thing we know there'll be male models swanning about in their pants riding highly fashionable bikes. Sounds awful I know, but let's face it, anything would look better than cycling shorts. Anyway, it would be a good idea to go over these pages a couple of times — you might not feel as if you can fit any more information in your head, but you can, I promise.

Limiting Factors in Photosynthesis

I'd love to tell you that you'd finished photosynthesis... but I'd be lying.

There are Optimum Conditions for Photosynthesis

The **ideal conditions** for photosynthesis vary from one plant species to another, but the conditions below would be ideal for **most** plant species in temperate climates like the UK.

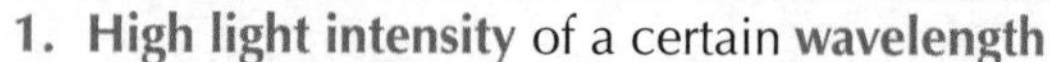

1. High light intensity of a certain wavelength

- Light is needed to provide the **energy** for the **light-dependent reaction** — the **higher** the **intensity** of the light, the **more energy** it provides.
- Only certain **wavelengths** of light are used for photosynthesis. The photosynthetic pigments chlorophyll a, chlorophyll b and carotene only **absorb** the **red** and **blue** light in sunlight. (**Green** light is **reflected**, which is why plants look green.)

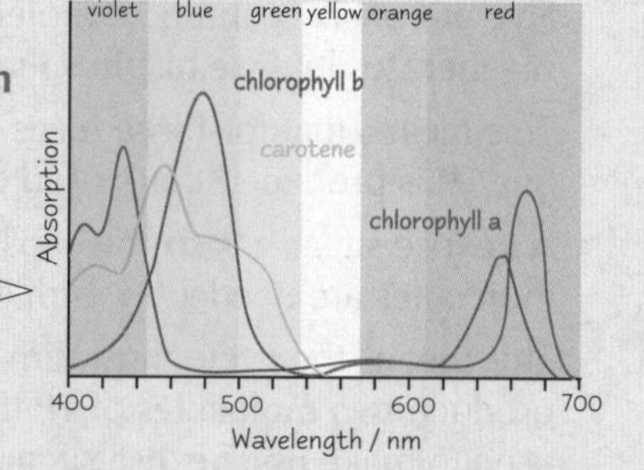

2. Temperature around 25 °C

- Photosynthesis involves **enzymes** (e.g. ATP synthase, rubisco). If the temperature falls **below 10 °C** the enzymes become **inactive**, but if the temperature is **more than 45 °C** they may start to **denature**.
- Also, at **high** temperatures **stomata close** to avoid losing too much water. This causes photosynthesis to slow down because **less CO_2** enters the leaf when the stomata are closed.

3. Carbon dioxide at 0.4%

- Carbon dioxide makes up **0.04%** of the gases in the atmosphere.
- Increasing this to **0.4%** gives a **higher rate** of photosynthesis, but any higher and the stomata start to **close**.

Plants also need a **constant supply** of **water** — **too little** and photosynthesis has to **stop** but **too much** and the soil becomes **waterlogged** (**reducing** the uptake of **minerals** such as **magnesium**, which is needed to make **chlorophyll a**).

Light, Temperature and CO_2 can all Limit Photosynthesis

1) **All three** of these things need to be at the **right level** to allow a plant to photosynthesise as quickly as possible.
2) If any **one** of these factors is **too low** or **too high**, it will **limit photosynthesis** (slow it down). Even if the other two factors are at the perfect level, it won't make **any difference** to the speed of photosynthesis as long as that factor is at the wrong level.
3) On a warm, sunny, windless day, it's usually **CO_2** that's the limiting factor, and at night it's the **light intensity**.
4) However, **any** of these factors could become the limiting factor, depending on the **environmental conditions**.

Light intensity

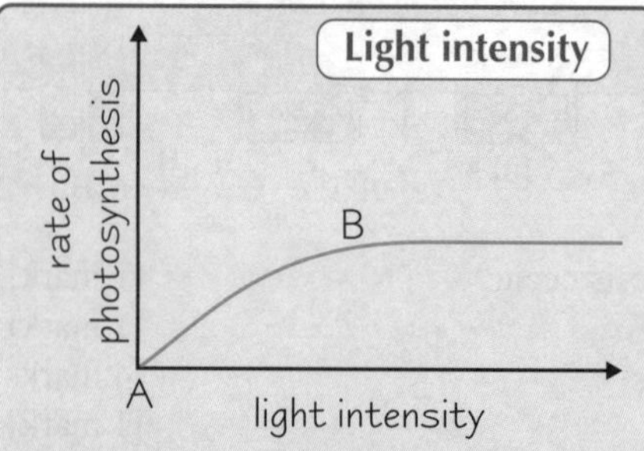

Between points A and B, the rate of photosynthesis is limited by the **light intensity**. So as the light intensity **increases**, so can the rate of photosynthesis. Point B is the **saturation point** — increasing light intensity after this point makes no difference, because **something else** has become the limiting factor. The graph now **levels off**.

Temperature

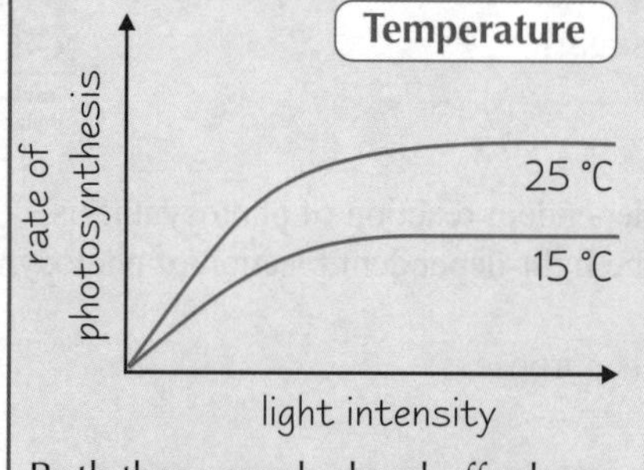

Both these graphs level off when **light intensity** is no longer the limiting factor. The graph at **25 °C** levels off at a **higher point** than the one at **15 °C**, showing that **temperature** must have been a limiting factor at **15 °C**.

CO_2 concentration

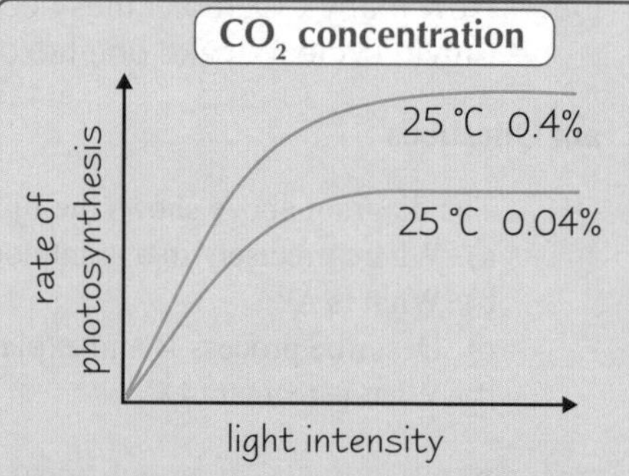

Again, both these graphs level off when **light intensity** is no longer the limiting factor. The graph at **0.4% CO_2** levels off at a **higher point** than the one at **0.04%**, so **CO_2 concentration** must have been a limiting factor at **0.04% CO_2**. The limiting factor here **isn't temperature** because it's the **same** for both graphs (25 °C).

The saturation point is where a factor is no longer limiting the reaction — something else has become the limiting factor.

Limiting Factors in Photosynthesis

Growers Use Information About Limiting Factors to Increase Plant Growth

Commercial growers (e.g. farmers) know the **factors** that **limit photosynthesis** and therefore limit **plant growth**. This means they can create an **environment** where plants get the **right amount** of everything that they need, which **increases growth** and so **increases yield**. Growers create optimum conditions in **glasshouses**, in the following ways:

Limiting Factor	Management in Glasshouse
Carbon dioxide concentration	CO_2 is added to the air, e.g. by burning a small amount of propane in a CO_2 generator.
Light	Light can get in through the glass. Lamps provide light at night-time.
Temperature	Glasshouses trap heat energy from sunlight, which warms the air. Heaters and cooling systems can also be used to keep a constant optimum temperature, and air circulation systems make sure the temperature is even throughout the glasshouse.

You Need to be Able to Interpret Data on Limiting Factors

Here are some **examples** of the kind of **data** you might get in the exam:

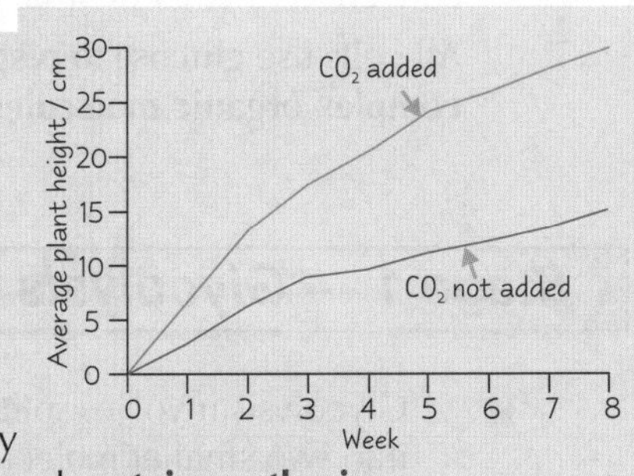

The graph on the **right** shows the effect on plant growth of **adding carbon dioxide** to a greenhouse.

1) In the greenhouse **with added CO_2** plant **growth** was **faster** (the line is steeper) and on average the plants were **larger** after 8 weeks than they were in the control greenhouse (30 cm compared to only 15 cm in the greenhouse where no CO_2 was added).
2) This is because the plants use CO_2 to produce **glucose** by photosynthesis. The more CO_2 they have, the more glucose they can produce, meaning they can **respire more** and so have **more ATP** for **DNA replication, cell division** and **protein synthesis**.

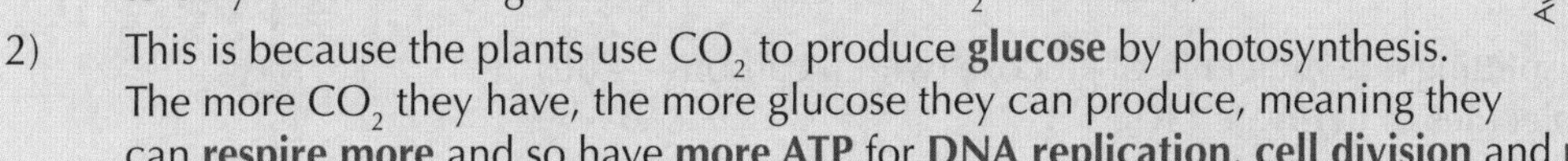

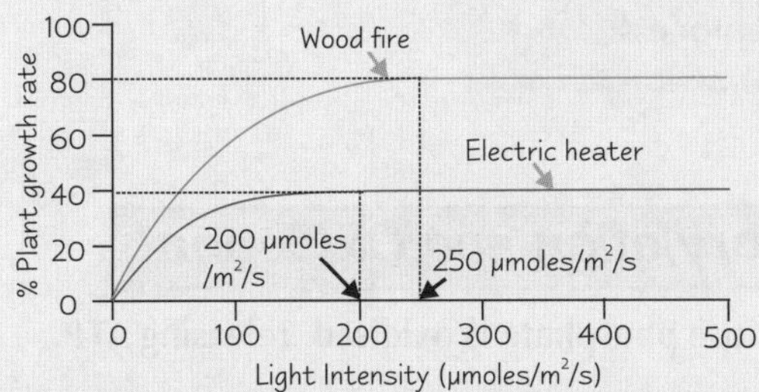

The graph on the **left** shows the effect of **light intensity** on plant growth, and the effect of two **different types** of **heater**.

1) At the start of the graph, the **greater** the **light intensity** the **greater** the **plant growth**.
2) At **200 µmoles/m²/s** of light the **bottom** graph flattens out, showing that **CO_2 concentration** or **temperature** is **limiting growth** in these plants.
3) At **250 µmoles/m²/s** of light the **top** graph flattens out.

The difference between the two graphs could be because the **wood fire increases** the **temperature more** than the electric heater or because it's **increasing** the **concentration** of **CO_2** in the air (an electric heater **doesn't** release CO_2).

Practice Questions

Q1 Name two factors that can limit plant growth.

Q2 How do commercial growers create an optimum level of CO_2 in a glasshouse?

Crop	Yield in glasshouse / kg	Yield grown outdoors / kg
Tomato	1000	200
Lettuce	750	230
Potato	850	680
Wheat	780	550

Exam Question

Q1 The table above shows the yields of various crops when they are grown in glasshouses and when grown outdoors.

a) Yields are usually higher overall in glasshouses.
Describe four ways in which conditions can be controlled in glasshouses to increase yields. [4 marks]

b) Glasshouses are not always financially viable for all crops.
Which crop above benefits the least from being grown in glasshouses? Explain your answer. [2 marks]

I'm a whizz at the factors that limit revision...

... watching Hollyoaks, making tea, watching EastEnders, walking the dog... not to mention staring into space (one of my favourites). Anyway, an interpreting data question could well come up in the exam — it could be any kind of data, but don't panic if it's not like the graphs above — as long as you understand limiting factors you'll be able to interpret it.

Respiration

From the last gazillion pages you know that plants make their own glucose. Unfortunately, that means now you need to learn how plant and animal cells release energy from glucose. It's not the easiest thing in the world to understand, but it'll make sense once you've gone through it a couple of times.

There are Four Stages in Aerobic Respiration

1) The four stages in aerobic respiration are **glycolysis**, the **link reaction**, the **Krebs cycle** and **oxidative phosphorylation**.
2) The **first three** stages are a **series of reactions**. The **products** from these reactions are **used** in the **final stage** to produce loads of ATP.
3) The **first** stage happens in the **cytoplasm** of cells and the **other three** stages take place in the **mitochondria**. You might want to refresh your memory of mitochondrion structure before you start.
4) **Anaerobic** respiration **doesn't involve** the **link reaction**, the **Krebs cycle** or **oxidative phosphorylation**. The **products** of **glycolysis** are converted to ethanol or lactate instead (see the next page for more).
5) All cells use **glucose** to **respire**, but organisms can also **break down** other **complex organic molecules** (e.g. fatty acids, amino acids), which can then be respired.

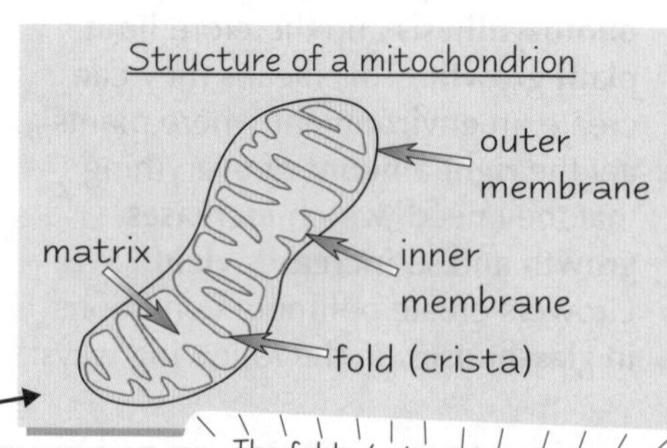

The folds (cristae) in the inner membrane of the mitochondrion provide a large surface area to maximise respiration.

Stage 1 — Glycolysis Makes Pyruvate from Glucose

1) Glycolysis involves splitting **one molecule** of glucose (with 6 carbons — 6C) into **two** smaller molecules of **pyruvate** (3C).
2) The process happens in the **cytoplasm** of cells.
3) Glycolysis is the **first stage** of both aerobic and anaerobic respiration and **doesn't need oxygen** to take place — so it's an **anaerobic** process.

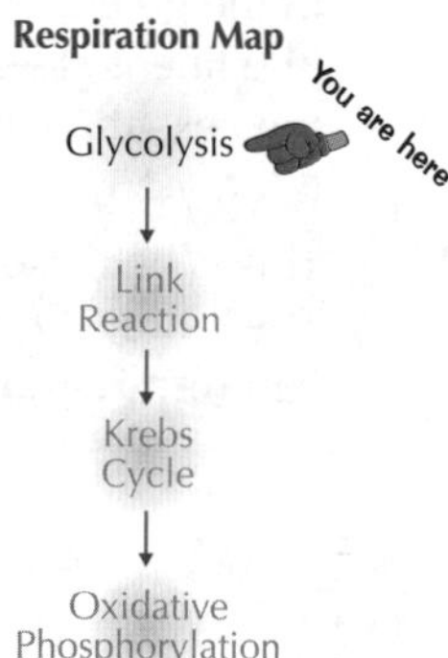

There are Two Stages in Glycolysis — Phosphorylation and Oxidation

First, **ATP** is **used** to **phosphorylate glucose** to triose phosphate. Then **triose phosphate** is **oxidised**, **releasing ATP**. Overall there's a **net gain** of **2 ATP**.

1 Stage One — Phosphorylation

1) Glucose is **phosphorylated** by adding **2 phosphates** from **2** molecules of **ATP**.
2) This creates **2** molecules of **triose phosphate** and **2** molecules of **ADP**.

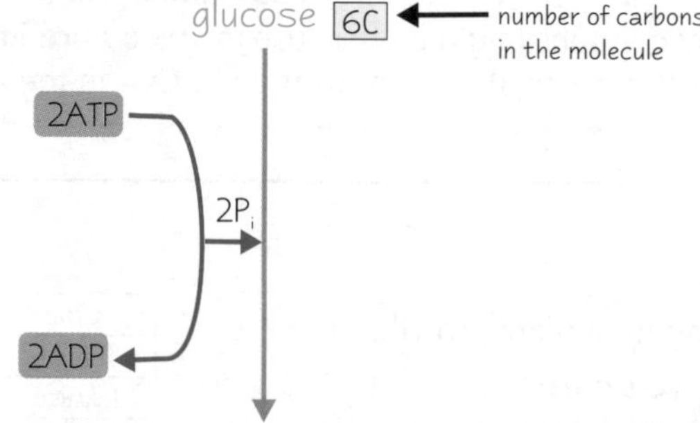

2 Stage Two — Oxidation

1) Triose phosphate is **oxidised** (loses hydrogen), forming **2** molecules of **pyruvate**.
2) **NAD** collects the hydrogen ions, forming **2 reduced NAD**.
3) **4 ATP** are **produced**, but 2 were used up in stage one, so there's a **net gain** of **2 ATP**.

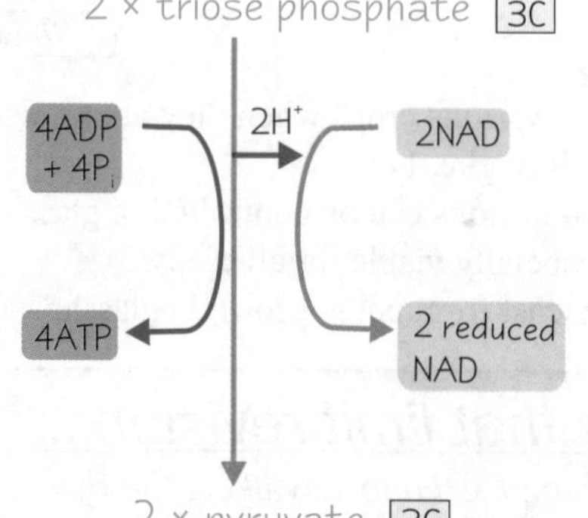

You're probably wondering what now happens to all the products of glycolysis...

1) The **two** molecules of **reduced NAD** go to the **last stage** (oxidative phosphorylation — see page 114).
2) The **two pyruvate** molecules go into the **matrix** of the **mitochondria** for the **link reaction** (see the next page).

Respiration

*Stage 2 — the **Link Reaction** converts **Pyruvate** to **Acetyl Coenzyme A***

1) **Pyruvate** is **decarboxylated** — **one carbon atom** is **removed** from pyruvate in the form of **CO_2**.
2) **NAD** is **reduced** — it collects **hydrogen** from pyruvate, changing pyruvate into **acetate**.
3) **Acetate** is combined with **coenzyme A** (CoA) to form **acetyl coenzyme A** (**acetyl CoA**).
4) **No ATP** is produced in this reaction.

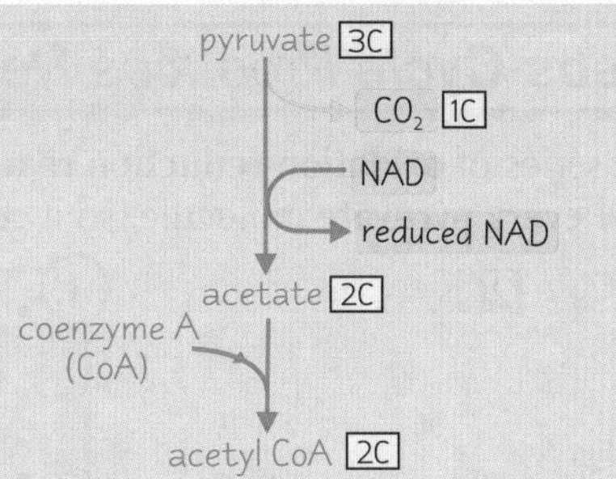

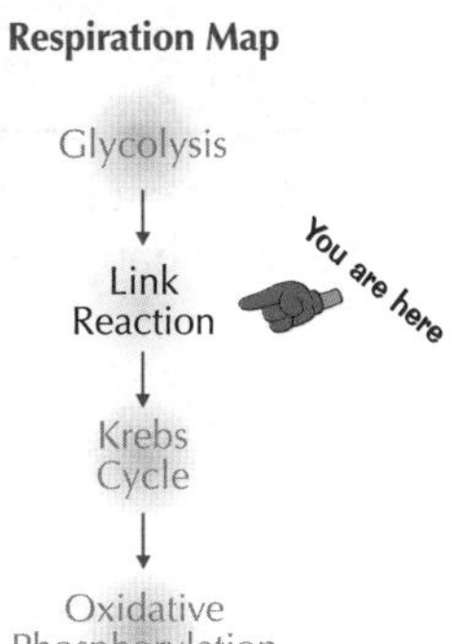

*The **Link Reaction** occurs **Twice** for every **Glucose Molecule***

Two pyruvate molecules are made for **every glucose molecule** that enters glycolysis. This means the **link reaction** and the third stage (the **Krebs cycle**) happen **twice** for every glucose molecule. So for each glucose molecule:

- **Two** molecules of **acetyl coenzyme A** go into the Krebs cycle (see the next page).
- **Two CO_2 molecules** are released as a waste product of respiration.
- **Two** molecules of **reduced NAD** are formed and go to the last stage (oxidative phosphorylation, see page 114).

*In **Anaerobic Respiration Pyruvate** is Converted to **Ethanol** or **Lactate***

1) In **aerobic** respiration (where there's **lots** of oxygen) **pyruvate** goes on to the third stage of respiration, the **Krebs cycle** (via the link reaction). In the Krebs cycle, **more ATP** is made and **NAD** is **reduced** (see next page).
2) However, in **anaerobic** respiration (where there's **no** oxygen) **pyruvate** is **converted** into **ethanol** (in **plants** and **yeast**) or **lactate** (in **animal** cells and some **bacteria**):

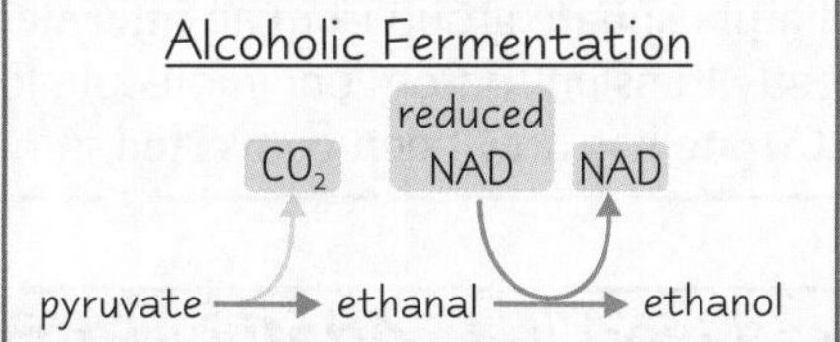

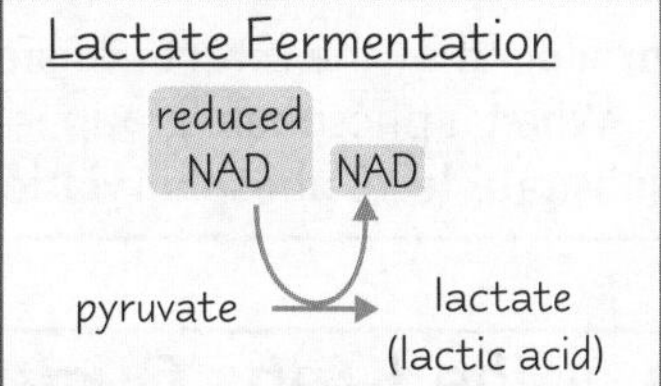

The production of lactate or ethanol **regenerates NAD**. This means **glycolysis** can **continue** even when there **isn't** much oxygen around, so a **small amount of ATP** can still be **produced** to keep some biological process going... clever.

Practice Questions

Q1 Where in the cell does glycolysis occur?

Q2 Is glycolysis an anaerobic or aerobic process?

Q3 How many ATP molecules are used up in glycolysis?

Q4 What is the product of the link reaction?

Exam Questions

Q1 Describe how a 6-carbon molecule of glucose is converted to pyruvate. [6 marks]

Q2 At the end of a 100 m sprint runners will have built up lactate in their muscle cells.
a) Write down an equation to show how lactate is produced. [2 marks]
b) What is the advantage of producing lactate in anaerobic respiration? [2 marks]

No ATP was harmed during this reaction...

Ahhhh... too many reactions. I'm sure your head hurts now, 'cause mine certainly does. Just think of revision as like doing exercise — it can be a pain while you're doing it (and maybe afterwards too), but it's worth it for the well-toned brain you'll have. Just keep going over and over it, until you get the first two stages of respiration straight in your head. Then relax.

Respiration

As you've seen, glycolysis produces a net gain of two ATP. Pah, we can do better than that. The Krebs cycle and oxidative phosphorylation are where it all happens — ATP galore.

*Stage 3 — the **Krebs Cycle** Produces **Reduced Coenzymes** and **ATP***

The Krebs cycle involves a series of **oxidation-reduction reactions**, which take place in the **matrix** of the **mitochondria**. The cycle happens **once** for **every pyruvate** molecule, so it goes round **twice** for **every glucose** molecule.

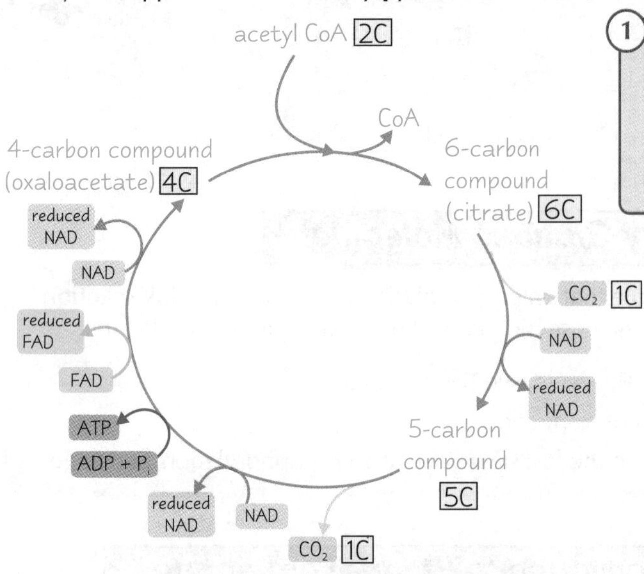

1
- **Acetyl CoA** from the link reaction combines with **oxaloacetate** to form **citrate**.
- **Coenzyme A** goes back to the **link reaction** to be used again.

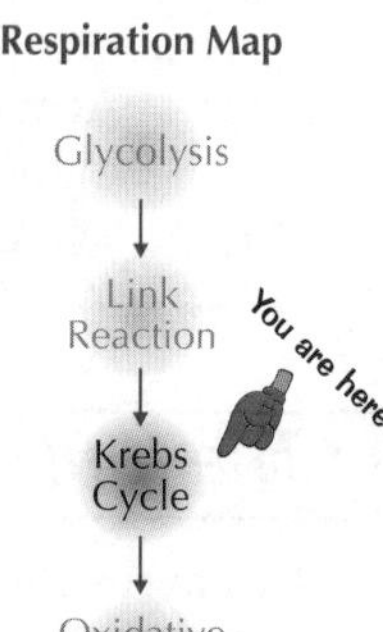

2
- The **6C citrate molecule** is converted to a **5C molecule.**
- **Decarboxylation** occurs, where **CO_2** is **removed.**
- **Dehydrogenation** also occurs, where **hydrogen** is **removed.**
- The hydrogen is used to **produce reduced NAD** from NAD.

- The **5C molecule** is then converted to a **4C molecule.** (There are some intermediate compounds formed during this conversion, but you don't need to know about them.)
- **Decarboxylation** and **dehydrogenation** occur, producing **one** molecule of **reduced FAD** and **two** of **reduced NAD.**
- **ATP** is **produced** by the **direct transfer** of a **phosphate** group from an **intermediate** compound **to ADP.** When a phosphate group is directly transferred from one molecule to another it's called **substrate-level phosphorylation. Citrate** has now been **converted** into **oxaloacetate.**

*Some **Products** of the **Krebs Cycle** are Used in **Oxidative Phosphorylation***

Some products are **reused**, some are **released** and others are used for the **next stage** of respiration:

Product from one Krebs cycle	Where it goes
1 coenzyme A	Reused in the next link reaction
Oxaloacetate	Regenerated for use in the next Krebs cycle
2 CO_2	Released as a waste product
1 ATP	Used for energy
3 reduced NAD	To oxidative phosphorylation
1 reduced FAD	To oxidative phosphorylation

Talking about oxidative phosphorylation was always a big hit with the ladies...

*Stage 4 — **Oxidative Phosphorylation Produces** Lots of **ATP***

1) Oxidative phosphorylation is the process where the **energy** carried by **electrons**, from **reduced coenzymes** (reduced NAD and reduced FAD), is used to **make ATP.** (The whole point of the previous stages is to make reduced NAD and reduced FAD for the final stage.)
2) Oxidative phosphorylation involves two processes — the **electron transport chain** and **chemiosmosis** (see the next page).

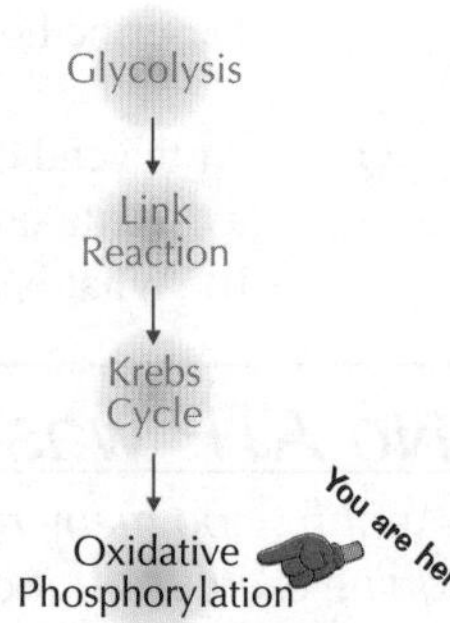

Respiration

Protons are Pumped Across the Inner Mitochondrial Membrane

So now on to how **oxidative phosphorylation** actually **works:**

1) **Hydrogen atoms** are released from **reduced NAD** and **reduced FAD** as they're oxidised to NAD and FAD. The H atoms **split** into **protons (H^+)** and **electrons (e^-)**.
2) The **electrons** move along the **electron transport chain** (made up of three **electron carriers**), **losing energy** at each carrier.
3) This energy is used by the electron carriers to **pump protons** from the **mitochondrial matrix into** the **intermembrane space** (the space **between** the inner and outer **mitochondrial membranes**).
4) The **concentration** of **protons** is now **higher** in the **intermembrane space** than in the mitochondrial matrix — this forms an **electrochemical gradient** (a **concentration gradient** of **ions**).
5) Protons **move down** the **electrochemical gradient**, back into the mitochondrial matrix, via **ATP synthase**. This **movement** drives the synthesis of **ATP** from **ADP** and **inorganic phosphate** (P_i).
6) The movement of H^+ ions across a membrane, which generates ATP, is called **chemiosmosis**.
7) In the mitochondrial matrix, at the end of the transport chain, the **protons**, **electrons** and **O_2** (from the blood) combine to form **water**. Oxygen is said to be the final **electron acceptor**.

The regenerated coenzymes are reused in the Krebs cycle.

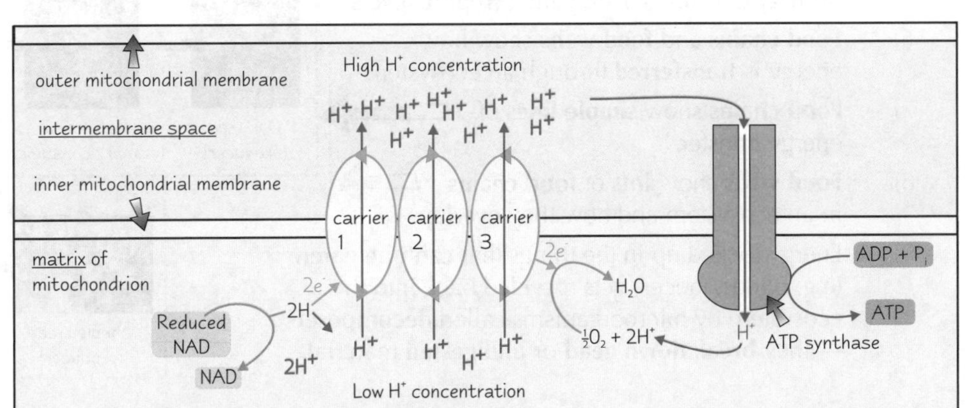

32 ATP Can be Made from One Glucose Molecule

As you know, **oxidative phosphorylation makes ATP** using energy from the reduced coenzymes — **2.5 ATP** are made from each **reduced NAD** and **1.5 ATP** are made from each **reduced FAD**. The table on the right shows **how much** ATP a cell can make from **one molecule** of **glucose** in **aerobic respiration**. (Remember, one molecule of glucose produces 2 pyruvate, so the link reaction and Krebs cycle happen twice.)

Stage of respiration	Molecules produced	Number of ATP molecules
Glycolysis	2 ATP	2
Glycolysis	2 reduced NAD	2 × 2.5 = 5
Link Reaction (×2)	2 reduced NAD	2 × 2.5 = 5
Krebs cycle (×2)	2 ATP	2
Krebs cycle (×2)	6 reduced NAD	6 × 2.5 = 15
Krebs cycle (×2)	2 reduced FAD	2 × 1.5 = 3
		Total ATP = 32

The number of ATP produced per reduced NAD or reduced FAD was thought to be 3 and 2, but new research has shown that the figures are nearer 2.5 and 1.5.

Practice Questions

Q1 Where in the cell does the Krebs cycle occur?

Q2 How many times does decarboxylation happen during one turn of the Krebs cycle?

Q3 What do the electrons lose as they move along the electron transport chain in oxidative phosphorylation?

Exam Question

Q1 Carbon monoxide inhibits the final electron carrier in the electron transport chain.

a) Explain how this affects ATP production via the electron transport chain. [2 marks]

b) Explain how this affects ATP production via the Krebs cycle. [2 marks]

The electron transport chain isn't just a FAD with the examiners...

Oh my gosh, I didn't think it could get any worse... You may be wondering how to learn these pages of crazy chemistry, but basically you have to put in the time and go over and over it. Don't worry though, it WILL pay off, and before you know it you'll be set for the exam. And once you know this section you'll be able to do anything, e.g. world domination...

Energy Transfer and Productivity

Some organisms get their energy from the sun and some get it from other organisms, and it's all very friendly. Yeah right.

Energy is Transferred Through Ecosystems

1) An **ecosystem** includes all the **organisms** living in a particular area and all the **non-living** (abiotic) conditions.
2) The **main route** by which energy **enters** an ecosystem is **photosynthesis** (e.g. by plants, see p. 104). (Some energy enters sea ecosystems when bacteria respire chemicals from deep sea vents.)
3) During photosynthesis plants **convert sunlight energy** into a form that can be **used** by other organisms — plants are called **producers** (even though they're only converting the energy, not producing it).
4) Energy is **transferred** through the **living organisms** of an ecosystem when organisms **eat** other organisms, e.g. producers are eaten by organisms called **primary consumers**. Primary consumers are then eaten by **secondary consumers** and secondary consumers are eaten by **tertiary consumers**.
5) Each of the stages (e.g. producers, primary consumers) are called **trophic levels**.
6) **Food chains** and **food webs** show how energy is **transferred** through an ecosystem.
7) **Food chains** show **simple lines** of energy transfer.
8) **Food webs** show **lots** of **food chains** in an ecosystem and how they **overlap**.
9) Energy locked up in the things that **can't be eaten** (e.g. bones, faeces) gets recycled back into the ecosystem by microorganisms called **decomposers** — they **break down dead** or **undigested** material.

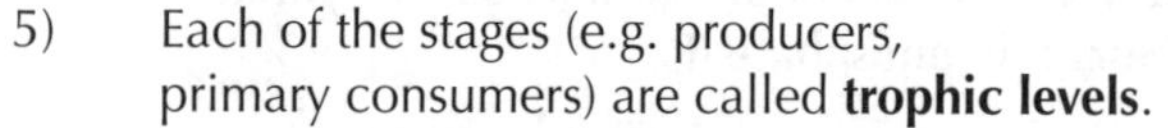

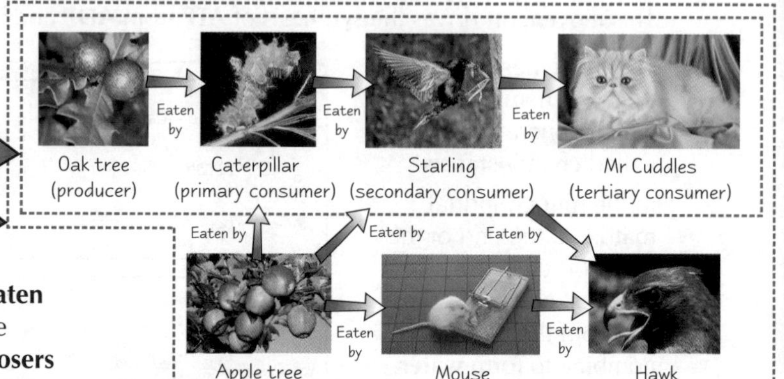

Not All Energy gets Transferred to the Next Trophic Level

1) **Not all** the energy (e.g. from sunlight or food) that's available to the organisms in a trophic level is **transferred** to the **next** trophic level — around **90%** of the **total available energy** is **lost** in various ways.
2) Some of the available energy (**60%**) is **never taken in** by the organisms in the first place. For example:
   - Plants **can't use** all the light energy that reaches their leaves (e.g. some is the **wrong wavelength**).
   - Some **parts** of food, e.g. **roots** or **bones**, **aren't eaten** by organisms so the energy isn't taken in.
   - Some parts of food are **indigestible** so **pass through** organisms and come out as **waste**, e.g. **faeces**.
3) The rest of the available energy (**40%**) is **taken in** (**absorbed**) — this is called the **gross productivity**. But not all of this is available to the next trophic level either.
   - **30%** of the **total energy** available (75% of the gross productivity) is **lost to the environment** when organisms use energy produced from **respiration** for **movement** or body **heat**. This is called **respiratory loss**.
   - **10%** of the **total energy** available (25% of the gross productivity) becomes **biomass** (e.g. it's **stored** or used for **growth**) — this is called the **net productivity**.
4) **Net productivity** is the amount of energy that's **available** to the **next trophic level**. Here's how it's **calculated**:

100% available energy
60% not taken in
40% gross productivity
10% net productivity (available to the next trophic level)
30% respiratory loss

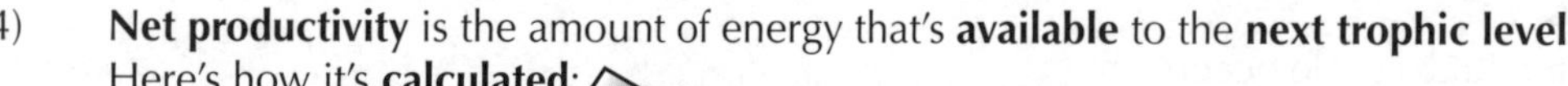

net productivity = gross productivity – respiratory loss

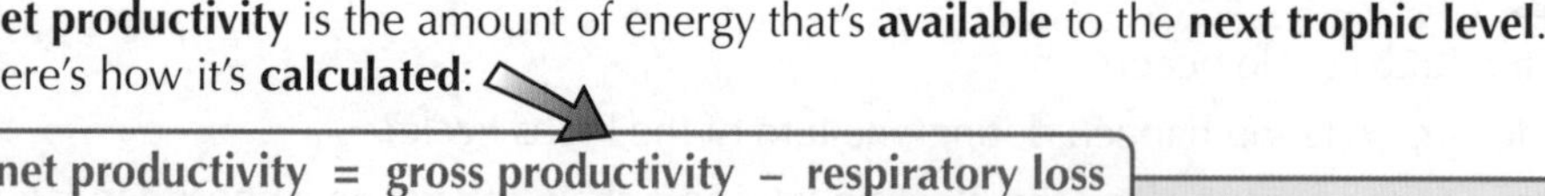

EXAMPLE: The rabbits in an ecosystem receive **20 000 $kJm^{-2}yr^{-1}$** of energy, but don't take in **12 000 $kJm^{-2}yr^{-1}$** of it, so their gross productivity is **8000 $kJm^{-2}yr^{-1}$** (20 000 – 12 000). They lose **6000 $kJm^{-2}yr^{-1}$** using energy from **respiration**. You can use this to **calculate** the **net productivity** of the rabbits:

net productivity = 8000 – 6000 = 2000 $kJm^{-2}yr^{-1}$

5) You might be asked to **calculate** how **efficient energy transfer** from one trophic level to another is:

The rabbits receive **20 000 $kJm^{-2}yr^{-1}$**, and their **net productivity** is **2000 $kJm^{-2}yr^{-1}$**. So the **percentage efficiency** of **energy transfer** is:

(2000 ÷ 20 000) × 100 = 10%

Energy Transfer and Productivity

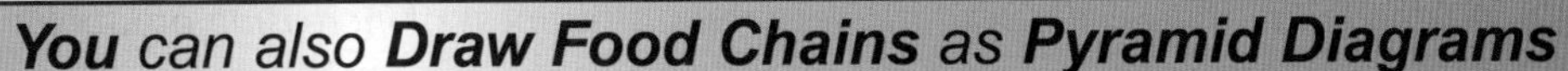

You can also Draw Food Chains as Pyramid Diagrams

1) **Food chains** can be shown by drawing **pyramids** with each block representing a **trophic level**.
2) **Producers** are always on the **bottom**, then **primary consumers** are above them, followed by **secondary consumers** then **tertiary consumers**.
3) The **area** of each block tells you about the **size** of the trophic level.
4) There are **three** types of pyramid — pyramids of **number**, **biomass** and **energy**:

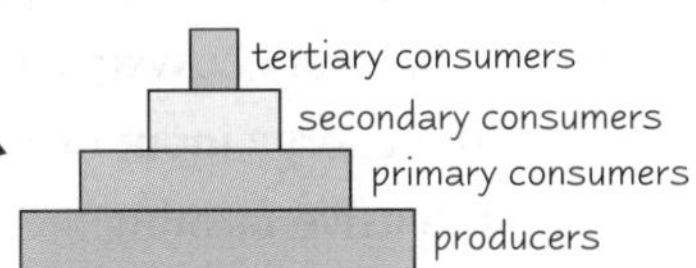

Pyramids of Numbers

- Pyramids of numbers show the **number** of organisms in each trophic level.
- They're not always **pyramid shaped** though — **small numbers** of **big organisms** (like trees) or **large numbers** of **small organisms** (like parasites) change the shape.

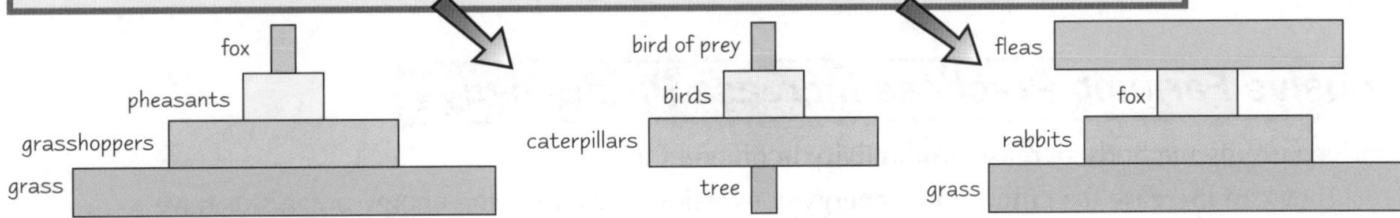

Pyramids of Biomass

- Pyramids of biomass show the **amount** of **biomass** in each trophic level (the **dry mass** of the organisms in **kgm**$^{-2}$) at a **single moment** in time.
- They **nearly** always come out pyramid-shaped. An exception is when they're based on **plant plankton** (microorganisms that photosynthesise) — the amount of plant plankton is quite **small** at any **given moment**, but because they have a **short life span** and **reproduce very quickly** there's **a lot** around over a **period of time**.

Pyramids of Energy

- Pyramids of energy show the **amount** of **energy** available in each trophic level in **kilojoules** per **square metre** per **year** (**kJm**$^{-2}$ **yr**$^{-1}$) — the **net productivity** of each trophic level (see previous page).
- Pyramids of energy are **always** pyramid shaped.

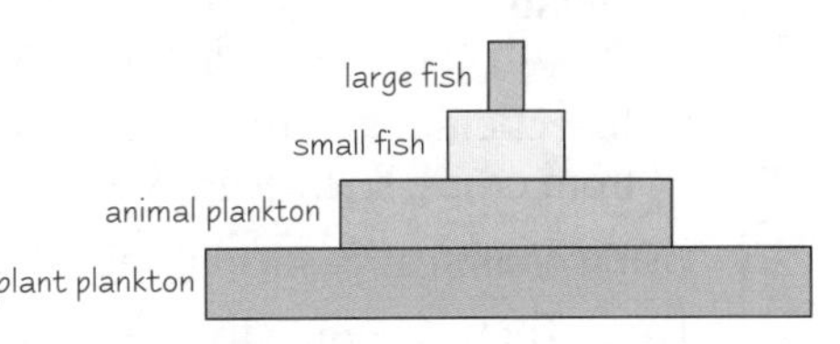

Practice Questions

Q1 State the main way that energy enters an ecosystem.

Q2 What do food webs show?

Q3 What do pyramids of biomass show?

Exam Question

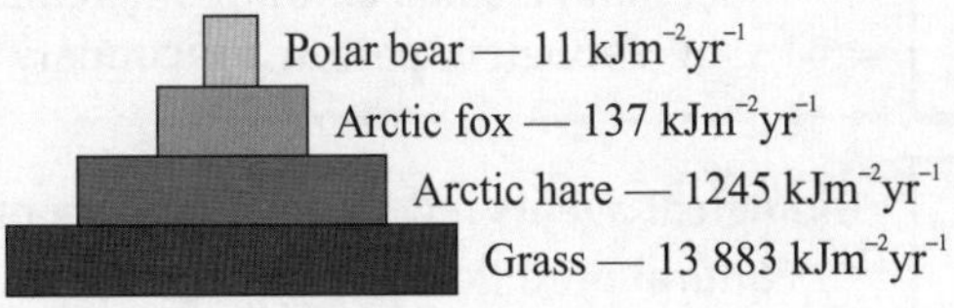

Q1 The pyramid of energy for a food chain is shown above.

a) The respiratory loss of the Arctic hare is 4165 kJm^{-2} yr^{-1}.
Calculate the gross productivity of the Arctic hare, showing your working. [2 marks]

b) Explain why the gross productivity of the Arctic hare is less than the net productivity of the grass. [3 marks]

c) Calculate the percentage efficiency of energy transfer from the Arctic fox to the polar bear. [2 marks]

Boy, do I need an energy transfer this morning...

It's really important to remember that energy transfer through an ecosystem isn't 100% efficient — most gets lost along the way so the next organisms don't get all the energy. Food chains and pyramids are a nice simple way of picturing what happens, but you need to remember that real ecosystems are a bit more complicated, so food webs are needed too.

Farming Practices and Productivity

Farmers may still wear wellies and say ooh-ar, but farming's all about the very serious business of increasing productivity.

Intensive Farming Systems are More Productive than Natural Ecosystems

1) A **natural ecosystem** is an ecosystem that **hasn't been changed** by **human activity**.
2) The **energy input** of a natural ecosystem is the **amount of sunlight** captured by the producers in the ecosystem.
3) **Intensive farming** involves changing an ecosystem by **controlling** the **biotic** or **abiotic conditions**, e.g. the presence of pests or the amount of nutrients available, to make it **more favourable** for crops or livestock.
4) This means intensively farmed **crops** or **livestock** can have **greater net productivity** (a greater **amount** of **biomass**) than **organisms** in **natural ecosystems**.
5) The **energy input** might be **greater** in an intensively farmed area than in a natural ecosystem, e.g. cattle may be given food that's **higher in energy** than their natural food. Or it might be the **same** as a natural ecosystem, e.g. a field of crops still receives the **same** amount of **sunlight** as a natural field.

Intensive Farming Practices Increase Productivity

Intensive farming methods **increase productivity** in different ways:

1) They can **increase** the **efficiency** of **energy conversion** — more of the energy organisms **have** is used for **growth** and less is used for **other activities**, e.g. recovering from **disease** or **movement**.
2) They can remove growth **limiting factors** — **more** of the energy **available** can be used for **growth**.
3) They can **increase energy input** — more energy is **added** to the ecosystem so there's **more energy** for **growth**.

Here are **three** of the main intensive farming practices used:

1 Killing Pest Species

Pests are organisms that **reduce** the **productivity** of **crops** by reducing the amount of energy available for **growth**. This means the crops are **less efficient** at **converting energy**. Here are **three** ways that farmers reduce pest numbers:

Using chemical pesticides

- **Herbicides** kill **weeds** that **compete** with agricultural crops for **energy**. Reducing competition means crops receive **more energy**, so they grow **faster** and become **larger**, **increasing** productivity.
- **Fungicides** kill **fungal infections** that **damage** agricultural crops. The crops **use more** energy for **growth** and **less** for fighting infection, so they grow **faster** and become **larger**, **increasing** productivity.
- **Insecticides** kill **insect** pests that **eat** and **damage** crops. Killing insect pests means **less biomass** is **lost** from crops, so they grow to be **larger**, which means productivity is **greater**.

Using **chemical pesticides** raises **environmental issues**:

1) They may **directly** affect (**damage** or **kill**) other **non-pest species**, e.g. butterflies.
2) They may **indirectly** affect other **non-pest species**, e.g. eating a lot of **primary consumers** that each contain a **small amount** of **chemical pesticide** can be enough to **poison** a **secondary consumer**.

There are also **economic issues**:

Chemical pesticides can be **expensive**. It may not be **profitable** for some farmers to use chemical pesticides — their **cost** may be **greater** than the **extra money** made from **increased productivity**.

Using biological agents

Biological agents reduce the **numbers of pests**, so crops lose **less energy** and **biomass**, **increasing** productivity.

- **Natural predators** introduced to the ecosystem **eat** the pest species, e.g. ladybirds eat greenfly.
- **Parasites** live in or lay their **eggs** on a **pest insect**. Parasites either **kill** the insect or **reduce** its ability to **function**, e.g. some species of wasps lay their eggs inside caterpillars — the eggs hatch and **kill** the caterpillars.
- **Pathogenic** (disease-causing) **bacteria** and **viruses** are used to kill pests, e.g. the bacterium ***Bacillus thuringiensis*** produces a **toxin** that kills a wide range of **caterpillars**.

Using biological agents raises **environmental issues**:

1) Natural predators introduced to an ecosystem may **become** a **pest species** themselves.
2) Biological agents can **affect** (damage or kill) other **non-pest species**.

There are also **economic issues**:

Biological agents may be less **cost-effective** than chemical pesticides, i.e. they may increase productivity **less** in the **short term** for the **same amount** of money invested.

Farming Practices and Productivity

Using integrated systems

Integrated systems use **both chemical pesticides** (e.g. insecticides) and **biological agents** (e.g. parasites).

1) The **combined effect** of using both can reduce pest numbers **even more** than either method **alone**, meaning **productivity** is **increased** even more.
2) Integrated systems can **reduce costs** if one method is **particularly expensive** — the expensive method can be used **less** because the two methods are used **together**.
3) Integrated systems can **reduce** the **environmental impact** of things like pesticides, because **less** is used.

2 Using Fertilisers

Fertilisers are chemicals that provide crops with **minerals** needed **for growth**, e.g. **nitrates**. Crops **use up** minerals in the soil as they **grow**, so their growth is **limited** when there **aren't enough** minerals. Adding fertiliser **replaces** the lost minerals, so **more energy** from the ecosystem can be used to grow, **increasing** the **efficiency** of energy conversion.

1) **Natural** fertilisers are **organic** matter — they include **manure** and **sewage sludge** (that's "muck" to you and me).
2) **Artificial** fertilisers are **inorganic** — they contain **pure chemicals** (e.g. ammonium nitrate) as powders or pellets.

Using fertilisers raises **environmental issues**:

1) Fertiliser can be washed into **rivers** and **ponds**, **killing fish** and **plant life** because of **eutrophication** (see p. 124).
2) Using fertilisers changes the **balance** of **nutrients** in the soil — **too much** of a particular nutrient can cause crops and other plants to **die**.

There are also **economic issues**:

Farmers need to get the **amount** of fertiliser they apply **just right**. **Too much** and money is **wasted** as excess fertiliser is **washed away** (causing **eutrophication**). **Too little** and productivity **won't** be increased, so **less money** can be made from **selling** the crop.

3 Rearing Livestock Intensively

Rearing livestock **intensively** involves **controlling** the **conditions** they live in, so **more** of their **energy** is used for **growth** and **less** is used for **other activities** — the **efficiency** of energy conversion is increased so **more biomass** is produced and productivity is **increased**.

1) Animals may be kept in **warm, indoor** pens where their **movement** is **restricted**. **Less energy** is **wasted** keeping **warm** and **moving around**.
2) Animals may be given **feed** that's **higher in energy** than their natural food. This **increases** the **energy input**, so **more energy** is available for **growth**.

The benefits are that **more food** can be produced in a **shorter** space of time, often at **lower cost**. However, enhancing productivity by intensive rearing raises **ethical issues**. For example, some people think the **conditions** intensively reared animals are kept in cause the animals **pain**, **distress** or restricts their **natural behaviour**, so it **shouldn't be done**.

Practice Questions

Q1 Name two types of pesticide.

Q2 What are fertilisers?

Q3 What does intensive farming involve?

Exam Question

Q1 Organic farmers don't use artificial chemicals on their land. Describe and explain how an organic farmer might increase productivity by reducing pest numbers on their farm. [5 marks]

Farming practices — baa-aa-aa-rmy...

Crikey, so farming's not just about getting up early to feed the chooks then — farmers want to produce as much food as they can so some of them use intensive methods. Make sure you pay attention to all the environmental, economic and ethical issues — you need to be able to talk about them to evaluate the consequences of using intensive methods.

The Carbon Cycle and Global Warming

Carbon — found in plants, animals, your petrol tank and on your burnt toast.

*The **Carbon Cycle** shows how **Carbon** is **Passed On** and **Recycled***

All organisms need carbon to make **essential compounds**, e.g. plants use CO_2 in photosynthesis to make glucose. The **carbon cycle** is how carbon **moves** through **living organisms** and the **non-living environment**. It involves four processes — **photosynthesis**, **respiration**, **decomposition** and **combustion**:

1) **Carbon** (in the form of CO_2 from **air** and **water**) is **absorbed** by plants when they carry out **photosynthesis** — it becomes carbon compounds in **plant tissues**.
2) Carbon is **passed on** to **primary consumers** when they **eat** the plants. It's passed on to **secondary** and **tertiary consumers** when they eat other consumers.
3) All living organisms **die** and the carbon compounds in the **dead organisms** are digested by **microorganisms** called **decomposers**, e.g. bacteria and fungi. Feeding on dead organic matter is called **saprobiontic nutrition**.
4) Carbon is **returned** to the air (and water) as **all living organisms** (including the decomposers) carry out **respiration**, which **produces** CO_2.
5) If dead organic matter ends up in places where there **aren't any** decomposers, e.g. deep oceans or bogs, their carbon compounds can be turned into **fossil fuels** over **millions of years** (by heat and pressure).
6) The carbon in fossil fuels is **released** when they're **burnt** — this is called **combustion**.

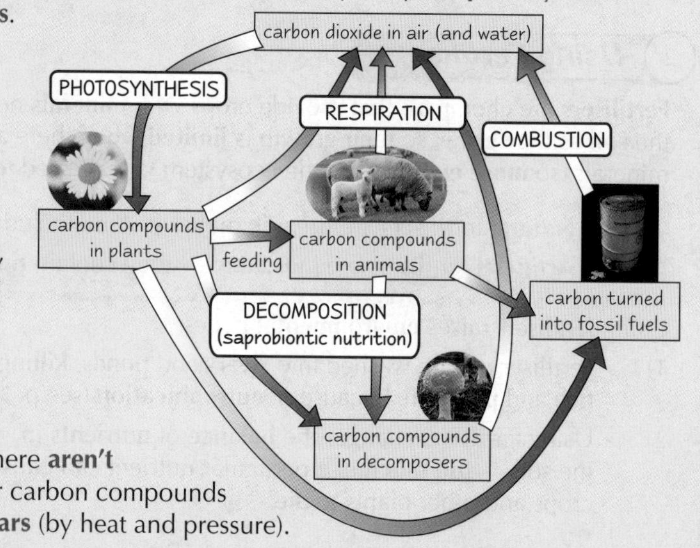

Respiration and Photosynthesis Cause Fluctuations in CO_2 Concentration

Respiration (which is carried out by **all** organisms) **adds** CO_2 to the atmosphere. **Photosynthesis removes** CO_2 from the atmosphere. The **amount** of respiration and photosynthesis going on **varies** on a **daily** and a **yearly** basis, so the amount of **atmospheric CO_2 changes**.

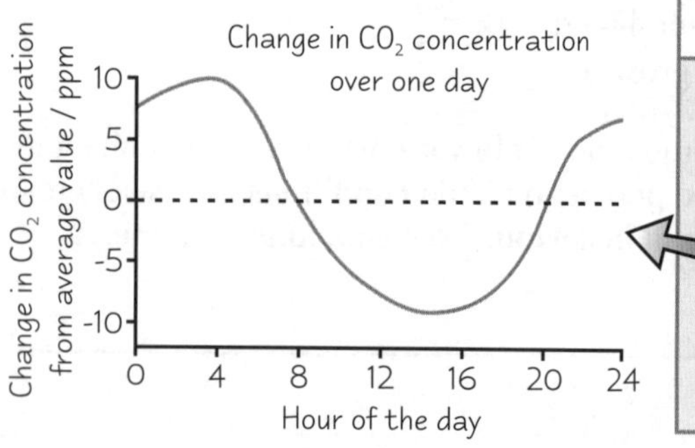

Daily change in CO_2 concentration

1) **Respiration** is carried out **constantly** through the **day and night**.
2) **Photosynthesis** only takes place during the **daylight** hours.
3) CO_2 concentration **falls** during the **day** because it's being **removed** by plants as they carry out photosynthesis.
4) CO_2 concentration **increases** at **night** because it's **no longer** being removed (**no** photosynthesis is happening), but all organisms are **still respiring** and **adding** CO_2 to the atmosphere.

Yearly change in CO_2 concentration

1) Most **plant life** exists in the **northern hemisphere** because that's where **most land** is.
2) Most plant **growth** occurs in the **summer** (June-Aug in the northern hemisphere) because that's when the **light intensity** is greatest — **more photosynthesis** can occur, which means there's **more energy** to grow.
3) CO_2 concentration **falls** during the **summer** because **more** is being **removed** from the atmosphere as **more plants** are **photosynthesising**.
4) CO_2 concentration **increases** throughout **autumn** and **winter** (Sep-April in the northern hemisphere) because **less** is being **removed** from the atmosphere, as **fewer plants** are **photosynthesising**.

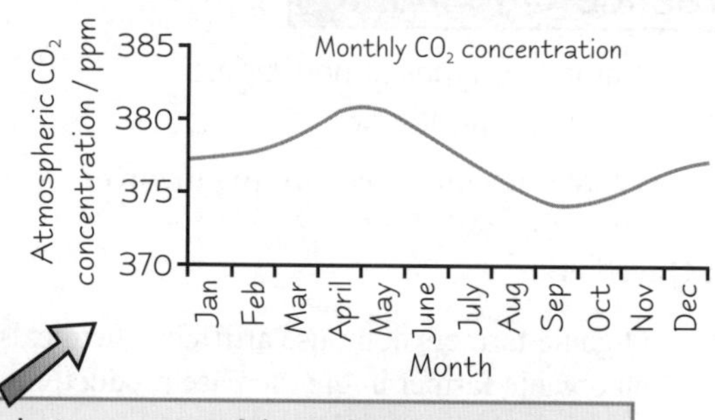

The Carbon Cycle and Global Warming

Global Warming is Caused by Increasing CO_2 and Methane Concentrations

1) **Global warming** is the term for the **increase** in **average global temperature** over the last century.
2) There's a **scientific consensus** that this increase has been **caused** by **human activity**. (It can't be explained by **natural causes**, which happen **more slowly**.)
3) Human activity has caused global warming by **enhancing** the **greenhouse effect** — the effect of greenhouse gases absorbing outgoing **energy**, so that less is **lost** to space.
4) The greenhouse effect is **essential** to keep the planet warm, but **too much** greenhouse gas in the atmosphere means the planet **warms up**.
5) **Two** of the main greenhouse gases are CO_2 and **methane**.

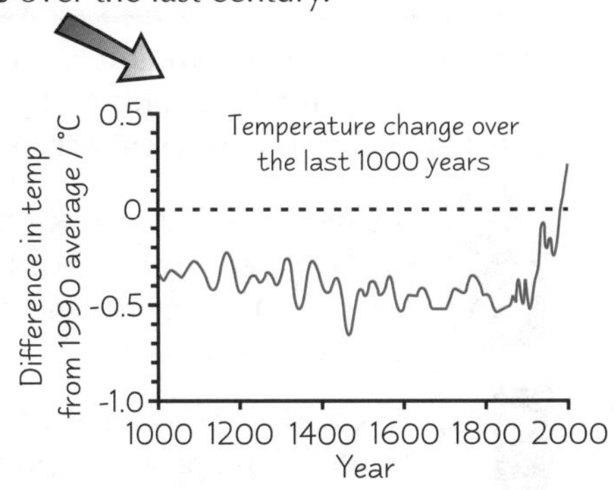

Carbon dioxide (CO_2)

- **Atmospheric CO_2** concentration has **increased rapidly** since the **mid-19th century** from **280 ppm** (parts per million) to nearly **380 ppm**. The concentration had been **stable** for the previous **10 000 years**.
- CO_2 concentration is **increasing** as more **fossil fuels** like coal, oil, natural gas and petrol are **burnt**, e.g. in power stations or in cars. Burning fossil fuels **releases CO_2**.
- CO_2 concentration is also **increased** by the **destruction** of **natural sinks** (things that keep CO_2 **out** of the atmosphere by storing **carbon**). E.g. trees are a big CO_2 sink — they store the carbon as **organic compounds**. CO_2 is **released** when trees are **burnt**, or when **decomposers break down** the organic compounds and **respire**.

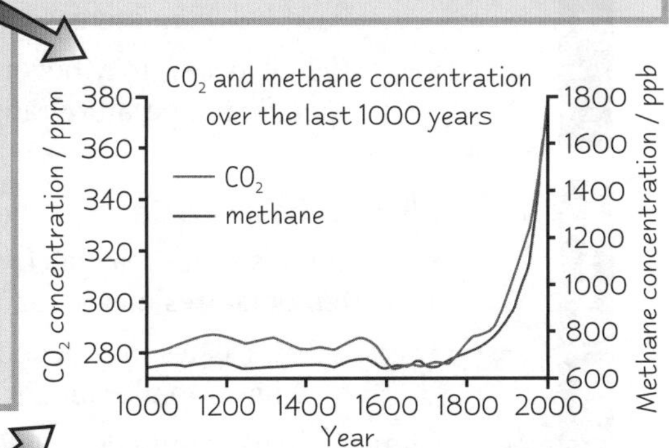

Methane

- **Atmospheric methane** concentration has **increased rapidly** since the **mid-19th century** from **700 ppb** (parts per billion) to **1700 ppb** in **2000**. The level had been **stable** for the previous **850 years**.
- Methane concentration is **increasing** because **more** methane is being **released** into the atmosphere, e.g. because **more fossil fuels** are being **extracted**, there's more **decaying waste** and there are **more cattle** which give off methane as a **waste gas**.
- Methane can also be released from **natural stores**, e.g. **frozen ground** (permafrost). As temperatures **increase** it's thought these stores will **thaw** and release **large amounts** of methane into the atmosphere.

An increase in **human activities** like **burning fossil fuels** (for industry and in cars), **farming** and **deforestation** has **increased** atmospheric concentrations of CO_2 and methane. This has **enhanced** the greenhouse effect and **caused** a rise in average global temperature — **global warming**.

Practice Questions

Q1 What is saprobiontic nutrition?

Q2 Why does CO_2 concentration decrease during daylight hours?

Exam Questions

Q1 Describe how carbon is cycled through living organisms and the non-living environment. [6 marks]

Q2 a) Explain why atmospheric CO_2 and methane concentrations have increased since the mid-19th century. [6 marks]

b) What is global warming? How are rising atmospheric CO_2 and methane concentrations causing it? [4 marks]

Daily pattern of my concentration — low during lessons and revision...

I know, I know, you might think these pages seem a bit too geographical, but I say it's the other way round — those boring geographers have nicked our biology. The carbon cycle might look a bit messy, but it isn't as complicated as it looks. You just need to break it down into the four processes of photosynthesis, respiration, combustion and decomposition.

Effects of Global Warming

Global warming might mean you can wear a bikini in Scotland, but it's bad news for some organisms...

Global Warming Could Affect all Organisms

Increasing **CO_2 concentration** is causing **global warming**, which is leading to other climate changes, e.g. different **rainfall patterns** and changes to **seasonal weather patterns**. All organisms could be **affected** by this, but **different organisms** could be affected in **different ways**:

Crop yield

The **increasing CO_2 concentration** that's **causing** global warming could **also** be **causing** an **increase** in **crop yields** (the **amount** of crops produced from an area). CO_2 concentration is a **limiting factor** for photosynthesis (see p. 110), so increasing global CO_2 concentration could mean crops grow **faster**, **increasing** crop yields.

Insect pests

1) Climate change may affect the **life cycle** of some insect species. For example, it's thought that increasing global temperature (**global warming**) means some insects go through their **larval stage** quicker and emerge as **adults earlier**, e.g. some butterflies may spend **10** fewer days as larvae for every **1 °C** rise in temperature.
2) Climate change may also affect the **numbers** of some insect species:
   - Some species are becoming **more** abundant, e.g. **warmer** and **wetter** summers in some places have led to an **increase** in the number of **mosquitoes**.
   - Other species may become **less** abundant, e.g. some **tropical** insect species can only thrive in **specific temperature ranges**, so if it gets **too hot** fewer insects may be able to **reproduce successfully**.

Wild animals and plants

1) Climate change could affect the **distribution** of many wild **animal** and **plant** species:
   - Some species may become **more** widely distributed, e.g. species that need **warmer temperatures** may spread **further** as the conditions they **thrive** in exist over a **wider** area.
   - Other species may become **less** widely distributed, e.g. species that need **cooler temperatures** may have **smaller** ranges as the conditions they **thrive** in exist over a **smaller** area.
2) Climate change could also affect the **number** of wild animals and plants:
   - Some species are becoming **more** abundant, e.g. **boarfish** are increasing in number in parts of the Atlantic Ocean where sea temperature is **rising**.
   - Other species are becoming **less** abundant, e.g. **polar bears** need frozen sea ice to hunt and **global warming** is causing more sea ice to **melt**. It's thought that the number of polar bears is **decreasing** because there isn't enough sea ice for them to hunt on.

You Need to be able to Analyse Data on the Effects of Global Warming

Analysing data's pretty important when looking at the **effects** of global warming. Here are a few examples:

1 Example 1 — Temperature and Crop Yield

A study was carried out to investigate whether **rising growing season temperature** is affecting **crop yields**. Some of the data from the study is shown on the graph. You might be asked to:

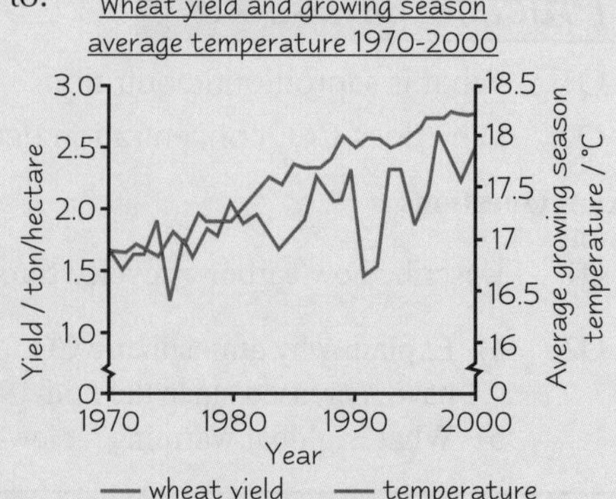

1) **Describe the data:**

   The **temperature fluctuated** between **1970** and **2000**, but the general trend was a **steady increase** from just under **17 °C** to just under **18 °C**.

   The **wheat yield** also showed a trend of **increasing** from around **1.6 tons** per hectare in **1970** to around **2.7 tons** per hectare in **2000**.

2) **Draw a conclusion:**

   The graph shows a **positive correlation** between **temperature** and **wheat yield**. The increasing growing season temperature could be **linked** to the increasing wheat yields.

   Even though the graph shows a **correlation**, you can't conclude that the increase in temperature **caused** the increase in wheat yield — there could have been **other factors** involved. This study actually found that the rising growing season temperature had a **negative effect** on wheat yields, but **improvements in technology** during the **same period** meant that crop yields **increased overall**.

Effects of Global Warming

2 Example 2 — Temperature and Insect Numbers

Temperature and number of greenfly caught

Average global temperature / °C: 13.7, 13.9, 14.1, 14.3, 14.5
Number of greenfly caught: 0, 100, 200, 300, 400, 500, 600
Year: 1960, 1970, 1980, 1990, 2000
— temperature — greenfly numbers

A study counted the **number** of **greenfly** in an area from 1960 to 2000.
A separate study collected data on **global temperature** at the same time.
The results are shown on the graph. You might be asked to:

1) **Describe the data:**

   The **temperature fluctuated** between **1960** and **2000**, but the general trend was a **steady increase** from just over **13.8 °C** to just over **14.4 °C**.

   The **number of greenfly** also **fluctuated** with a generally **increasing** trend from around **110** in **1960** to just around **480** in **2000**.

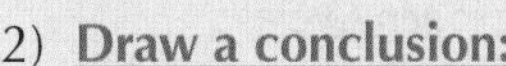

2) **Draw a conclusion:**

   There's a **positive correlation** between **temperature** and **numbers of greenfly**.
   The increasing global temperature could be **linked to** the increasing greenfly numbers.

3) **Suggest an explanation for your conclusion:**

   Greenfly numbers **could** be increasing because higher temperatures may **increase** their **food supply**, e.g. the rate of **photosynthesis** may **increase** at higher temperatures, allowing plants to **grow faster** and **become larger**.

3 Example 3 — Temperature and the Distribution of Organisms

A study was carried out to investigate the changing **distribution** of **subtropical plankton** species in the north Atlantic.
The results are shown below, along with data that's been collected on **global sea surface temperature**.
You might be asked to:

1) **Describe the data:**

   Sea surface temperature fluctuated around the average between **1950** and **1978**, then there was a **steady increase** between 1978 and 2000, up to just over **0.3 °C** greater than the average.

   Subtropical plankton species were found in the sea **south of the UK** in 1958-1981. By 2000-2002 their distribution had moved **further north** along the west coast of the UK and Ireland to the **Arctic Ocean**.

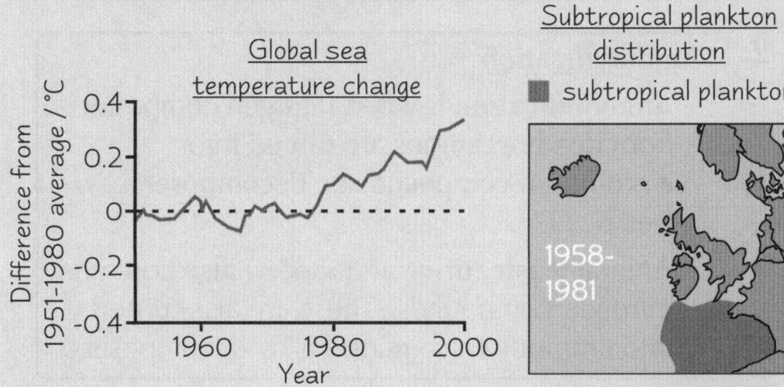

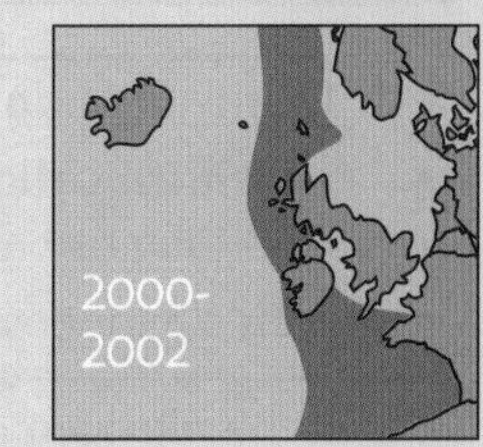

2) **Draw a conclusion:**

   There's a link between **rising global sea surface temperature** and the **northward** change in **distribution** of subtropical plankton.

   The data shows a **link**, but you can't say that the increase in temperature **caused** the change in distribution — there could have been **other factors** involved, e.g. **overfishing** could have removed plankton **predator species**.

Practice Questions

Q1 Give one way that climate change is affecting populations of insect pests.

Q2 Give one way that climate change is affecting wild animal species.

Exam Question

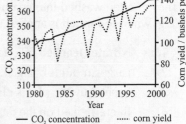

CO_2 concentration and corn yield

CO_2 concentration / ppm: 310, 320, 330, 340, 350, 360, 370, 380, 390
Corn yield / bushels per acre: 60, 80, 100, 120, 140, 160
Year: 1980, 1985, 1990, 1995, 2000
— CO_2 concentration ····· corn yield

Q1 The graph on the right shows CO_2 concentration and corn yield.

a) Describe what the graph is showing. [4 marks]

b) Draw a conclusion. [1 mark]

c) Use your knowledge to suggest an explanation for your conclusion. [2 marks]

Global warming effects — not as much fun as special effects...

Boy, that wasn't fun, but the business of analysing data is an important one if you want to profit in your exam. There could be lots of questions on data, so have a good read through these examples, and never mix up correlation and cause.

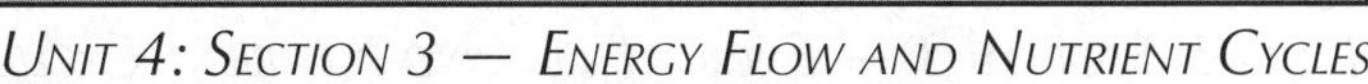

The Nitrogen Cycle and Eutrophication

Sorry, there's some more cycling to do here — sadly, the nitrogen cycle's a little bit more tiring than the carbon cycle.

The Nitrogen Cycle shows how Nitrogen is Passed on and Recycled

Plants and animals **need** nitrogen to make **proteins** and **nucleic acids** (DNA and RNA). The atmosphere's made up of about 78% nitrogen, but plants and animals **can't use it** in that form — they need **bacteria** to **convert** it into **nitrogen compounds** first. The **nitrogen cycle** shows how nitrogen is **converted** into a useable form and then **passed** on between different **living** organisms and the **non-living** environment.

The nitrogen cycle includes **food chains** (nitrogen is passed on when organisms are eaten), and four different processes that involve bacteria — **nitrogen fixation**, **ammonification**, **nitrification** and **denitrification**:

1 Nitrogen fixation

- **Nitrogen fixation** is when nitrogen **gas** in the atmosphere is turned into **ammonia** by **bacteria** called *Rhizobium*. The ammonia can then be **used** by plants.
- *Rhizobium* are found inside **root nodules** (growths on the roots) of **leguminous** plants (e.g. peas, beans and clover).
- They form a **mutualistic** relationship with the plants — they provide the plant with **nitrogen compounds** and the plant provides them with **carbohydrates**.

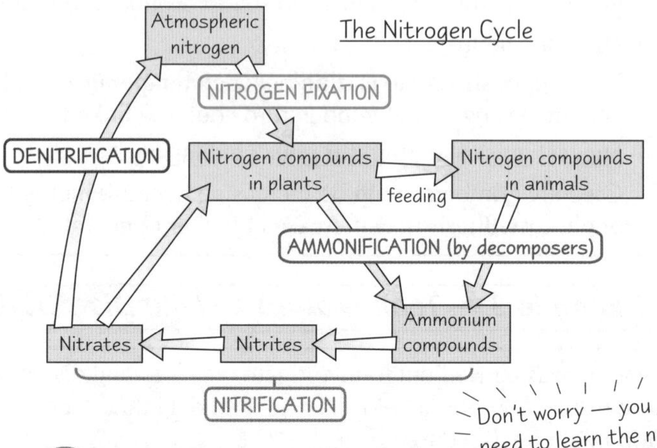

Don't worry — you don't need to learn the names of the microorganisms.

2 Ammonification

- **Ammonification** is when nitrogen compounds from **dead organisms** are turned into **ammonium compounds** by **decomposers** (see p. 116).
- Animal **waste** (**urine** and **faeces**) also contains nitrogen compounds. These are also turned into ammonium compounds by decomposers.

3 Nitrification

- **Nitrification** is when **ammonium compounds** in the soil are **changed** into **nitrogen compounds** that can then be **used** by plants.
- First **nitrifying bacteria** (e.g. *Nitrosomonas*) change **ammonium compounds** into **nitrites**.
- Then other nitrifying bacteria called *Nitrobacter* change **nitrites** into **nitrates**.

4 Denitrification

- **Denitrification** is when nitrates in the soil are **converted** into **nitrogen gas** by **denitrifying bacteria** — they use nitrates in the soil to carry out **respiration** and produce nitrogen gas.
- This happens under **anaerobic conditions** (where there's **no** oxygen), e.g. in **waterlogged** soils.

Parts of the nitrogen cycle can also be carried out **artificially** and on an **industrial** scale. The **Haber process** produces **ammonia** from **atmospheric nitrogen** — it's used to make things like **fertilisers**.

Nitrogen Fertilisers can Leach into Water and Cause Eutrophication

Leaching is when **water-soluble** compounds in the soil are **washed away**, e.g. by rain or irrigation systems. They're often **washed** into nearby **ponds** and **rivers**. If **nitrogen fertiliser** is leached into waterways (e.g. when **too much** is applied to a field) it can cause **eutrophication**:

1) **Nitrates leached** from fertilised fields stimulate the **growth** of **algae** in ponds and rivers.
2) Large amounts of algae **block light** from reaching the plants below.
3) Eventually the **plants die** because they're **unable** to photosynthesise enough.
4) **Bacteria** feed on the dead plant matter.
5) The **increased** numbers of **bacteria reduce** the **oxygen** concentration in the water by carrying out **aerobic respiration**.
6) **Fish** and other aquatic organisms **die** because there **isn't enough dissolved oxygen**.

Hey, who turned out the lights?

The Nitrogen Cycle and Eutrophication

You Need to be able to Analyse, Interpret and Evaluate Data on Eutrophication

You've got to know how to **analyse data**, so here's an example of the kind of thing you might get in your exam:

A study was conducted to investigate the effect, on a nearby **river**, of adding **fertiliser** to **farmland**.

The **oxygen** and **algal** content of a river that runs past a field where **nitrate fertiliser** had been applied, was measured **at the field** and up to a distance of **180 m** away. A similar **control river** next to an **unfertilised** field was also studied. The results are shown in the graphs on the right.

1) **Describe the data:**

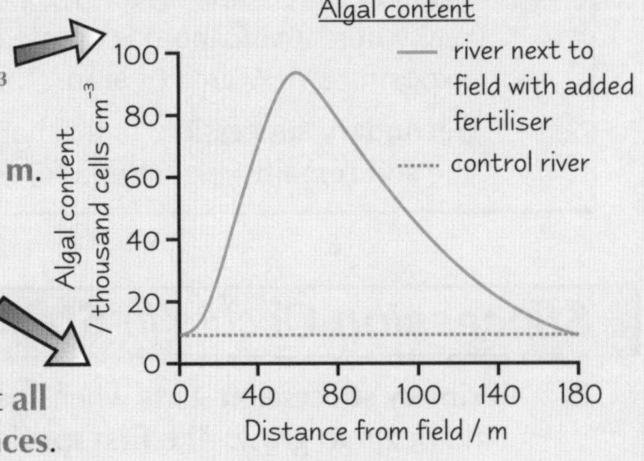

The **algal content** of the water **increases** sharply from **10 000 cells cm^{-3}** at the field to **95 000 cells cm^{-3}** at a distance of **60 m** from the field. Algal content then **decreases** beyond **60 m** to **10 000 cells cm^{-3} at 180 m.**

The **oxygen content** of the water **decreases** from **8 $mgdm^{-3}$** at the field to **2 $mgdm^{-3}$** at a distance of **80 m** from the field. The oxygen content then **increases** beyond **80 m** up to **13 $mgdm^{-3}$** at 180 m, where it begins to **level off**.

The control river showed a **steady algal content** of **10 000 cells cm^{-3}** at **all distances**, as well as a **steady oxygen content** of **8 $mgdm^{-3}$** at **all distances.**

Oxygen content

Oxygen content of water / $mgdm^{-3}$ (axis 0–16); Distance from field / m (0, 40, 80, 100, 140, 180); — river next to field with added fertiliser; ······ control river

2) **Draw a conclusion:**

There's a **negative correlation** between the algal content and the oxygen content of the water — as the algal content **increases**, the oxygen content **decreases**, and vice versa.

3) **Evaluate the methodology:**

A control river was used which helps to control the effect of **some variables**, e.g. water temperature. But it doesn't remove the effect of **all** variables, e.g. **different organisms** may live in the control river, which could **affect** the algal or oxygen content.

The experiment only looked at **two rivers**, which means the **sample size** was **small**. Studying other rivers may have produced **different results** and a **different conclusion. More experiments** and results would be needed to make the data **more reliable.**

4) **Suggest an explanation for your conclusion:**

The results **suggest leaching** of the fertiliser and **eutrophication** have occurred. **Nitrate fertilisers** from the field could have **leached** into the river and caused the algal content of the river to **increase** by **stimulating** algal growth. The increased algal content could have **prevented light** from reaching plants **below**, causing them to die and be decomposed by **bacteria**. The bacteria **use up** the oxygen in the river when carrying out **aerobic respiration**, resulting in **decreased** dissolved oxygen levels.

Practice Questions

Q1 What is denitrification?

Q2 What is leaching?

Q3 Briefly describe eutrophication.

Exam Question

Q1 The diagram on the right shows the nitrogen cycle.

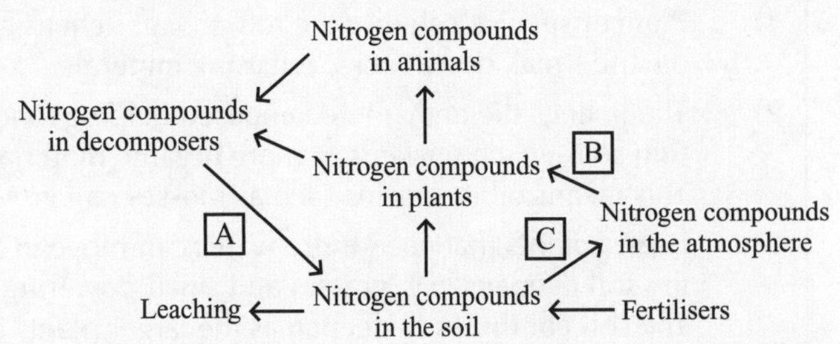

a) Name the processes labelled A and C in the diagram. [2 marks]

b) Name and describe process B in detail. [3 marks]

Nitrogen fixation — cheaper than a shoe fixation...

The nitrogen cycle's not as bad as it seems. Divide up the four processes of nitrogen fixation, ammonification, nitrification and denitrification and learn them separately, then hey presto — you've learnt the whole cycle. Eutrophication's alright too, it's just got a scary name. Watch out though, they like to combine all this stuff and get you to interpret data in the exam.

Succession

Repeat after me: successful succession involves several simple successive seral stages.

Succession is the Process of Ecosystem Change

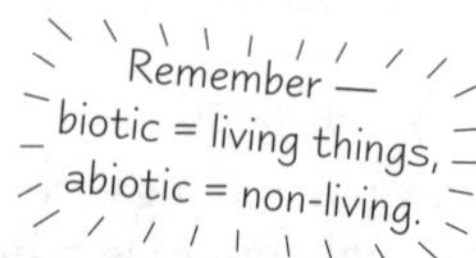

Succession is the process by which an **ecosystem** (see p. 94) **changes** over **time**. The **biotic conditions** (e.g. **plant** and **animal communities**) change as the **abiotic conditions** change (e.g. **water** availability). There are **two** types of succession:

1) **Primary succession** — this happens on land that's been **newly formed** or **exposed**, e.g. where a **volcano** has erupted to form a **new rock surface**, or where **sea level** has **dropped** exposing a new area of land. There's **no soil** or **organic material** to start with, e.g. just bare rock.
2) **Secondary succession** — this happens on land that's been **cleared** of all the **plants**, but where the **soil remains**, e.g. after a **forest fire** or where a forest has been **cut down by humans**.

Succession Occurs in Stages called Seral Stages

1) **Primary succession** starts when species **colonise** a new land surface. **Seeds** and **spores** are blown in by the **wind** and begin to **grow**. The **first species** to colonise the area are called **pioneer species** — this is the **first seral stage**.
   - The **abiotic conditions** are **hostile** (**harsh**), e.g. there's no soil to **retain water**. Only pioneer species **grow** because they're **specialised** to cope with the harsh conditions, e.g. **marram grass** can grow on sand dunes near the sea because it has **deep roots** to get water and can **tolerate** the salty environment.
   - The pioneer species **change** the **abiotic conditions** — they **die** and **microorganisms decompose** the dead **organic material** (**humus**). This forms a **basic soil**.
   - This makes conditions **less hostile**, e.g. the basic soil helps to **retain water**, which means **new organisms** can move in and grow. These then die and are decomposed, adding **more** organic material, making the soil **deeper** and **richer in minerals**. This means **larger plants** like **shrubs** can start to grow in the deeper soil, which retains **even more** water.
2) **Secondary succession** happens in the **same way**, but because there's already a **soil layer** succession starts at a **later seral stage** — the pioneer species in secondary succession are **larger plants**, e.g. shrubs.
3) At each stage, **different** plants and animals that are **better adapted** for the improved conditions move in, **out-compete** the plants and animals that are already there, and become the **dominant species** in the ecosystem.
4) As succession goes on, the ecosystem becomes **more complex**. New species move in **alongside** existing species, which means the **species diversity** (the number of **different species** and the **abundance** of each species) **increases**.
5) The **final seral stage** is called the **climax community** — the ecosystem is supporting the **largest** and **most complex** community of plants and animals it can. It **won't change** much more — it's in a **steady state**.

This example shows primary succession on bare rock, but succession also happens on sand dunes, salt marshes and even on lakes.

Example of primary succession — bare rock to woodland

1) **Pioneer species colonise** the rocks. E.g. **lichens** grow **on** and **break down** rocks, **releasing minerals**.
2) The lichens **die** and are **decomposed** helping to form a **thin soil**, which thickens as more **organic material** is formed. This means other species such as **mosses** can **grow**.
3) **Larger plants** that need **more water** can move in as the soil **deepens**, e.g. **grasses** and **small flowering plants**. The soil **continues to deepen** as the larger plants die and are decomposed.
4) **Shrubs**, **ferns** and **small trees** begin to grow, **out-competing** the grasses and smaller plants to become the **dominant** species. **Diversity increases**.
5) Finally, the soil is **deep** and **rich** enough in **nutrients** to support **large trees**. These become the dominant species, and the **climax community** is formed.

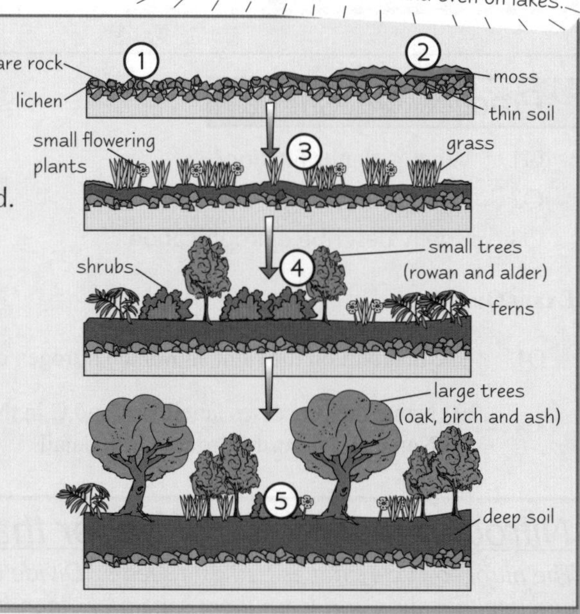

Succession

Different Ecosystems have Different Climax Communities

Which species make up the climax community depends on what the **climate's** like in an ecosystem. The climax community for a **particular** climate is called its **climatic climax**. For example:

- In a **temperate climate** there's **plenty** of **available water, mild temperatures** and not much **change** between the seasons. The climatic climax will contain **large trees** because they **can grow** in these conditions once **deep soils** have developed.
- In a **polar climate** there's **not much available water**, temperatures are **low** and there are **massive changes** between the seasons. Large trees **won't ever** be able to grow in these conditions, so the climatic climax contains only **herbs** or **shrubs**, but it's still the **climax community**.

Conservation Often Involves Managing Succession

Human activities can **prevent succession**, stopping a climax community from **developing**. When succession is stopped **artificially** like this the climax community is called a **plagioclimax**. For example:

A **regularly mown** grassy field **won't develop** shrubs and trees (**woody plants**), even if the climate of the ecosystem could support them. The **growing points** of the woody plants are **cut off** by the lawnmower, so larger plants **can't establish** themselves. The **longer** the interval between mowing, the **further** succession can progress and the more **diversity increases**. But with **more frequent** mowing, succession can't progress and diversity will be **lower** — only the grasses can **survive** being mowed.

Man had been given a mighty weapon with which they would tame the forces of nature.

Conservation (the **protection** and **management** of ecosystems) sometimes involves preventing succession in order to **preserve** an ecosystem in its **current** seral stage. For example, there are large areas of **moorland** in **Scotland** that provide **habitats** for many species of plants and animals. If the moorland was left to **natural processes**, succession would lead to a **climax community** of **spruce forest**. This would mean the **loss** of the moorland habitat and could lead to the loss of some of the plants and animals that **currently** live there. Preventing succession keeps the moorland ecosystem **intact**. There are a couple of ways to **manage succession** to **conserve** the moorland ecosystem:

1) **Animals** are allowed to **graze** on the land. This is similar to **mowing** — the animals eat the **growing points** of the shrubs and trees, which **stops** them from establishing themselves and helps to keep vegetation **low**.
2) **Managed fires** are lit. After the fires, **secondary succession** will occur on the moorland — the species that grow back **first** (**pioneer species**) are the species that are being **conserved**, e.g. heather. Larger species will take **longer** to grow back and will be **removed again** the next time the moor's burnt.

Practice Questions

Q1 What is the difference between primary and secondary succession?

Q2 What is the name given to species that are the first to colonise an area during succession?

Q3 What is meant by a climax community?

Exam Question

Q1 A farmer has a field where he plants crops every year. When the crops are fully grown he removes them all and then ploughs the field (churns up all the plants and soil so the field is left as bare soil). The farmer has decided not to plant crops or plough the field for several years.

a) Describe, in terms of succession, what will happen in the field over time. [6 marks]

b) Explain why succession doesn't usually take place in the farmer's field. [2 marks]

Revision succession — bare brain to a woodland of knowledge...

When answering questions on succession, examiners are pretty keen on you using the right terminology — that means saying "pioneer species" instead of "the first plants to grow there". This stuff's all quite wordy, but the concept of succession is simple enough — some plants start growing, change the environment so it's less hostile, then others can move in.

Conservation

Conservation is important for us and the environment — won't somebody think of the polar bears...

Conserving Species and Habitats is Important for Many Reasons

Conservation is the **protection** and **management** of **species** and **habitats** (**ecosystems**).
It's **important** for **many reasons**:

1) **Species** are **resources** for lots of things that **humans need**, e.g. **rainforests** contain species that provide things like **drugs**, **clothes** and **food**. If the species and their habitats **aren't** conserved, the resources that we use now will be **lost**. Resources that **may be useful** in the **future** could also be **lost**.
2) Some people think we should conserve species simply because it's the **right thing to do**, e.g. most people think organisms have a **right to exist**, so they shouldn't become extinct as a result of **human activity**.
3) Many species and habitats bring **joy** to lots of people because they're **attractive** to **look at**. The species and habitats may be **lost** if they **aren't** conserved, so **future generations** won't be able to enjoy them.
4) Conserving species and habitats can help to prevent **climate change**. E.g. when trees are **burnt**, CO_2 is **released** into the atmosphere, which contributes to global warming. If they're conserved, this **doesn't happen**.
5) Conserving species and habitats helps to **prevent** the **disruption** of **food chains**. Disruption of food chains could mean the **loss** of **resources**. E.g. some species of **bear feed** on **salmon**, which feed on **herring** — if the number of herring **decreases** it can affect **both** the salmon and the bear populations.

Not everyone agrees with every conservation measure though — there's often **conflict** when conservation **affects people's livelihoods**, e.g. conservation of the Siberian tiger in Russia affects people who make money from killing the tigers and selling their fur (there's conflict between the conservationists and the hunters).

There are Many Different Ways to Conserve Species and Habitats

Different species and habitats need to be conserved in **different ways**.
Here are a few examples of **some** of the different **conservation methods** that can be used.

1 *Plants can be Conserved using Seedbanks*

The seedbank — 0% APR on branch transfers.

1) A **seedbank** is a **store** of lots of **seeds** from lots of **different plant species**.
2) They help to conserve species by storing the seeds of **endangered** plants.
3) They also help to conserve **different varieties** of each species by storing a **range** of seeds from plants with **different characteristics**, e.g. seeds from tall sunflowers and seeds from short sunflowers.
4) If the plants become **extinct** in the wild the stored seeds can be used to **grow new plants**.
5) Seedbanks are a **good way** of conserving plant species — **large numbers** of species can be conserved because seeds don't need **much space**. Seeds can also be **stored anywhere** and for a **long time**, as long as it's **cool** and **dry**.
6) But there are **disadvantages** — the seeds have to be regularly tested to see if they're still **viable** (whether they can grow into a plant), which can be **expensive** and **time-consuming**.

2 *Fish species can be Conserved using Fishing Quotas*

1) **Fishing quotas** are **limits** to the **amount** of certain fish species that fishermen are **allowed** to **catch**.
2) **Scientists** study different species and decide **how big** their populations need to be for them to **maintain** their numbers. Then they decide **how many** it's **safe** for fishermen to take without reducing the population **too much**.
3) **International agreements** are made (e.g. the Common Fisheries Policy in the EU) that state the **amount** of fish **each country** can take, and **where** they're allowed to take them from.
4) Fishing quotas help to **conserve** fish species by **reducing** the numbers that are **caught** and **killed**, so the populations aren't **reduced** too much and the species aren't at risk from becoming **extinct**.
5) There are **problems** with fishing quotas though — many fishermen **don't agree** with the scientists who say that the fish numbers are **low**. Some also think introducing quotas will cause **job losses**.

Conservation

3 Animals can be Conserved using Captive Breeding Programmes

1) Captive breeding programmes involve breeding animals in **controlled environments**.
2) Species that are **endangered**, or already **extinct in the wild**, can be **bred** in captivity to help **increase their numbers**, e.g. pandas are bred in captivity because their numbers are **critically low** in the wild.
3) There are some **problems** with captive breeding programmes though, e.g. animals can have **problems breeding** outside their **natural habitat**, which can be hard to **recreate** in a zoo. For example, pandas don't reproduce as **successfully** in captivity as they do in the wild.
4) Animals bred in captivity can be **reintroduced to the wild**. This **increases** their **numbers** in the wild, which can help to conserve their **numbers** or bring them **back** from the **brink of extinction**.
5) Reintroducing animals into the wild can cause **problems** though, e.g. reintroduced animals could bring **new diseases** to habitats, **harming** other species **living there**.

No way, I'm not breeding with him. He's ugly and his breath smells of bamboo.

4 Any organism can be Conserved by Relocation

1) **Relocating** a species means **moving** a population of a species to a **new location** because they're directly under **threat**, e.g. from poaching, or the **habitat** they're living in is under threat, e.g. from rising sea levels.
2) The species is moved to an area where it's **not at risk** (e.g. a protected national park, see below), but with a **similar environment** to where it's come from, so the species is still able to **survive**.
3) It's often used for species that only exist in **one place** (if that population **dies out**, the species will be **extinct**).
4) It helps to **conserve** species because they're relocated to a place where they're **more likely** to **survive**, so their numbers may **increase**.
5) Relocating species can cause **problems** though, e.g. native species in the new area may be **out-competed** by the species that's moved in and become **endangered** themselves.

5 Habitats can be Conserved using Protected Areas

1) **Protected areas** such as **national parks** and **nature reserves** protect habitats (and so protect the **species** in them) by **restricting urban development**, **industrial development** and **farming**.
2) Habitats in **protected areas** can be **managed** to conserve them, e.g. by **coppicing** — **cutting** down trees in a way that lets them **grow back**, so they don't need to be **replanted**. This helps to conserve the woodland, but allows some wood to be **harvested**.
3) There are **problems** with using protected areas to conserve habitats though, e.g. national parks are also used as **tourist destinations** (many are **funded** by **revenue** from the tourists that visit). This means there's conflict between the need to **conserve** the habitats and the need to allow people to **visit** and **use** them.

Practice Questions

Q1 Suggest why conservation of species and habitats is important for humans.

Q2 What is a seedbank?

Exam Question

Q1 A conservationist has argued that the deforestation of tropical rainforests will have terrible consequences for human beings and the environment.

a) Use your knowledge to outline the reasons for the conservation of tropical rainforests. [6 marks]

b) Name two suitable methods that could be used to conserve species from a tropical rainforest. [2 marks]

Captive breeding — you will procreate, or else...

There's lots of debate about conservation — what should be conserved and what's the best way to do it. That means there's a chance to get top marks in your exam — examiners love it when you talk about both sides of something. For example, even if you don't agree with captive breeding you need to say that it's got both positive and negative points.

Conservation Evidence and Data

And now my pretties, it's time for some data. Mwa ha ha...

You May Have to Evaluate Evidence *and* Data *About* Conservation Issues

You need to be able to **evaluate** any **evidence** or **data** about **conservation** projects and research that the examiners throw at you — so here's an example I made earlier:

In recent years, **native British bluebells** have become **less common** in woodland areas. It's thought that this is due to the presence of **non-native Spanish bluebells**, which compete with the native species for a **similar niche**. An experiment was carried out to see if **removing** the invasive Spanish species would help to **conserve** the native species. Each year for 15 years the **percentage cover** of native species was estimated in a **50 m by 50 m** area of **woodland** using random sampling and 250, **1 m² quadrats**. After five years, **all** the Spanish bluebells were **removed**. A **similar sized control woodland** in which the Spanish bluebells remained **untouched** was also studied. The results are shown on the right. You might be asked to:

% cover of native British bluebells in a woodland

% cover of native bluebell (0–100); years (0–15); Spanish species removed

1) Describe the data:
   - For the first **five years**, the **percentage cover** of **native bluebells fell** from **50%** to around **25%**. After the Spanish species was **removed**, it **increased** from around **25%** to around **45%** in **ten years**.

   - The **control experiment** shows a fairly **steady drop** in native bluebell percentage cover from **60%** to **20%** over the 15 years.

Control experiment

% cover of native bluebell (0–100); years (0–15)

2) Draw conclusions:
   The removal of Spanish bluebells **resulted** in an **increase** in the percentage cover of **native bluebells** over a **ten year period**. This suggests that the **recent decrease** in native British bluebells is due to **competition** with the Spanish bluebells.
3) Evaluate the method:
   - The effects of some **other variables** (e.g. **changing weather**) were **removed** by the **control experiment**, where the percentage cover of native bluebells continued to fall throughout the 15 year study. This makes the data **more reliable**.
   - The **study area** and **sample size** were quite **large**, giving **more accurate** data.
   - **Random sampling** removed bias — the data's **more likely** to be an **accurate estimate** of the **whole area**.

There's more about interpreting data on pages 90-92, and 196-198.

You Need to be Able to Consider Conflicting Evidence

1) The **evidence** from **one study** alone **wouldn't usually be enough** to conclude that there's a **link** between decreasing percentage cover of native bluebells, and the presence of Spanish bluebells.
2) **Similar studies** would be carried out to **investigate** the link. If these studies came to the **same conclusion**, the conclusion would become **increasingly accepted**.
3) Sometimes studies come up with **conflicting evidence** though — evidence that leads to a **different conclusion** than other studies. For example:

Another study was carried out to **investigate** the effect on native bluebells of **removing** Spanish bluebells. It was **similar** to the study above except a **20 m by 20 m** area was sampled using a random sample of **20 quadrats**, and **no control** woodland was used. You might be asked to:

1) Describe the data:
   In the first five years, the **percentage cover** of **native bluebells fell** from **50%** to around **25%**. After the Spanish species was **removed**, it **kept decreasing** to around **15%** after the **full 15** years.

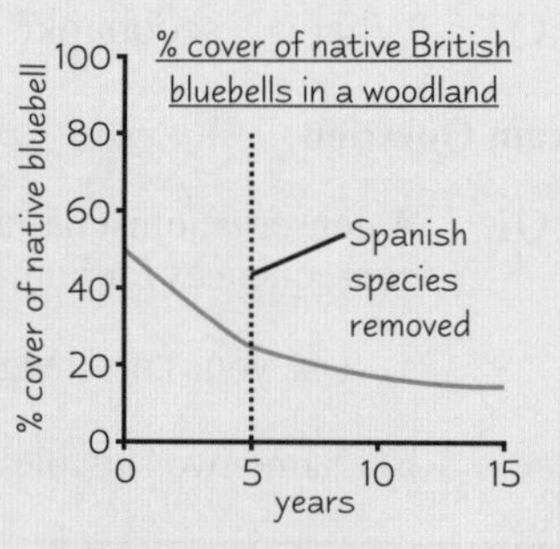

2) Draw conclusions:
   The **removal** of the Spanish bluebells had **no effect** on the **decreasing** percentage cover of native bluebells — which **conflicts** with the study above.
3) Evaluate the method:
   - There **wasn't** a **control** woodland, so the **continuing decrease** in native bluebell cover after the removal of the Spanish bluebells could be due to **another factor**, e.g. cold weather in years 5-10.
   - The **study area** and **sample size** were quite small, giving a **less accurate** total percentage cover.

Conservation Evidence and Data

Conservation Relies on Science to Make Informed Decisions

1) Scientists carry out **research** to provide **information** about conservation issues.
2) This information can then be used to make **informed decisions** about **which** species and habitats **need** to be conserved, and the **best way** to conserve them.
3) For example, the study at the **top of the previous page** showed that **native bluebell** coverage increased after the removal of **Spanish bluebells**, which **suggests** that the decrease in native bluebell coverage is due to **competition** with the Spanish species. It provides evidence that there's a **conservation issue** (native bluebells are decreasing) and a way to **solve it** (remove the Spanish species).

Many conservation **decisions** have been made using the results of **scientific research** — take a look at these examples:

Scientific results	Decision
Between 1970 and 1989 the number of African elephants dropped from around 3 million to around 50 000 because they were being hunted for their ivory tusks.	In 1989, the Convention on International Trade in Endangered Species banned ivory trade to end the demand for elephant tusks, so that fewer elephants would be killed for their tusks.
The commonly used pesticide DDT was found to have contributed to the loss of half the peregrine falcon population in the UK in the 1950s and 1960s. DDT built up in the food chain and caused the falcon eggs to have thin shells. This meant the eggs were crushed and the chicks weren't hatched.	The use of DDT as a pesticide was banned in the UK in 1984 to try to conserve and increase peregrine falcon numbers.
The numbers of some species of sea turtle have dropped so low that they're now endangered. Many eggs are removed from the beaches by poachers before the turtles hatch and reach the sea.	Conservation agencies have set up hatching programmes where eggs are taken away from beaches and looked after until they hatch. The young turtles are then released into the sea.
A reduction in the size of hedgerows in farmers' fields was found to cause a decrease in biodiversity in the British countryside.	The government provides subsidies to encourage farmers to plant hedgerows and leave margins of ground unharvested around fields. This increases the size of hedgerows and conserves biodiversity.
Whale numbers were found to have dropped massively due to whale hunting.	Commercial whaling was banned in 1986 by the International Whaling Commission in order to conserve whale numbers.

Practice Questions

Q1 What is conflicting evidence?

Q2 Give one example of scientific evidence that has informed decision-making about conservation issues.

Q3 Give one example of a decision that has been made as a result of scientific evidence about conservation issues.

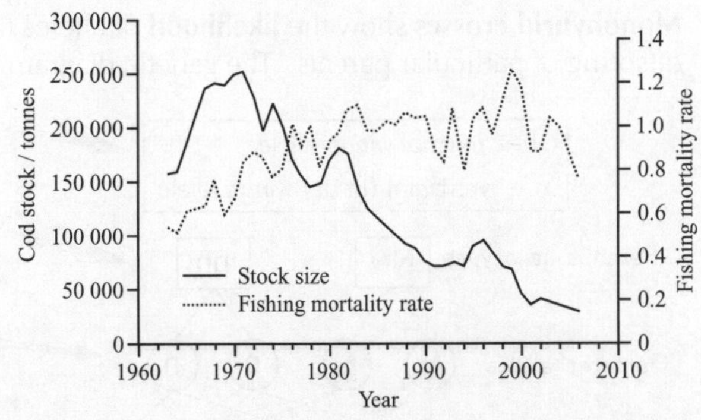

Exam Question

Q1 The graph shows the stock of spawning cod in the North Sea and the rate of mortality caused by fishing since 1960.

a) Describe the results shown by the graph. [4 marks]

b) Suggest a conclusion that could be drawn from the graph. [2 marks]

c) Scientists have stated that 150 000 tonnes is the minimum stock needed to preserve a cod population. In which year did cod stocks first fall below this level? [1 mark]

d) How might this data be used to make informed decisions about the conservation of cod stocks? [1 mark]

I'm considering conflict after these pages, I tell you...

Ah hah ha, aaaaah ha ha ha... oh, I think I need to stop my evil laugh now. I quite enjoyed that. Evaluating evidence and data's an important nut to crack — you might have to do it in your exam for conservation or for another topic altogether.

Inheritance

If you've ever wondered what causes colour blindness, how gender is controlled or how genetic diseases are passed on, then this is the section for you. If you've never wondered this and don't really care — tough. You still need to know it.

You Need to Know These Genetic Terms

'Codes for' means 'contains the instructions for'.

TERM	DESCRIPTION
Gene	A sequence of bases on a DNA molecule that codes for a protein (polypeptide), which results in a characteristic, e.g. the gene for eye colour.
Alleles	One or more alternative versions of the same gene. Most plants and animals, including humans, have two alleles of each gene, one from each parent. The order of bases in each allele is slightly different — they code for different versions of the same characteristic. They're represented using letters, e.g. the allele for brown eyes (B) and the allele for blue eyes (b).
Genotype	The genetic constitution of an organism — the alleles an organism has, e.g. BB, Bb or bb for eye colour.
Phenotype	The expression of the genetic constitution and its interaction with the environment — an organism's characteristics, e.g. brown eyes.
Dominant	An allele whose characteristic appears in the phenotype even when there's only one copy. Dominant alleles are shown by a capital letter. E.g. the allele for brown eyes (B) is dominant — if a person's genotype is Bb or BB, they'll have brown eyes.
Recessive	An allele whose characteristic only appears in the phenotype if two copies are present. Recessive alleles are shown by a lower case letter. E.g. the allele for blue eyes (b) is recessive — if a person's genotype is bb, they'll have blue eyes.
Codominant	Alleles that are both expressed in the phenotype — neither one is recessive, e.g. the alleles for haemoglobin (see page 133).
Locus	The fixed position of a gene on a chromosome. Alleles of a gene are found at the same locus on each chromosome in a pair.
Homozygote	An organism that carries two copies of the same allele, e.g. BB or bb.
Heterozygote	An organism that carries two different alleles, e.g. Bb.

Genetic Diagrams Show the Possible Genotypes of Offspring

Individuals have **two alleles** for **each gene**. **Gametes** (sex cells) contain only **one allele** for each gene. When gametes from two parents fuse together, the alleles they contain form the **genotype** of the **offspring** produced. **Genetic diagrams** can be used to **predict** the **genotypes** and **phenotypes** of the offspring produced if two parents are **crossed** (bred). You need to know how to use genetic diagrams to predict the results of various crosses, including **monohybrid crosses**.

Monohybrid inheritance is the inheritance of a **single characteristic** (gene) controlled by **different alleles**. **Monohybrid crosses** show the **likelihood** of alleles (and so different versions of the characteristic) being **inherited** by offspring of particular parents. The genetic diagram below shows how **wing length** is inherited in fruit flies:

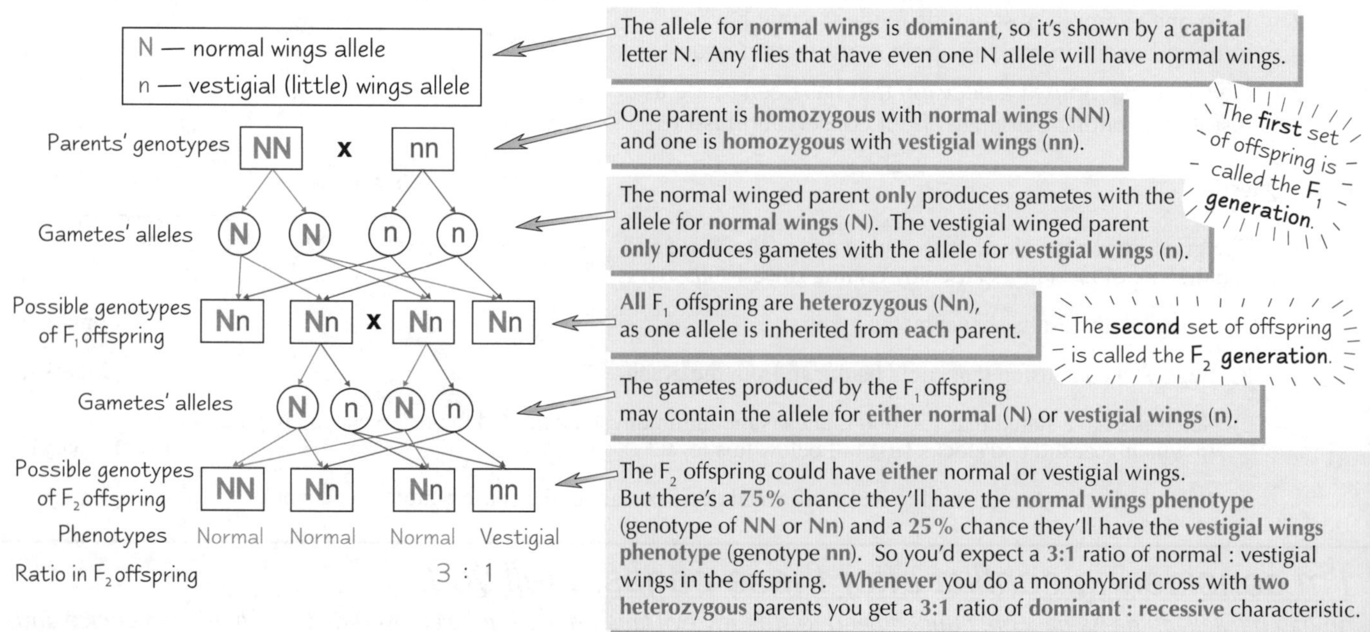
N — normal wings allele
n — vestigial (little) wings allele
Parents' genotypes
NN x nn
Gametes' alleles
N N n n
Possible genotypes of F1 offspring
Nn Nn x Nn Nn
Gametes' alleles
N n N n
Possible genotypes of F2 offspring
NN Nn Nn nn
Phenotypes
Normal Normal Normal Vestigial
Ratio in F2 offspring
3 : 1
The allele for normal wings is dominant, so it's shown by a capital letter N. Any flies that have even one N allele will have normal wings.
One parent is homozygous with normal wings (NN) and one is homozygous with vestigial wings (nn).
The first set of offspring is called the F1 generation.
The normal winged parent only produces gametes with the allele for normal wings (N). The vestigial winged parent only produces gametes with the allele for vestigial wings (n).
All F1 offspring are heterozygous (Nn), as one allele is inherited from each parent.
The second set of offspring is called the F2 generation.
The gametes produced by the F1 offspring may contain the allele for either normal (N) or vestigial wings (n).
The F2 offspring could have either normal or vestigial wings. But there's a 75% chance they'll have the normal wings phenotype (genotype of NN or Nn) and a 25% chance they'll have the vestigial wings phenotype (genotype nn). So you'd expect a 3:1 ratio of normal : vestigial wings in the offspring. Whenever you do a monohybrid cross with two heterozygous parents you get a 3:1 ratio of dominant : recessive characteristic.

Inheritance

A **Punnett square** is just another way of showing a **genetic diagram** — they're also used to predict the **genotypes** and **phenotypes** of offspring. The Punnett squares below show the same crosses from the previous page:

1) First work out the alleles the **gametes** would have.

Parents' genotypes: NN, nn
Gametes' alleles: N, N, n, n

2) Next **cross** the **parents' gametes** to show the possible genotypes of the F_1 **generation** — **all heterozygous**, Nn.

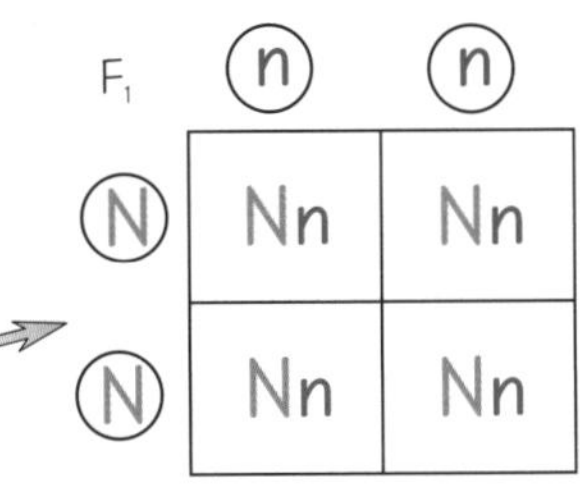

3) Then **cross** the **gametes' alleles** of the F_1 **generation** to show the possible **genotypes** of the F_2 **generation**. The Punnett square shows a **75%** chance that offspring will have **normal wings** and a **25%** chance that they'll have **vestigial wings**, i.e. a **3:1 ratio**.

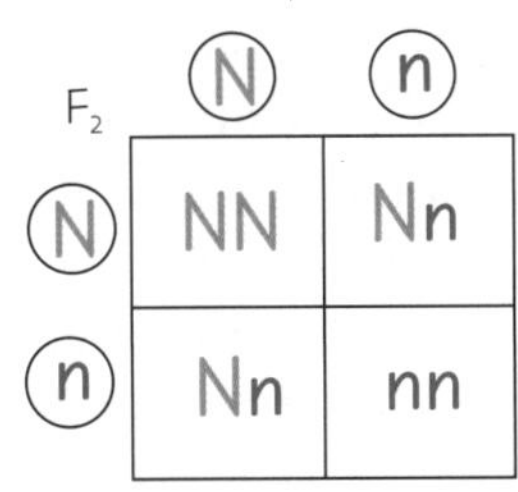

- 1 in 4 chance of offspring having the genotype NN (normal wings)
- 2 in 4 chance of offspring having the genotype Nn (normal wings)
- 1 in 4 chance of offspring having the genotype nn (vestigial wings)
- So, phenotype ratio normal:vestigial = 3:1

Some Genes Have Codominant Alleles

You need to be able to work out genetic diagrams for codominant alleles too.

Occasionally, alleles show **codominance** — **both alleles** are expressed in the **phenotype**, **neither one** is recessive. One example in humans is the allele for **sickle-cell anaemia**:

1) People who are **homozygous** for **normal haemoglobin** (**H^NH^N**) don't have the disease.
2) People who are **homozygous** for **sickle haemoglobin** (**H^SH^S**) have **sickle-cell anaemia** — all their **blood cells** are **sickle-shaped** (crescent-shaped).
3) People who are **heterozygous** (**H^NH^S**) have an **in-between** phenotype, called the **sickle-cell trait** — they have **some** normal haemoglobin and some sickle haemoglobin. The two alleles are **codominant** because they're **both** expressed in the **phenotype**.
4) The **genetic diagram** on the right shows the possible offspring from **crossing** two parents with **sickle-cell trait** (**heterozygous**).

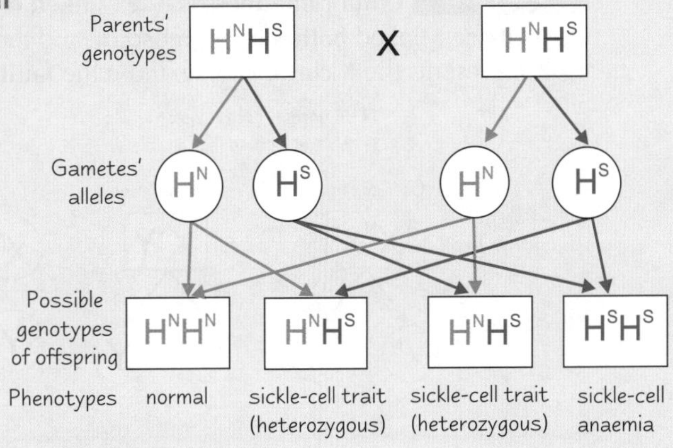

Practice Questions

Q1 What is meant by the term genotype?

Q2 What is meant by the term phenotype?

Q3 What is meant by the term codominance?

Exam Question

Q1 In pea plants, seed texture (round or wrinkled) is passed from parent to offspring by monohybrid inheritance. The allele for round seeds is represented by R and the allele for wrinkled seeds is represented by r.

a) Draw a genetic diagram to show the possible genotypes of F_1 offspring produced by crossing a homozygous round seed pea plant with a homozygous wrinkled seed pea plant. [3 marks]

b) What ratio of round to wrinkled seeds would you expect to see in the F_2 generation? [3 marks]

If there's a dominant revision allele I'm definitely homozygous recessive...

OK, so there are a lot of fancy words on these pages and yes, you do need to know them all. Sorry about that. But don't despair — once you've learnt what the words mean and know how genetic diagrams work it'll all just fall into place.

Inheritance

Now you know how these genetic diagram thingies work, you can use them to work out all kinds of clever stuff — even cleverer than the stuff you can already do. The crosses on these pages are a bit trickier, but nothing you can't handle.

Some **Characteristics** are **Sex-linked**

1) The genetic information for **gender** (**sex**) is carried on two **sex chromosomes**.
2) In mammals, **females** have **two X** chromosomes (XX) and **males** have **one X** chromosome and **one Y** chromosome (XY). The genetic diagram on the right shows how gender is **inherited**. The probability of having **male offspring** is **50%** and the probability of having **female offspring** is **50%**.

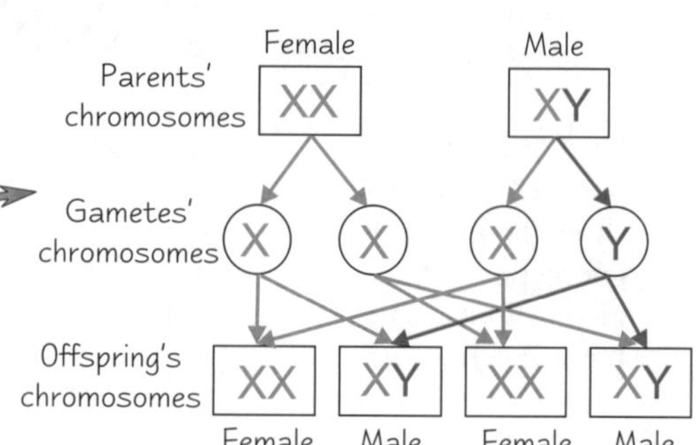

3) A **characteristic** is said to be **sex-linked** when the allele that codes for it is located on a **sex chromosome**.
4) The **Y chromosome** is **smaller** than the X chromosome and carries **fewer genes**. So most genes on the sex chromosomes are **only carried** on the X chromosome (called **X-linked** genes).

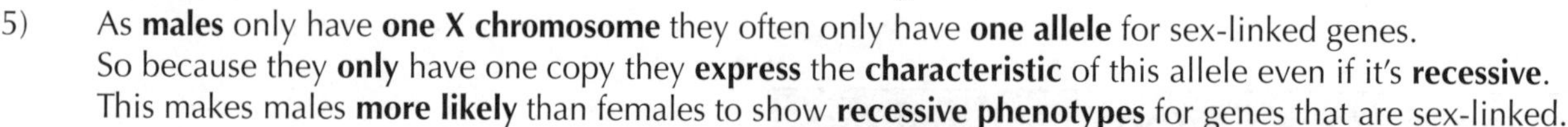

5) As **males** only have **one X chromosome** they often only have **one allele** for sex-linked genes. So because they **only** have one copy they **express** the **characteristic** of this allele even if it's **recessive**. This makes males **more likely** than females to show **recessive phenotypes** for genes that are sex-linked.
6) Genetic disorders caused by **faulty alleles** located on sex chromosomes include **colour blindness** and **haemophilia**. The faulty alleles for both of these disorders are carried on the X chromosome and so are called **X-linked disorders**. **Y-linked disorders** do exist but are **less common**.

Example **Colour blindness** is a **sex-linked disorder** caused by a faulty allele carried on the **X** chromosome. As it's sex-linked **both** the chromosome and the allele are **represented** in the **genetic diagram**, e.g. X^n, where **X** represents the **X chromosome** and **n** the **faulty allele** for **colour vision**. The **Y chromosome** doesn't have an allele for colour vision so is **just** represented by **Y**. **Females** would need **two copies** of the **recessive allele** to be colour blind, while **males** only need **one copy**. This means colour blindness is **much rarer** in **women** than **men**.

N — normal colour vision allele
n — faulty colour vision allele

Carrier female — Parents' genotypes: X^NX^n — Gametes' alleles: X^N, X^n

Normal male — Parents' genotypes: X^NY — Gametes' alleles: X^N, Y

Possible offspring

	X^N	X^n
X^N	X^NX^N Normal female	X^NX^n Carrier female
Y	X^NY Normal male	X^nY Colour-blind male

A carrier is a person carrying an allele which is not expressed in the phenotype but that can be passed on to offspring.

Some **Genes** Have **Multiple Alleles**

Inheritance is **more complicated** when there are **more than two** alleles of the same gene (**multiple alleles**).

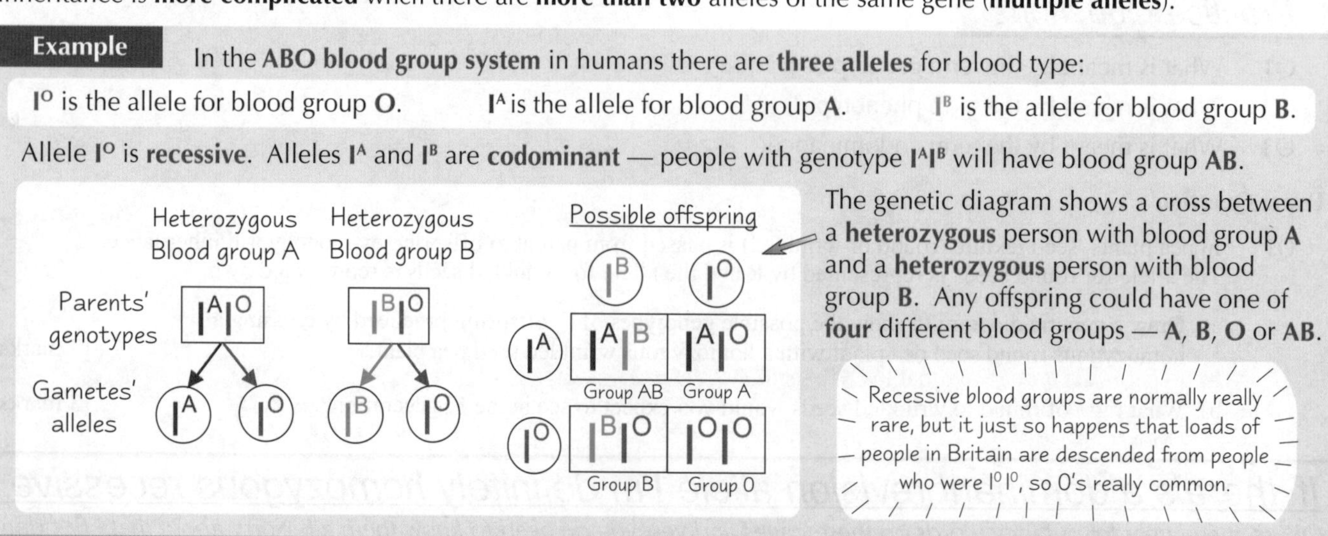

Example In the **ABO blood group system** in humans there are **three alleles** for blood type:

I^O is the allele for blood group **O**. I^A is the allele for blood group **A**. I^B is the allele for blood group **B**.

Allele I^O is **recessive**. Alleles I^A and I^B are **codominant** — people with genotype I^AI^B will have blood group **AB**.

The genetic diagram shows a cross between a **heterozygous** person with blood group **A** and a **heterozygous** person with blood group **B**. Any offspring could have one of **four** different blood groups — **A, B, O** or **AB**.

Recessive blood groups are normally really rare, but it just so happens that loads of people in Britain are descended from people who were I^OI^O, so O's really common.

Inheritance

Genetic Pedigree Diagrams Show How Traits Run in Families

Genetic pedigree diagrams show an **inherited trait** (characteristic) in a group of **related individuals**. You might have to **interpret** genetic pedigree diagrams to work out the **genotypes** or **potential phenotypes** of individuals:

Example **Cystic fibrosis (CF)** is an inherited disorder that's caused by a faulty **recessive** allele (f) — it codes for a **faulty chloride ion channel**. A person will only have the disorder if they're **homozygous** for the allele (ff) — they must inherit one recessive allele **from each parent**. If a person is **heterozygous** (Ff), they **won't** have CF but they'll be a **carrier**.

A key will show what the shapes represent:
- Unaffected male
- Unaffected female
- Male with CF
- Female with CF

Two parents are joined by a horizontal line.

A vertical line goes from parents to children.

Children have a vertical line above them.

This female has CF (ff), but neither of her parents do. She must have inherited an f allele from each parent, so both parents must be carriers (Ff).

This male has CF (ff), like his mother (ff). He's inherited an f allele from each parent, so his father must be a carrier (Ff).

From the **information** in the diagram you could do **genetic crosses** to work out the probability that further children would have CF or be a carrier. E.g. to work out the **chances of the next child** born to individuals 4 and 5 having CF you would cross **Ff** (individual 4) and **ff** (individual 5).

Parents' genotypes: Ff X ff

Gametes' alleles: F, f, f, f

Possible genotypes of offspring	Ff	Ff	ff	ff
Phenotypes	Carrier	Carrier	CF	CF

With a face this cute, Dillon knew he'd never have to worry about pedigree diagrams.

Practice Questions

Q1 What is a sex-linked gene?

Q2 What is a carrier?

Q3 What do pedigree diagrams show?

Key: Unaffected male; Unaffected female; Male with ADA deficiency; Female with ADA deficiency

Exam Questions

Q1 Haemophilia A is a sex-linked genetic disorder caused by a recessive allele carried on the X chromosome. Explain why haemophilia A is more common in males than females. [3 marks]

Q2 Using a genetic diagram, show the probability of a heterozygous person with blood group A and a homozygous person with blood group B having a child with blood group B. [4 marks]

Q3 ADA deficiency is an inherited metabolic disorder caused by a recessive allele (a). Use the genetic pedigree diagram above to answer the following questions:

a) Give the possible genotype(s) of individual 2. [1 mark]

b) What is the genotype of individual 6? Explain your answer. [2 marks]

c) What is the probability that the next child born to individuals 5 and 6 will have ADA deficiency? Show your working. [4 marks]

Sex-linkage and multiple alleles — it's all starting to sound a little bit kinky...

So sex-linked characteristics are characteristics linked to your sex, more than two alleles for one gene are called multiple alleles, and genetic pedigree diagrams are just family trees with a few genotypes thrown in. Another two pages ticked off.

The Hardy-Weinberg Principle

Sometimes you need to look at the genetics of a whole population, rather than a cross between just two individuals. And that's where those spiffing fellows Hardy and Weinberg come in...

Members of a Population Share a Gene Pool

1) A **species** is defined as a group of **similar organisms** that can **reproduce** to give **fertile offspring**.
2) A **population** is a group of organisms of the **same species** living in a **particular area**.
3) Species can exist as **one** or **more populations**, e.g. there are populations of the American black bear (*Ursus americanus*) in parts of America and in parts of Canada.
4) The **gene pool** is the complete range of **alleles** present in a **population**.
5) How **often** an **allele occurs** in a population is called the **allele frequency**. It's usually given as a **percentage** of the total population, e.g. 35%, or a **number**, e.g. 0.35.

Yogi wanted everyone to know what population he was in.

The Hardy-Weinberg Principle Predicts That Allele Frequencies Won't Change

1) The **Hardy-Weinberg principle** predicts that the **frequencies** of **alleles** in a population **won't change** from **one generation** to the **next**.
2) But this prediction is **only true** under **certain conditions** — it has to be a **large population** where there's **no immigration**, **emigration**, **mutations** or **natural selection** (see p. 138). There also needs to be **random mating** — all possible genotypes can breed with all others.
3) The **Hardy-Weinberg equations** (see below) are based on this principle. They can be used to **estimate the frequency** of particular **alleles** and **genotypes** within populations.
4) The Hardy-Weinberg equations can also be used to test whether or not the Hardy-Weinberg principle **applies** to **particular alleles** in **particular populations**, i.e. to test whether **selection** or any **other factors** are **influencing** allele frequencies — if frequencies **do change** between generations in a large population then there's a pressure of some kind (see next page).

The Hardy-Weinberg Equations Can be Used to...

...Predict Allele Frequency...

1) You can **figure out** the frequency of one allele if you **know the frequency of the other**, using this equation:

$$p + q = 1$$

Where: **p** = the **frequency** of the **dominant** allele
q = the **frequency** of the **recessive** allele

The **total frequency** of **all possible alleles** for a characteristic in a certain population is **1.0**. So the frequencies of the **individual alleles** (the dominant one and the recessive one) must **add up to 1.0**.

2) E.g. a species of plant has either **red** or **white** flowers. Allele **R** (red) is **dominant** and allele **r** (white) is **recessive**. If the frequency of **R** is **0.4**, then the frequency of **r** is 1 – 0.4 = **0.6**.

...Predict Genotype Frequency...

Make sure you **learn** both Hardy-Weinberg equations.

1) You can **figure out** the frequency of one genotype if you **know the frequencies of the others**, using this equation:

$$p^2 + 2pq + q^2 = 1$$

Where p^2 = the **frequency** of the **homozygous dominant genotype**
2pq = the **frequency** of the **heterozygous genotype**
q^2 = the **frequency** of the **homozygous recessive genotype**

The **total frequency** of **all possible genotypes** for one characteristic in a certain population is **1.0**. So the frequencies of the **individual genotypes** must **add up to 1.0**.

2) E.g. If there are **two alleles** for **flower colour** (R and r), there are **three possible genotypes** — **RR**, **Rr** and **rr**. If the frequency of genotype **RR** (p^2) is **0.34** and the frequency of genotype **Rr** (2pq) is **0.27**, the frequency of genotype rr (q^2) must be 1 – 0.34 – 0.27 = **0.39**.

The Hardy-Weinberg Principle

...*Predict* the *Percentage* of a *Population* that has a *Certain Genotype...*

Example

The **frequency** of **cystic fibrosis** (genotype ff) in the UK is currently approximately **1 birth in 2000**.
From this information you can estimate the **proportion** of people in the UK that are cystic fibrosis **carriers (Ff)**.
To do this you need to find the **frequency** of **heterozygous genotype Ff**, i.e. **2pq**, using **both** equations:

First calculate q:

- Frequency of cystic fibrosis (homozygous recessive, ff) is 1 in 2000
- $ff = q^2 = 1 \div 2000 = 0.0005$
- So, $q = \sqrt{0.0005} = 0.022$

Next calculate p:

- using $p + q = 1$, $p = 1 - q$
- $p = 1 - 0.022$
- $p = 0.978$

Then calculate 2pq:

- $2pq = 2 \times 0.978 \times 0.022$
- $2pq = 0.043$

The **frequency** of **genotype Ff** is **0.043**, so the **percentage** of the UK population that are **carriers** is **4.3%**.

...and *Show if External Factors* are *Affecting Allele Frequency*

Example

If the **frequency** of **cystic fibrosis** is measured **50 years later** it might be found to be **1 birth in 3000**.
From this information you can estimate the **frequency** of the **recessive allele** (f) in the population, i.e. **q**.

To calculate q:

- Frequency of cystic fibrosis (homozygous recessive, ff) is 1 in 3000
- $ff = q^2 = 1 \div 3000 = 0.00033$
- So, $q = \sqrt{0.00033} = 0.018$

The frequency of the recessive allele is now **0.018**, compared to **0.022** currently (see above). As the frequency of the allele has **changed** between generations the **Hardy-Weinberg principle doesn't apply** so there must have been some **factors** affecting **allele frequency**, e.g. **immigration, migration, mutations** or **natural selection**.

Practice Questions

Q1 What is a population?

Q2 What is a gene pool?

Q3 What conditions are needed for the Hardy-Weinberg principle to apply?

Q4 Which term represents the frequency of the homozygous recessive genotype in the Hardy-Weinberg equations?

Q5 Which term represents the frequency of the heterozygous genotype in the Hardy-Weinberg equations?

Exam Question

Q1 Cleft chins are controlled by a single gene with two alleles. The allele coding for a cleft chin (C) is dominant over the allele coding for a non-cleft chin (c). In a particular population the frequency of the homozygous dominant genotype for cleft chin is 0.14.

a) What is the frequency of the recessive allele in the population? [3 marks]

b) What is the frequency of the homozygous recessive genotype in the population? [1 mark]

c) What percentage of the population have a cleft chin? [2 marks]

This stuff's surely not that bad — Hardly worth Weining about...

Two equations that you absolutely have to know — so learn 'em. And whilst you're at it make sure that you learn what each of the terms means as well. You'll feel like a right wally if you know that $p^2 + 2pq + q^2 = 1$ but haven't got a clue what p^2, 2pq and q^2 stand for. It's the kind of stuff that falls out of your head really easily so learn it, learn it, learn it.

Allele Frequency and Speciation

I'm sure my brother's from a totally different species, boom boom — it's an oldie but a goodie.

Allele Frequency is *Affected* by *Differential Reproductive Success*

1) Sometimes the **frequency** of an **allele** within a population **changes**. This can happen when the allele codes for a characteristic that **affects** the **chances** of an organism **surviving**.
2) Not all individuals are as likely to **reproduce** as each other. There's **differential reproductive success** in a population — individuals that have an allele that **increases** their **chance of survival** are **more likely** to **survive**, **reproduce** and **pass on** their genes (including the **beneficial** allele), than individuals with different alleles.
3) This means that a **greater proportion** of the next generation **inherit** the **beneficial allele**.
4) They, in turn, are **more likely** to **survive**, **reproduce** and **pass on** their genes.
5) So the **frequency** of the beneficial allele **increases** from generation to generation.
6) This process is called **natural selection**.

Different Types of *Natural Selection* Lead to *Different Frequency Patterns*

Stabilising selection and **directional selection** are **types** of **natural selection** that affect **allele frequency** in different ways:

Stabilising selection is where individuals with alleles for characteristics towards the **middle** of the range are more likely to **survive** and **reproduce**. It occurs when the environment **isn't changing**, and it **reduces the range** of possible **phenotypes**.

EXAMPLE In any **mammal population** there's a **range** of **fur length**. In a **stable climate**, having fur at the **extremes** of this range **reduces** the **chances** of **surviving** as it's harder to maintain the **right body temperature**. Animals with alleles for **average fur length** are the **most** likely to **survive**, **reproduce** and **pass on** their alleles. So these alleles **increase** in **frequency**. The **proportion** of the **population** with **average fur length increases** and the **range** of fur lengths **decreases**.

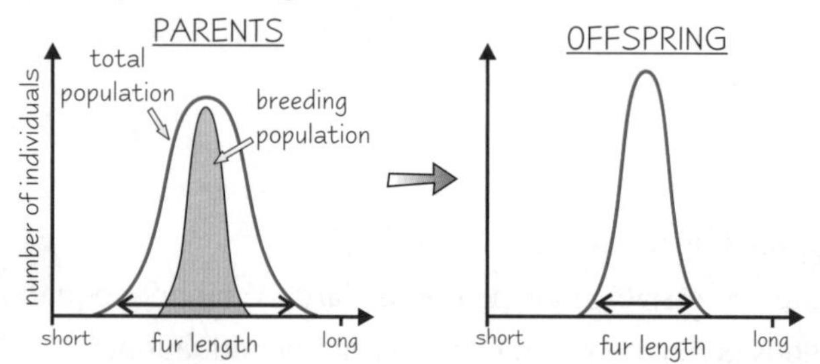

Directional selection is where individuals with alleles for characteristics of an **extreme type** are more likely to **survive** and **reproduce**. This could be in response to an **environmental change**.

EXAMPLE **Cheetahs** are the **fastest** animals on land. It's likely that this characteristic was developed through **directional selection**, as individuals that have **alleles** for **speed** are **more likely** to **catch prey** than slower individuals. So they're **more likely** to **survive**, **reproduce** and **pass on** their alleles. Over time the **frequency** of alleles for **high speed increases** and the population becomes **faster**.

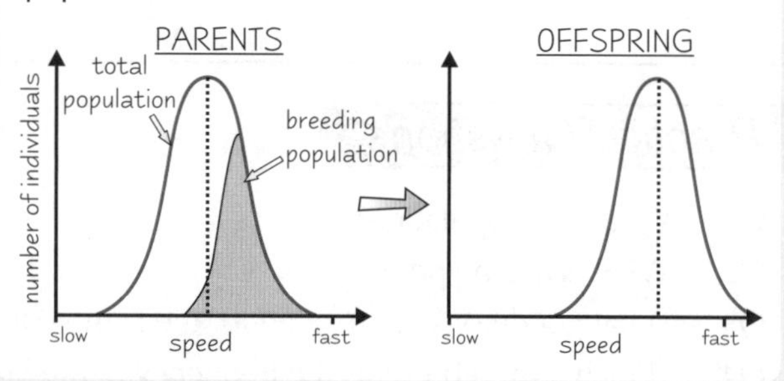

You Need to be able to Interpret Data Relating to the Effect of Selection

1) For example, there are **two common forms** of the peppered moth in the UK — one **dark coloured** and one **pale**.
2) The allele for **dark colouring** is **dominant** (**M**) over the allele for **pale colouring** (**m**).
3) The table shows how **allele frequency** in a population of **peppered moths** changed between 1852 and 1860 as the number of **coal-powered factories** in the area increased.
4) The frequency of the **m** allele **decreases** from **0.75** to **0.39** as the number of factories increases. So the frequency of the **M allele** must **increase** from **0.25** to **0.61** (remember: p + q = 1, see page 136).
5) As the allele frequencies are **changing**, it's likely there's selective pressure **for** dark colouring. This could be because of **pollution** — more factories means more pollution, which **darkens** buildings etc. The **dark coloured moths** would be better **camouflaged**, making them **more likely** to **survive**, **reproduce** and pass on **M**.

Year	Number of coal-powered factories	Frequency of m allele
1852	1	0.75
1854	3	0.70
1856	5	0.63
1858	7	0.47
1860	10	0.39

Allele Frequency and Speciation

Geographical Isolation and Natural Selection Lead to Speciation

1) **Speciation** is the development of a **new species**.
2) Speciation occurs when populations of the same species become **reproductively isolated**.
3) This can happen when a **physical barrier**, e.g. a flood or an earthquake, **divides** a population of a species, causing some individuals to become **separated** from the main population. This is known as **geographical isolation**.
4) Populations that are geographically separated will experience slightly **different conditions**. For example, there might be a **different climate** on each side of the physical barrier.
5) The populations will experience **different selective pressures** and so **different changes** in allele frequencies:
   - Different **alleles** will be **more advantageous** in the different populations. For example, if geographical separation places one population in a **colder climate** than before, **longer fur length** will be **beneficial**. **Directional selection** will then act on the **alleles** for fur length in this population, changing the frequency of the allele for **longer fur length**.
   - Allele frequencies will also change as **mutations** (see p. 174) will occur **independently** in each population.
6) The changes in allele frequency will lead to **differences** accumulating in the **gene pools** of the separated populations, causing changes in **phenotype frequencies**.
7) Eventually, individuals from the different populations will have changed so much that they won't be able to breed with one another to produce **fertile** offspring — they'll have become **reproductively isolated**.
8) The two groups will have become **separate species**.

Sandra's hair had caused her reproductive isolation.

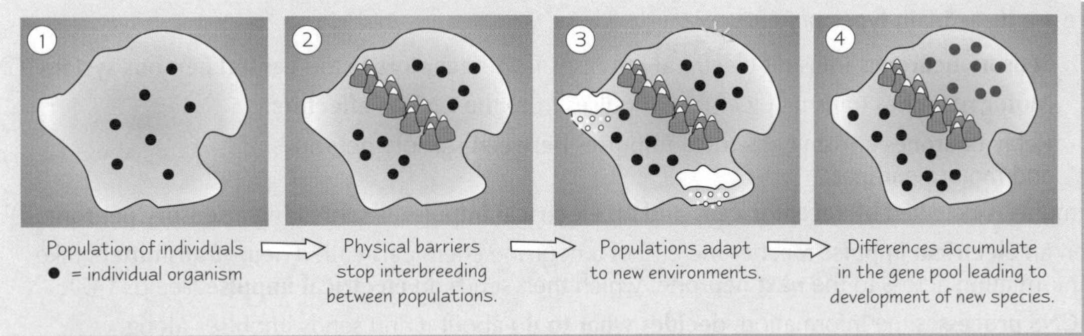
1
2
3
4
Population of individuals
● = individual organism
Physical barriers stop interbreeding between populations.
Populations adapt to new environments.
Differences accumulate in the gene pool leading to development of new species.

Practice Questions

Q1 What is stabilising selection?

Q2 What is directional selection?

Q3 What is speciation?

Exam Question

Average Temp / °C	Frequency of h allele
22	0.11
21	0.13
19	0.19
18	0.20
16	0.23

Q1 The table above shows the results of an investigation into hair length in golden hamsters in a climate where the temperature is decreasing. Hair length is controlled by a single gene with two alleles. H represents the allele for short hair, which is dominant over the allele for long hair, represented by h.

a) Describe the relationship between the frequency of the recessive long hair allele and temperature. Suggest an explanation for this relationship. [4 marks]

b) What type of selection is responsible for this change in allele frequency? [1 mark]

Differential reproductive success — not PC, but it sorts the hot from the not...

And that's the end of the section. Whoopdeedoo. All that's left for you to do is learn these two pages — no small feat I grant you, as they cover a lot of stuff. Maybe have another quick flick through the section before you sign off from genetic variation entirely. You know it makes sense, and who knows — you might enjoy it... OK maybe not, but do it anyway.

Nervous and Hormonal Communication

Right, it's time to get your brain cells fired up and get your teeth stuck into a mammoth — a mammoth section, that is...

Responding to their ***Environment*** Helps ***Organisms Survive***

1) **Animals increase** their **chances** of **survival** by **responding** to **changes** in their **external environment**, e.g. by **avoiding harmful environments** such as places that are too hot or too cold.
2) They also **respond** to **changes** in their **internal environment** to make sure that the **conditions** are always **optimal** for their **metabolism** (all the chemical reactions that go on inside them).
3) **Plants** also **increase** their **chances** of **survival** by **responding** to **changes** in their **environment** (see p. 156).
4) Any **change** in the internal or external **environment** is called a **stimulus**.

Receptors Detect ***Stimuli*** and ***Effectors*** Produce a ***Response***

1) **Receptors detect stimuli** — they can be **cells** or **proteins** on **cell surface membranes**. There are **loads** of **different types** of receptors that detect **different stimuli**.
2) **Effectors** are cells that bring about a **response** to a **stimulus**, to produce an **effect**. Effectors include **muscle cells** and cells found in **glands**, e.g. the **pancreas**.
3) Receptors **communicate** with effectors via the **nervous system** or the **hormonal system**, or sometimes using **both**.

The ***Nervous System*** Sends Information as ***Electrical Impulses***

1) The **nervous system** is made up of a **complex network** of cells called **neurones**. There are **three main types** of neurone:
   - **Sensory neurones** transmit electrical impulses from **receptors** to the **central nervous system** (**CNS**).
   - **Motor neurones** transmit electrical impulses from the **CNS** to **effectors**.
   - **Relay neurones** transmit electrical impulses **between** sensory neurones and motor neurones.

Electrical impulses are also called nerve impulses.

2) A stimulus is detected by **receptor cells** and an **electrical impulse** is sent along a **sensory neurone**.
3) When an **electrical impulse** reaches the end of a neurone chemicals called **neurotransmitters** take the information across to the **next neurone**, which then sends an **electrical impulse** (see p. 147).
4) The **CNS processes** the information, **decides what to do** about it and sends impulses along **motor neurones** to an **effector**.

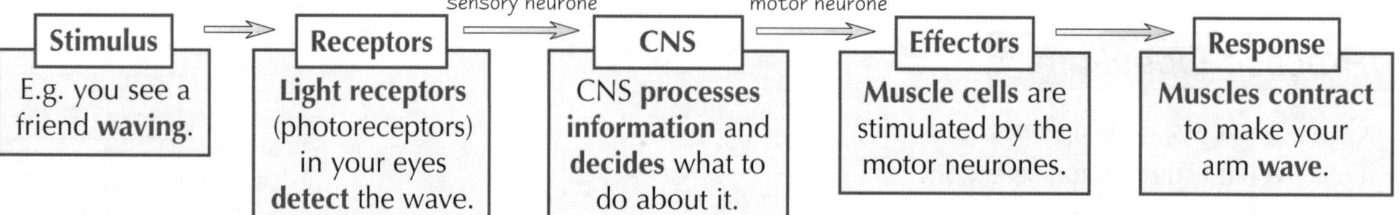

E.g. you see a friend waving.

5) The **nervous system** is split into two different systems:

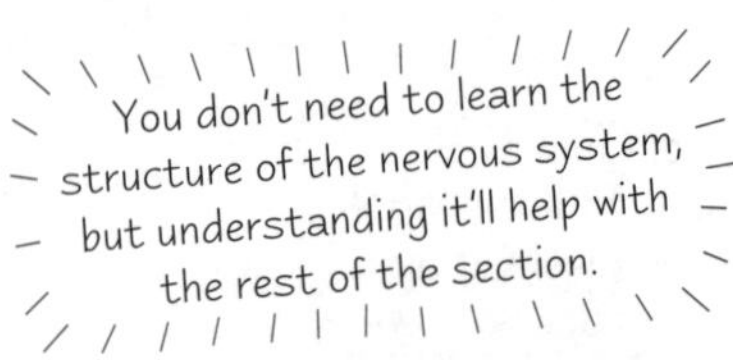

You don't need to learn the
structure of the nervous system,
but understanding it'll help with
the rest of the section.

The **central nervous system** (**CNS**) — made up of the **brain** and the **spinal cord**.

The **peripheral nervous system** — made up of the neurones that connect the CNS to the **rest** of the **body**. It also has two different systems:

- The **somatic nervous system** controls **conscious** activities, e.g. running and playing video games.
- The **autonomic nervous system** controls **unconscious** activities, e.g. digestion. It's got two divisions that have **opposite effects** on the body:
  - The **sympathetic** nervous system gets the body **ready for action**. It's the '**flight or fight**' system.
  - The **parasympathetic** nervous system **calms** the body down. It's the '**rest and digest**' system.

Harold thought it was about time his sympathetic nervous system took over.

Nervous and Hormonal Communication

Nervous System Communication is Localised, Short-lived and Rapid

1) When an **electrical impulse** reaches the end of a neurone, **neurotransmitters** are **secreted directly** onto **cells** (e.g. muscle cells) — so the nervous response is **localised**.
2) **Neurotransmitters** are **quickly removed** once they've done their job, so the response is **short-lived**.
3) Electrical impulses are **really fast**, so the response is **rapid** — this allows animals to **react quickly** to stimuli.

The Hormonal System Sends Information as Chemical Signals

1) The **hormonal system** is made up of **glands** and **hormones**:
   - A **gland** is a group of cells that are specialised to **secrete** a useful substance, such as a **hormone**. E.g. the **pancreas** secretes **insulin**.
   - **Hormones** are '**chemical messengers**'. Many hormones are **proteins** or **peptides**, e.g. **insulin**. Some hormones are **steroids**, e.g. **progesterone**.
2) **Hormones** are **secreted** when a **gland** is **stimulated**:
   - Glands can be **stimulated** by a **change** in **concentration** of a specific **substance** (sometimes **another hormone**).
   - They can also be **stimulated** by **electrical impulses**.
3) Hormones **diffuse directly into** the **blood**, then they're **taken** around the body by the **circulatory system**.
4) They **diffuse out** of the blood **all over** the **body** but each hormone will only **bind** to **specific receptors** for that hormone, found on the membranes of some cells (called **target cells**).
5) The hormones trigger a **response** in the **target cells** (the **effectors**).

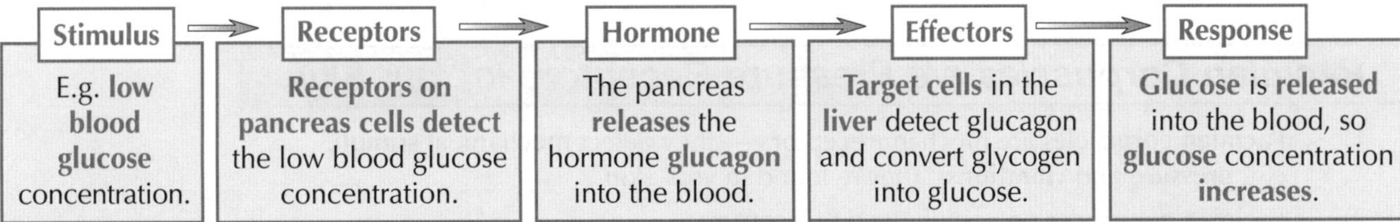

Stimulus
E.g. low blood glucose concentration.
Receptors
Receptors on pancreas cells detect the low blood glucose concentration.
Hormone
The pancreas releases the hormone glucagon into the blood.
Effectors
Target cells in the liver detect glucagon and convert glycogen into glucose.
Response
Glucose is released into the blood, so glucose concentration increases.

Hormonal System Communication is Slower, Long-lasting and Widespread

1) Hormones **aren't** released directly onto their target cells — they must **travel** in the **blood** to get there. This means that chemical communication (by hormones) is **slower** than electrical communication (by nerves).
2) They **aren't broken down as quickly** as neurotransmitters, so the **effects** of hormones can **last** for much **longer**.
3) Hormones are transported **all over** the **body**, so the response may be **widespread** if the target cells are widespread.

Practice Questions

Q1 Why do organisms respond to changes in their environment?

Q2 Give two types of effector.

Q3 How do hormones reach their target cells?

Exam Question

Q1 Bright light causes circular iris muscles in an animal's eyes to contract, which constricts the pupils and protects the eyes.

a) Suggest why this response uses nervous communication rather than hormonal communication. [1 mark]

b) Describe and explain the roles of receptors and effectors for this response. [5 marks]

Vacancy — talented gag writer required for boring biology topics...

Actually, this stuff is really quite fascinating once you realise just how much your body can do without you even knowing. Just sit back and relax, let your nerves and hormones do the work... Ah, apart from the whole revision thing — your body can't do that without you knowing, unfortunately. Get your head around these pages before you tackle the rest of the section.

Receptors

So now you know why organisms respond it's time for the (slightly less interesting but equally important) details... first up — receptors.

Receptors are Specific to One Kind of Stimulus

1) Receptors are **specific** — they only **detect one particular stimulus**, e.g. light, pressure or glucose concentration.
2) There are **many different types** of receptor that each detect a **different type of stimulus**.
3) Some receptors are **cells**, e.g. photoreceptors are receptor cells that connect to the nervous system. Some receptors are **proteins** on **cell surface membranes**, e.g. glucose receptors are proteins found in the cell membranes of some pancreatic cells.
4) Here's a bit more about how receptor cells that communicate information via the **nervous system** work:
   - When a nervous system receptor is in its **resting state** (not being stimulated), there's a **difference in charge** between the **inside** and the **outside** of the cell — this is generated by ion pumps and ion channels (see p. 144). This means that there's a **voltage** across the membrane. Voltage is also known as **potential difference**.
   - The **potential difference** when a cell is at **rest** is called its **resting potential**. When a stimulus is detected, the cell membrane is **excited** and becomes **more permeable**, allowing **more ions** to move **in** and **out** of the cell — **altering** the **potential difference**. The **change** in **potential difference** due to a stimulus is called the **generator potential**.
   - A **bigger stimulus** excites the membrane more, causing a **bigger movement** of ions and a **bigger change** in potential difference — so a **bigger generator potential** is produced.
   - If the **generator potential** is **big enough** it'll trigger an **action potential** — an electrical impulse along a neurone (see p. 144). An action potential is only triggered if the generator potential reaches a certain level called the **threshold** level. Action potentials are all one size, so the **strength** of the **stimulus** is measured by the **frequency** of **action potentials**.
   - If the stimulus is **too weak** the generator potential **won't reach** the **threshold**, so there's **no action potential**.

Pacinian Corpuscles are Pressure Receptors in Your Skin

1) **Pacinian corpuscles** are **mechanoreceptors** — they detect **mechanical stimuli**, e.g. **pressure** and **vibrations**. They're found in your **skin**.
2) Pacinian corpuscles contain the end of a **sensory neurone**, imaginatively called a **sensory nerve ending**.
3) The sensory nerve ending is **wrapped** in loads of layers of connective tissue called **lamellae**.
4) When a Pacinian corpuscle is **stimulated**, e.g. by a tap on the arm, the lamellae are **deformed** and **press** on the **sensory nerve ending**.
5) This causes **deformation** of **stretch-mediated sodium channels** in the sensory neurone's cell membrane. The sodium ion channels **open** and **sodium ions diffuse into** the cell, creating a **generator potential**.
6) If the **generator potential** reaches the **threshold**, it triggers an **action potential**.

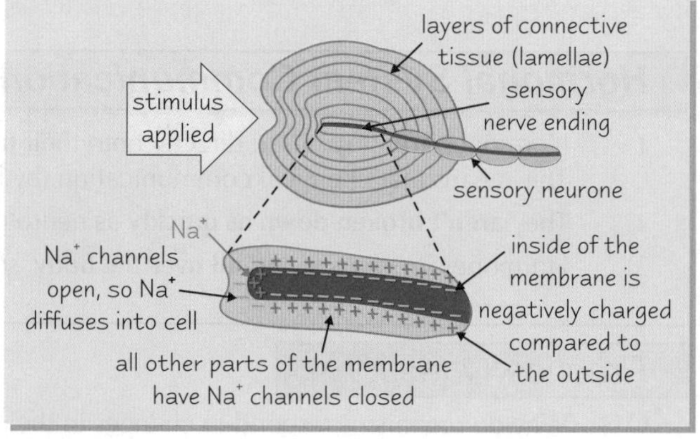

Photoreceptors are Light Receptors in Your Eye

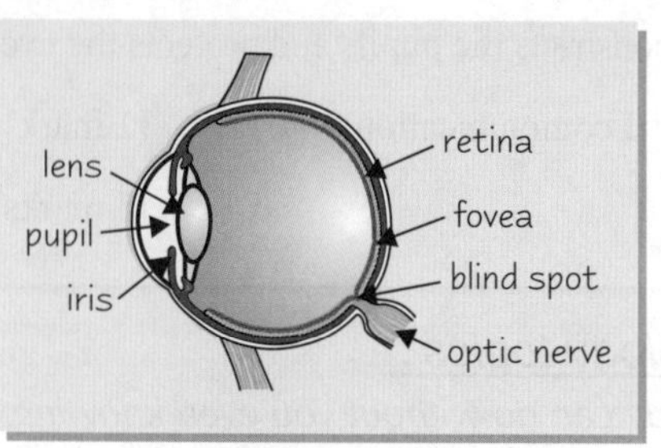

1) **Light** enters the eye through the **pupil**. The **amount** of light that enters is **controlled** by the muscles of the **iris**.
2) Light rays are **focused** by the **lens** onto the **retina**, which lines the inside of the eye. The retina contains **photoreceptor cells** — these **detect light**.
3) The **fovea** is an area of the retina where there are **lots** of **photoreceptors**.
4) **Nerve impulses** from the photoreceptor cells are carried from the **retina** to the **brain** by the **optic nerve**, which is a bundle of **neurones**. Where the optic nerve leaves the eye is called the **blind spot** — there **aren't** any **photoreceptor cells**, so it's **not sensitive** to **light**.

Receptors

Photoreceptors Convert Light into an Electrical Impulse

Light goes straight through the neurones to the photoreceptors.

1) **Light** enters the eye, hits the **photoreceptors** and is **absorbed** by **light-sensitive pigments**.
2) Light bleaches the pigments, causing a **chemical change** and altering the **membrane permeability** to **sodium**.
3) A **generator potential** is created and if it reaches the threshold, a nerve impulse is sent along a **bipolar neurone**.
4) Bipolar neurones connect **photoreceptors** to the **optic nerve**, which takes impulses to the **brain**.

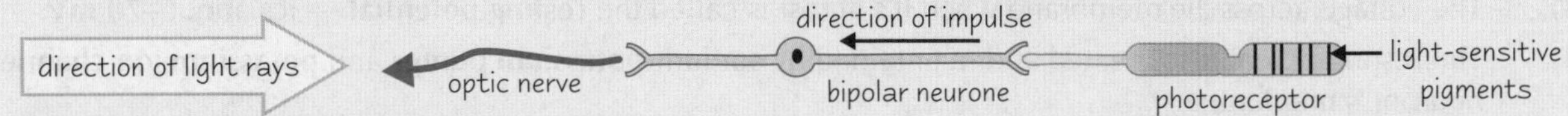

5) The human eye has **two types** of photoreceptor — **rods** and **cones**.
6) Rods are mainly found in the **peripheral** parts of the **retina**, and cones are found **packed together** in the **fovea**.
7) Rods only give information in **black and white** (monochromatic vision), but cones give information in **colour** (trichromatic vision). There are three types of cones — **red-sensitive**, **green-sensitive** and **blue-sensitive**. They're stimulated in **different proportions** so you see different colours.

Rods are More Sensitive, but Cones let you See More Detail

Sensitivity

- Rods are **very sensitive to light** (they fire action potentials in **dim light**). This is because **many rods** join **one neurone**, so many weak **generator potentials combine** to **reach** the **threshold** and trigger an action potential.
- Cones are **less sensitive** than rods (they only fire action potentials in **bright light**). This is because **one cone** joins **one neurone**, so it takes more light to reach the threshold and trigger an action potential.

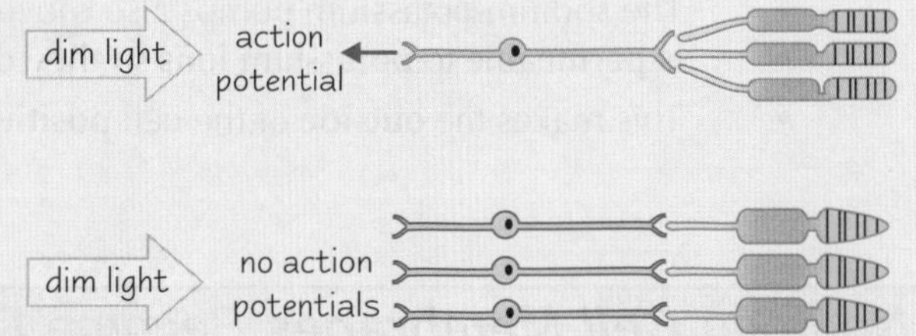

Visual acuity (the ability to tell apart points that are close together)

- Rods give **low visual acuity** because **many rods** join the **same neurone**, which means light from two objects close together **can't** be told apart.
- Cones give **high visual acuity** because cones are **close together** and **one cone** joins **one neurone**. When light from two points hits two cones, two action potentials (one from **each cone**) go to the brain — so you can distinguish two points that are close together as **two separate points**.

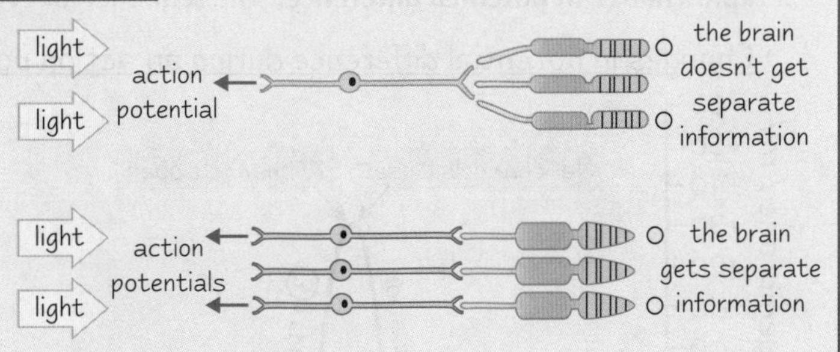

Practice Questions

Q1 What is a generator potential?

Q2 What is the threshold level of a receptor?

Exam Questions

Q1 Explain how a tap on the arm is converted into a nerve impulse. [6 marks]

Q2 Explain how the human eye has both high sensitivity and high acuity. [5 marks]

Pacinian corpuscles love deadlines — they work best under pressure...

Wow, loads of stuff here, so cone-gratulations if you manage to remember it all. Receptors are really important because without them you wouldn't be able to see this book, and without this book revision would be way trickier.

Nervous System — Neurones

Ah, on to the good stuff. Revision notepad at the ready, motor neurones fired up, OK — lights, camera, action potentials...

Neurone **Cell Membranes** are **Polarised** at **Rest**

1) In a neurone's **resting state** (when it's not being stimulated), the **outside** of the membrane is **positively charged** compared to the **inside**. This is because there are **more positive ions outside** the cell than inside.
2) So the membrane is **polarised** — there's a **difference in charge** (voltage).
3) The voltage across the membrane when it's at rest is called the **resting potential** — it's about **–70 mV**.
4) The resting potential is created and maintained by **sodium-potassium pumps** and **potassium ion channels** in a neurone's membrane:

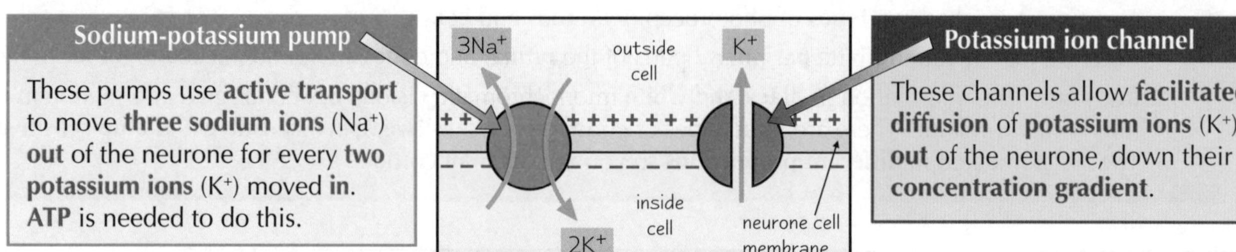

Sodium-potassium pump
These pumps use **active transport** to move **three sodium ions** (Na^+) **out** of the neurone for every **two potassium ions** (K^+) moved **in**. **ATP** is needed to do this.
3Na+
outside cell
K+
inside cell
2K+
neurone cell membrane
Potassium ion channel
These channels allow **facilitated diffusion** of **potassium ions** (K^+) **out** of the neurone, down their **concentration gradient**.

- The sodium-potassium pumps move **sodium ions out** of the neurone, but the membrane **isn't permeable** to **sodium ions**, so they **can't diffuse back in**. This creates a **sodium ion electrochemical gradient** (a **concentration gradient** of **ions**) because there are **more** positive sodium ions **outside** the cell than inside.
- The sodium-potassium pumps also move **potassium ions in** to the neurone, but the membrane **is permeable** to **potassium ions** so they **diffuse back out** through potassium ion channels.
- This makes the **outside** of the cell **positively charged** compared to the inside.

Neurone **Cell Membranes** Become **Depolarised** when they're **Stimulated**

A **stimulus** triggers other ion channels, called **sodium ion channels**, to **open**. If the stimulus is big enough, it'll trigger a **rapid change** in **potential difference**. The sequence of events that happen are known as an **action potential**:

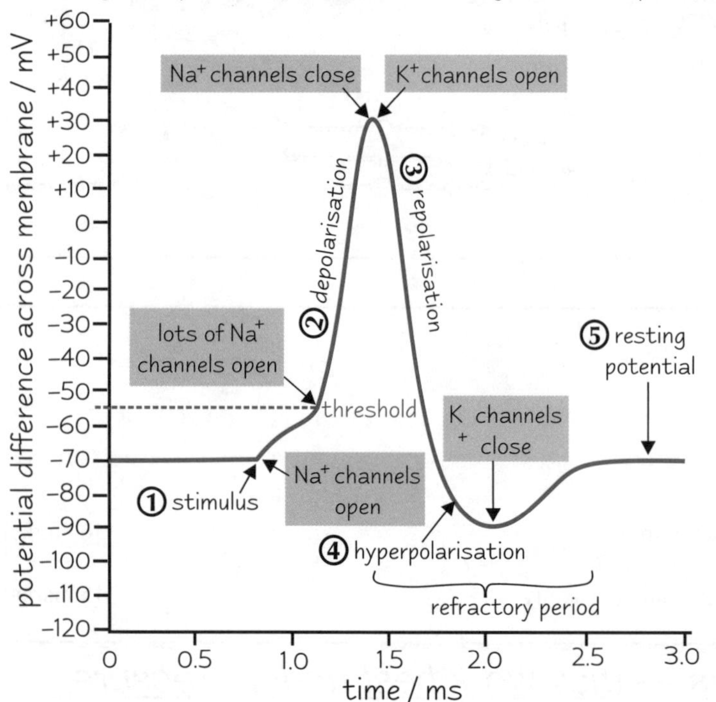

① **Stimulus** — this **excites** the neurone cell membrane, causing **sodium ion channels** to **open**. The membrane becomes **more permeable** to sodium, so **sodium ions diffuse into** the neurone down the sodium ion electrochemical gradient. This makes the **inside** of the neurone **less negative**.

② **Depolarisation** — if the potential difference reaches the **threshold** (around **–55 mV**), **more sodium ion channels open**. **More sodium ions diffuse into** the neurone.

Nervous System — Neurones

③ **Repolarisation** — at a potential difference of around **+30 mV** the **sodium ion channels close** and **potassium ion channels open**. The membrane is **more permeable** to potassium so **potassium ions diffuse out** of the neurone down the potassium ion concentration gradient. This starts to get the membrane **back** to its **resting potential**.

④ **Hyperpolarisation** — **potassium ion channels** are **slow to close** so there's a slight '**overshoot**' where too many potassium ions diffuse out of the neurone. The potential difference becomes **more negative** than the **resting potential** (i.e. less than –70 mV).

⑤ **Resting potential** — the ion channels are **reset**. The **sodium-potassium pump** returns the membrane to its **resting potential** and maintains it until the membrane's excited by another stimulus.

After an **action potential**, the neurone cell membrane **can't** be **excited** again straight away. This is because the ion channels are **recovering** and they **can't** be made to **open** — **sodium ion channels** are **closed** during repolarisation and **potassium ion channels** are **closed** during hyperpolarisation. This period of recovery is called the **refractory period**.

*The **Action Potential** Moves **Along** the **Neurone***

1) When an **action potential** happens, some of the **sodium ions** that enter the neurone **diffuse sideways**.
2) This causes **sodium ion channels** in the next region of the neurone to **open** and **sodium ions diffuse into** that part.
3) This causes a **wave of depolarisation** to travel along the neurone.
4) The **wave** moves **away** from the parts of the membrane in the **refractory period** because these parts **can't fire** an action potential.

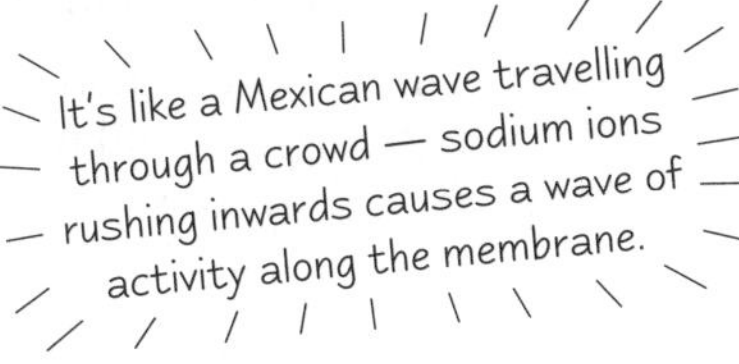

It's like a Mexican wave travelling through a crowd — sodium ions rushing inwards causes a wave of activity along the membrane.

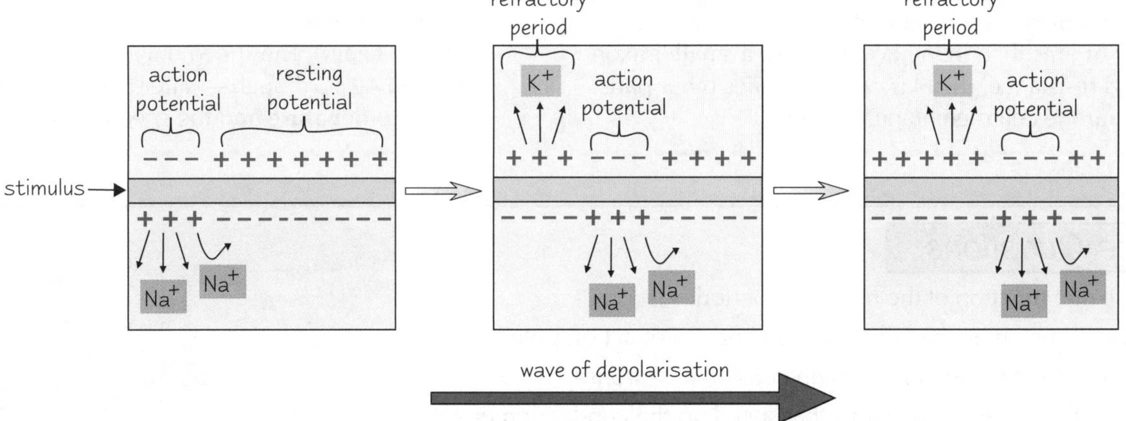

*The **Refractory Period** Produces **Discrete Impulses***

1) During the **refractory period**, **ion channels** are **recovering** and **can't** be **opened**.
2) So the refractory period acts as a **time delay** between one action potential and the next. This makes sure that **action potentials don't overlap** but pass along as **discrete** (separate) **impulses**.
3) The refractory period also makes sure **action potentials** are **unidirectional** (they only travel in **one direction**).

Action Potentials** have an **All-or-Nothing Nature

1) Once the threshold is reached, an action potential will **always fire** with the **same change in voltage**, no matter how big the stimulus is.
2) If **threshold isn't reached**, an action potential **won't fire** — this is the **all-or-nothing** nature of action potentials.
3) A **bigger stimulus** won't cause a bigger action potential but it will cause them to fire **more frequently**.

Nervous System — Neurones

Three Factors Affect the Speed of Conduction of Action Potentials

1) Myelination

1) Some neurones are **myelinated** — they have a **myelin sheath**.
2) The myelin sheath is an **electrical insulator**.
3) It's made of a type of cell called a **Schwann cell**.
4) Between the Schwann cells are tiny patches of **bare membrane** called the **nodes of Ranvier**. **Sodium ion channels** are **concentrated** at the nodes.
5) In a **myelinated** neurone, **depolarisation** only happens at the **nodes of Ranvier** (where sodium ions can get through the membrane).
6) The neurone's **cytoplasm conducts** enough electrical charge to **depolarise** the **next node**, so the impulse '**jumps**' from node to node.
7) This is called **saltatory conduction** and it's **really fast**.
8) In a **non-myelinated** neurone, the impulse travels as a **wave** along the **whole length** of the **axon membrane**.
9) This is **slower** than saltatory conduction (although it's still pretty quick).

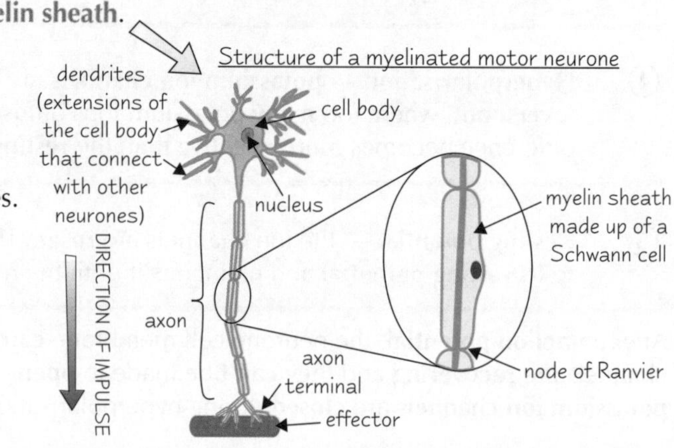

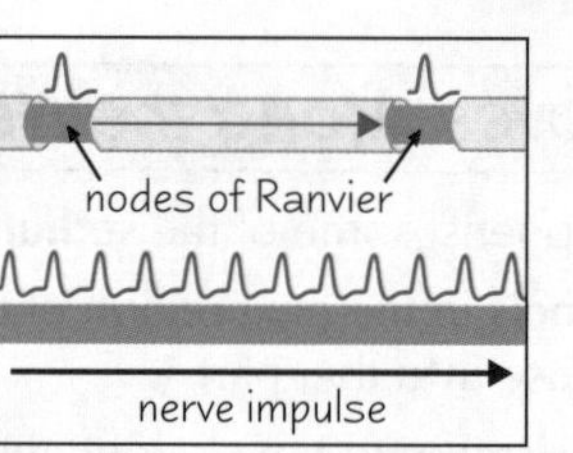

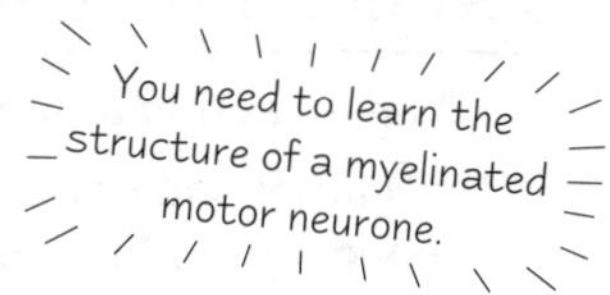

You need to learn the structure of a myelinated motor neurone.

2) Axon diameter

Action potentials are conducted **quicker** along axons with **bigger diameters** because there's **less resistance** to the **flow of ions** than in the cytoplasm of a smaller axon. With less resistance, **depolarisation reaches** other parts of the neurone cell membrane **quicker**.

3) Temperature

The speed of conduction increases as the **temperature increases** too, because **ions diffuse faster**. The speed only increases up to around **40 °C** though — after that the **proteins** begin to **denature** and the speed decreases.

Practice Questions

These questions cover pages 144-146.

Q1 Give one function of the refractory period.

Q2 What is meant by the 'all-or-nothing' nature of action potentials?

Q3 What is the function of Schwann cells on a neurone?

Q4 Name three features of axons that speed up the conduction of action potentials.

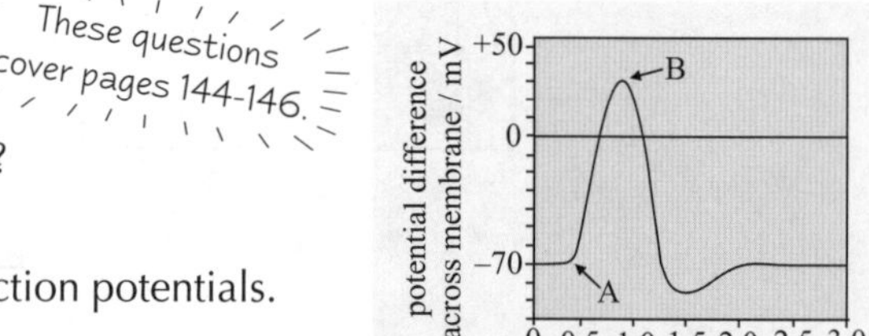

Exam Questions

Q1 The graph on the right shows an action potential across an axon membrane following the application of a stimulus.

a) What label should be added at point A? [1 mark]

b) Explain what causes the change in potential difference between point A and point B. [3 marks]

c) A stimulus was applied at 1.5 ms, but failed to produce an action potential. Suggest why. [2 marks]

Q2 Multiple sclerosis is a disease of the nervous system characterised by damage to the myelin sheaths of neurones. Explain how this will affect the transmission of action potentials. [5 marks]

I'm feeling a bit depolarised after all that...

All this stuff about neurones can be a bit tricky to get your head around at first. Take your time and try scribbling it all down a few times till it starts to make some kind of sense. Neurones work because there's an electrical charge across their membrane, which is set up by ion pumps and ion channels. It's a change in this charge that transmits an action potential.

Nervous System — Synaptic Transmission

When an action potential arrives at the end of a neurone the information has to be passed on to the next cell — this could be another neurone, a muscle cell or a gland cell.

A ***Synapse*** *is a* ***Junction*** *Between a* ***Neurone*** *and the* ***Next Cell***

1) A **synapse** is the junction between a **neurone** and another **neurone**, or between a **neurone** and an **effector cell**, e.g. a muscle or gland cell.
2) The **tiny gap** between the cells at a synapse is called the **synaptic cleft**.
3) The **presynaptic neurone** (the one before the synapse) has a **swelling** called a **synaptic knob**. This contains **synaptic vesicles** filled with **chemicals** called **neurotransmitters**.
4) When an **action potential** reaches the end of a neurone it causes **neurotransmitters** to be **released** into the synaptic cleft. They **diffuse across** to the **postsynaptic membrane** (the one after the synapse) and **bind** to **specific receptors**.
5) When neurotransmitters bind to receptors they might **trigger** an **action potential** (in a neurone), cause **muscle contraction** (in a muscle cell), or cause a **hormone** to be **secreted** (from a gland cell).

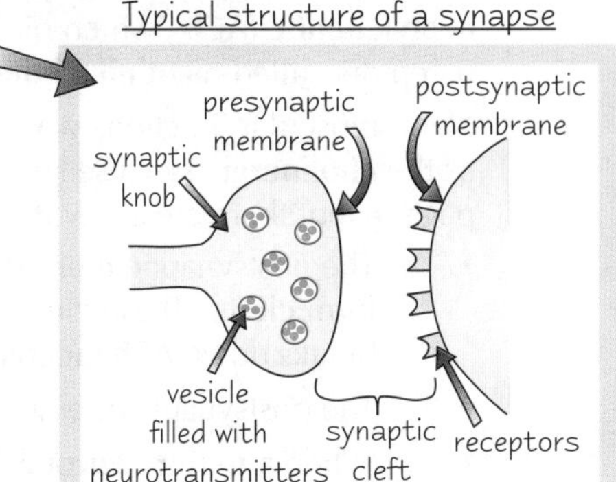

6) Because the receptors are **only** on the postsynaptic membranes, synapses make sure **impulses** are **unidirectional** — the impulse can only travel in **one direction**.
7) Neurotransmitters are **removed** from the **cleft** so the **response** doesn't keep happening, e.g. they're taken back into the **presynaptic neurone** or they're **broken down** by **enzymes** (and the products are taken into the neurone).
8) There are many **different** neurotransmitters. You need to know about one called **acetylcholine** (**ACh**), which binds to **cholinergic receptors**. Synapses that use acetylcholine are called **cholinergic synapses**.

ACh Transmits *the Nerve Impulse* ***Across*** *a* ***Cholinergic Synapse***

This is how a **nerve impulse** is transmitted across a **cholinergic synapse**:

1) An action potential (see p. 144) arrives at the **synaptic knob** of the **presynaptic neurone**.
2) The action potential stimulates **voltage-gated calcium ion channels** in the **presynaptic neurone** to **open**.
3) **Calcium ions diffuse into** the synaptic knob. (They're pumped out afterwards by active transport.)

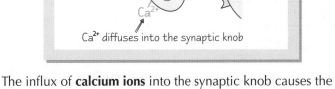

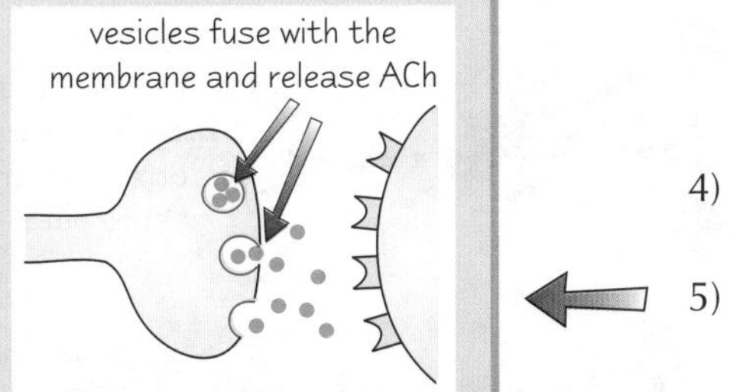

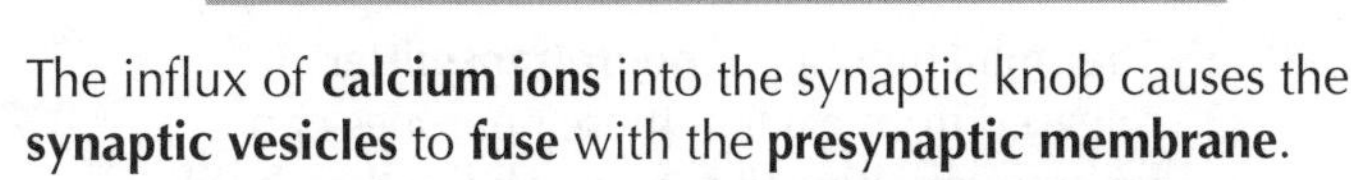

4) The influx of **calcium ions** into the synaptic knob causes the **synaptic vesicles** to **fuse** with the **presynaptic membrane**.
5) The **vesicles release** the neurotransmitter **acetylcholine** (**ACh**) into the **synaptic cleft** — this is called **exocytosis**.

6) ACh **diffuses** across the **synaptic cleft** and **binds** to specific **cholinergic receptors** on the **postsynaptic membrane**.
7) This causes **sodium ion channels** in the **postsynaptic neurone** to **open**.
8) The **influx** of **sodium ions** into the postsynaptic membrane causes an **action potential** on the **postsynaptic membrane** (if the **threshold** is reached).

9) ACh is **removed** from the **synaptic cleft** so the **response** doesn't keep happening. It's **broken down** by an **enzyme** called **acetylcholinesterase** (**AChE**) and the products are **re-absorbed** by the **presynaptic neurone** and used to make more ACh.

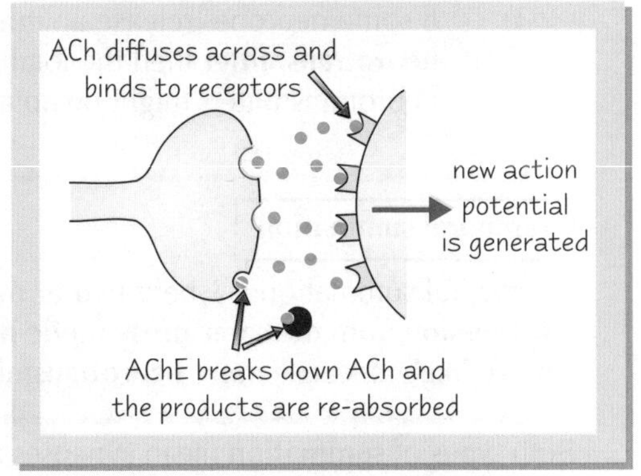

Nervous System — Synaptic Transmission

Neuromuscular Junctions are Synapses Between Neurones and Muscles

1) A **neuromuscular junction** is a **synapse** between a **motor neurone** and a **muscle cell**.
2) Neuromuscular junctions use the neurotransmitter **acetylcholine** (**ACh**), which binds to cholinergic receptors called **nicotinic cholinergic receptors**.
3) Neuromuscular junctions **work** in the **same way** as the **cholinergic synapse** shown on the previous page — but there are a few **differences**:
   - The postsynaptic membrane has lots of **folds** that form **clefts**. These clefts **store** the **enzyme** that breaks down **ACh** (**acetylcholinesterase** — **AChE**).
   - The postsynaptic membrane has **more receptors** than other synapses.
   - When a **motor neurone** fires an **action potential**, it **always triggers** a **response** in a muscle cell. This **isn't** always the case for a synapse between two neurones (see below).

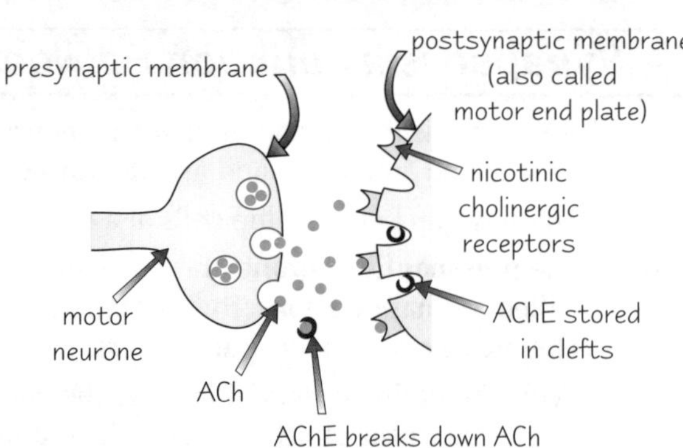

Neurotransmitters are Excitatory or Inhibitory

1) **Excitatory** neurotransmitters **depolarise** the postsynaptic membrane, making it fire an **action potential** if the **threshold** is reached. E.g. **acetylcholine** is an excitatory neurotransmitter — it binds to cholinergic receptors to cause an **action potential** in the postsynaptic membrane.
2) **Inhibitory** neurotransmitters **hyperpolarise** the postsynaptic membrane (make the potential difference more negative), **preventing** it from firing an action potential. E.g. **GABA** is an inhibitory neurotransmitter — when it binds to its receptors it causes **potassium ion channels** to open on the postsynaptic membrane, **hyperpolarising** the neurone.

Mum couldn't help wishing Johnny had a few more inhibitory neurotransmitters.

Summation at Synapses Finely Tunes the Nervous Response

If a stimulus is **weak**, only a **small amount** of **neurotransmitter** will be released from a neurone into the synaptic cleft. This might not be enough to **excite** the postsynaptic membrane to the **threshold** level and stimulate an action potential. **Summation** is where the effect of neurotransmitter released from many neurones (or one neurone that's stimulated a lot in a short period of time) is **added together**. There are two types of summation:

Spatial summation

1) Sometimes **many** neurones **connect** to **one** neurone.
2) The small amount of **neurotransmitter** released from **each** of these neurones can be enough **altogether** to **reach** the **threshold** in the postsynaptic neurone and **trigger** an **action potential**.
3) If some neurones release an **inhibitory neurotransmitter** then the total effect of all the neurotransmitters might be **no action potential**.

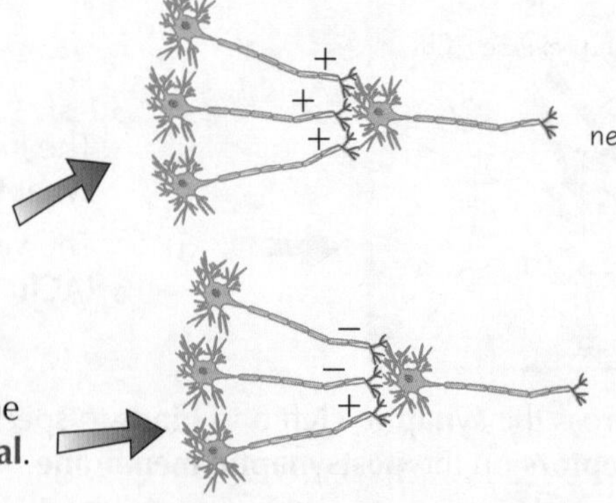

Many neurones release excitatory neurotransmitters (+) = action potential

More inhibitory neurotransmitters are released (-) than excitatory neurotransmitters (+) = no action potential

Temporal summation

Temporal summation is where **two or more** nerve impulses arrive in **quick succession** from the **same presynaptic neurone**. This makes an action potential **more likely** because **more neurotransmitter** is released into the **synaptic cleft**.

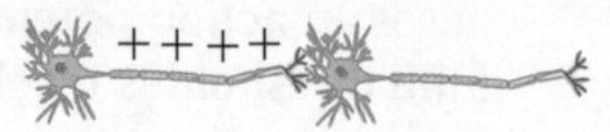

High frequency of weak excitatory inputs = action potential

Both types of **summation** mean synapses **accurately process information**, **finely tuning** the response.

Nervous System — Synaptic Transmission

Drugs Affect the *Action* of *Neurotransmitters* at *Synapses* in Various Ways

Some **drugs affect synaptic transmission**. You might have to **predict** the **effects** that a drug would have at a synapse in your exam. Here are some **examples** of how drugs can affect synaptic transmission:

1. Some drugs are the **same shape** as neurotransmitters so they **mimic** their action at receptors (these drugs are called **agonists**). This means **more receptors** are **activated**. E.g. **nicotine** mimics **acetylcholine** so binds to nicotinic cholinergic receptors in the brain.

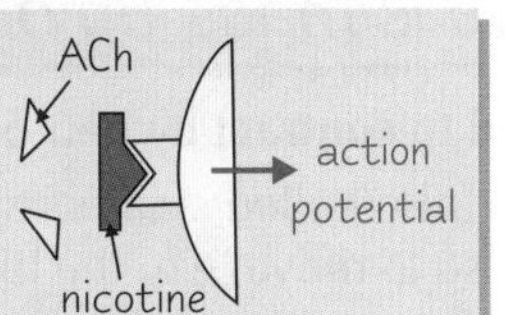

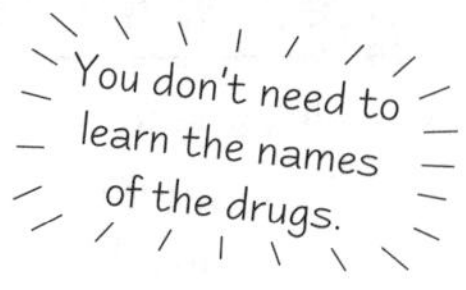

You don't need to learn the names of the drugs.

2. Some drugs **block receptors** so they **can't be activated** by neurotransmitters (these drugs are called **antagonists**). This means **fewer receptors** (if any) can be **activated**. E.g. **curare** blocks the effects of acetylcholine by blocking nicotinic cholinergic receptors at neuromuscular junctions, so muscle cells can't be stimulated. This results in the muscle being **paralysed**.

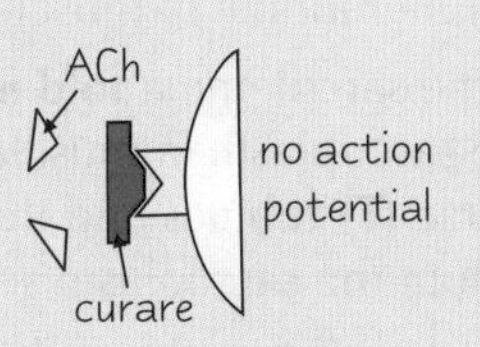

3. Some drugs **inhibit** the **enzyme** that breaks down neurotransmitters (they stop it from working). This means there are **more neurotransmitters** in the synaptic cleft to **bind** to **receptors** and they're there for **longer**. E.g. **nerve gases** stop acetylcholine from being broken down in the synaptic cleft. This can lead to **loss** of **muscle control**.

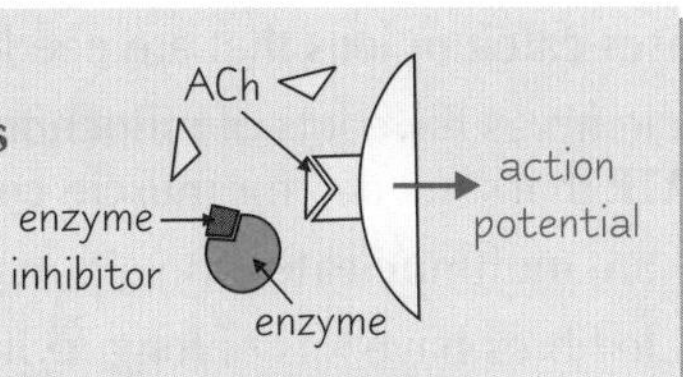

4. Some drugs **stimulate** the release of **neurotransmitter** from the presynaptic neurone so **more receptors** are activated, e.g. **amphetamines**.

5. Some drugs **inhibit** the release of neurotransmitters from the presynaptic neurone so **fewer receptors** are activated, e.g. **alcohol**.

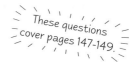

Practice Questions

These questions cover pages 147-149.

Q1 What neurotransmitter do you find at cholinergic synapses?

Q2 How do synapses ensure that nerve impulses are unidirectional?

Q3 Why are calcium ions important in synaptic transmission?

Q4 Give one way that neurotransmitters are removed from the synaptic cleft.

Q5 What kind of receptors are found at neuromuscular junctions?

Q6 What do inhibitory neurotransmitters do at synapses?

Q7 Name the two types of summation that can occur at synapses.

Exam Questions

Q1 Describe the sequence of events from the arrival of an action potential at the presynaptic membrane of a cholinergic synapse to the generation of a new action potential at the postsynaptic membrane. [6 marks]

Q2 Myasthenia gravis is a disease in which the body's immune system gradually destroys receptors at neuromuscular junctions. Suggest what symptoms a sufferer might have. Explain your answer. [4 marks]

Q3 Galantamine is a drug that inhibits the enzyme acetylcholinesterase (AChE). Predict the effect of galantamine at a neuromuscular junction and explain your answer. [3 marks]

Neurotransmitter revision inhibits any excitement...

Another three tough pages in a row, aren't I kind to you. And lots more diagrams to have a go at drawing and re-drawing. Don't worry if you're not the world's best artist, just make sure you add labels to your drawings to explain what's happening.

Effectors — Muscle Contraction

I reckon muscle cells are definitely the spoilt brats of the biology world. They're so special that everything muscly has to have its own special name — there's none of this "cell membrane" malarkey, oh no, it's "sarcolemma" if you please... So first get your head round all these silly posh names, and then you can concentrate on what's actually going on here.

Skeletal Muscle is made up of Long Muscle Fibres

Muscles are **stimulated** to **contract** by neurones and act as **effectors**.

1) **Skeletal muscle** (also called striated, striped or voluntary muscle) is the type of muscle you use to **move**, e.g. the biceps and triceps move the lower arm.
2) Skeletal muscle is made up of **large bundles** of **long cells**, called **muscle fibres**.
3) The cell membrane of muscle fibre cells is called the **sarcolemma**.
4) Bits of the sarcolemma **fold inwards** across the muscle fibre and stick into the **sarcoplasm** (a muscle cell's cytoplasm). These folds are called **transverse (T) tubules** and they help to **spread electrical impulses** throughout the sarcoplasm so they **reach** all parts of the **muscle fibre**.
5) A network of **internal membranes** called the **sarcoplasmic reticulum** runs through the sarcoplasm. The sarcoplasmic reticulum **stores** and **releases calcium ions** that are needed for muscle contraction (see p. 152).
6) Muscle fibres have lots of **mitochondria** to **provide** the **ATP** that's needed for **muscle contraction**.
7) They are **multinucleate** (contain many nuclei).
8) Muscle fibres have lots of **long**, **cylindrical organelles** called **myofibrils**. They're made up of proteins and are **highly specialised** for **contraction**.

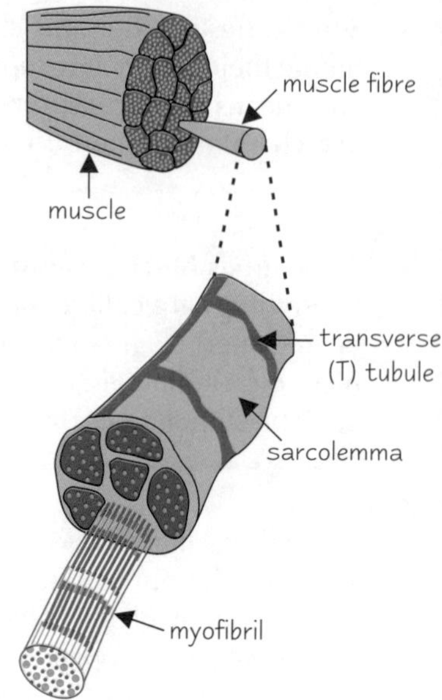

Myofibrils Contain Thick Myosin Filaments and Thin Actin Filaments

1) Myofibrils contain bundles of **thick** and **thin myofilaments** that **move past each other** to make muscles **contract**.
   - **Thick myofilaments** are made of the protein **myosin**.
   - **Thin myofilaments** are made of the protein **actin**.

There's more detail on actin and myosin on p. 152.

2) If you look at a **myofibril** under an **electron microscope**, you'll see a pattern of alternating **dark** and **light bands**:
   - **Dark** bands contain the **thick myosin filaments** and some overlapping thin actin filaments — these are called **A-bands**.
   - **Light** bands contain **thin actin filaments** only — these are called **I-bands**.

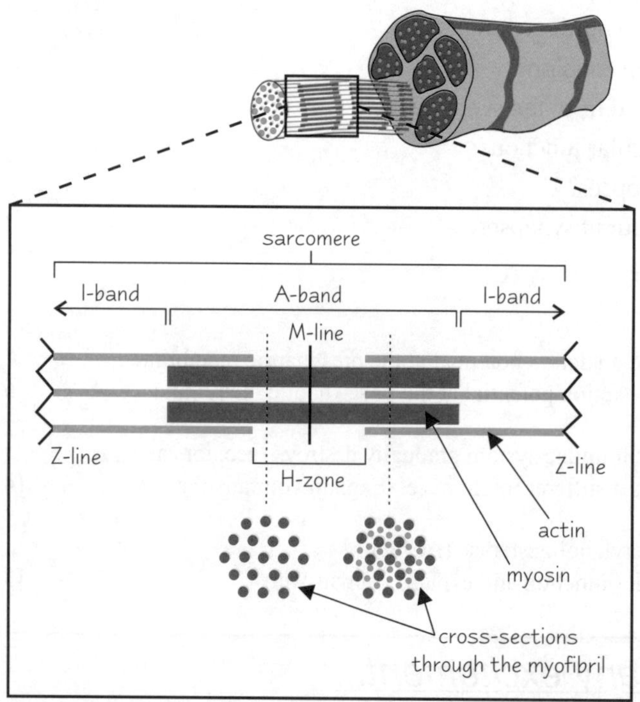

3) A myofibril is made up of many short units called **sarcomeres**.
4) The **ends** of each **sarcomere** are marked with a **Z-line**.
5) In the **middle** of each sarcomere is an **M-line**. The M-line is the **middle** of the **myosin** filaments.
6) **Around** the M-line is the **H-zone**. The H-zone **only** contains **myosin** filaments.

Derek was the proud winner of the biggest muscles AND the smallest pants.

Effectors — Muscle Contraction

Muscle Contraction is Explained by the Sliding Filament Theory

1) **Myosin** and **actin** filaments **slide** over one another to make the **sarcomeres contract** — the myofilaments themselves **don't** contract.
2) The **simultaneous contraction** of lots of **sarcomeres** means the **myofibrils** and **muscle fibres contract**.
3) Sarcomeres return to their **original length** as the muscle **relaxes**.

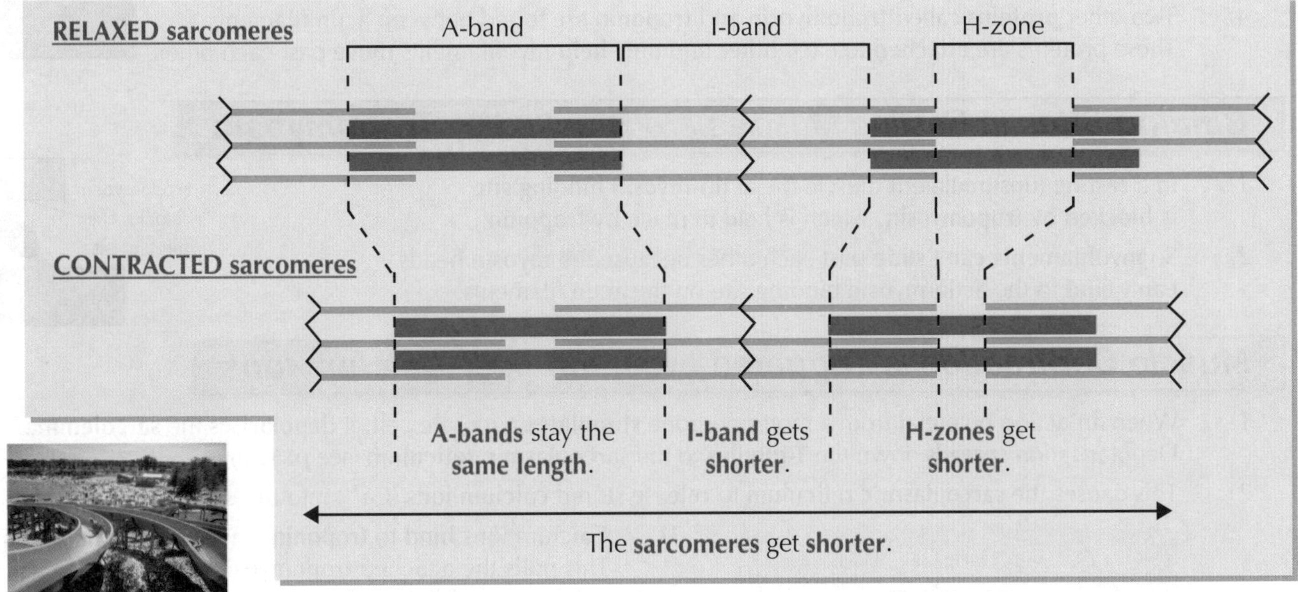

If only the sliding filament theory was as much fun...

Practice Questions

Q1 What are transverse (T) tubules?

Q2 Name the two proteins that make up myofibrils.

Q3 What are the dark bands called?

Q4 What are the light bands called?

Q5 How do myofilaments make sarcomeres contract?

Q6 What happens to sarcomeres as a muscle relaxes?

Exam Questions

Q1 Describe how myofilaments, muscle fibres, myofibrils and muscles are related to each other. [3 marks]

Q2 A muscle myofibril was examined under an electron microscope and a sketch was drawn (Figure 1).

Figure 1

A

B

C

a) What are the correct names for labels A, B and C? [3 marks]

b) Describe how the lengths of the different bands in a myofibril change during muscle contraction. [2 marks]

c) The myofibril was then cut through the M-line (Figure 2). State which of the cross-section drawings you would expect to see and explain why. [3 marks]

Figure 2

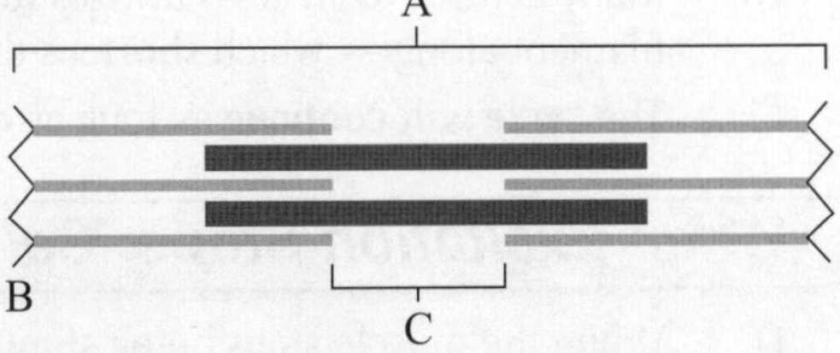

1 2 3

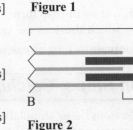

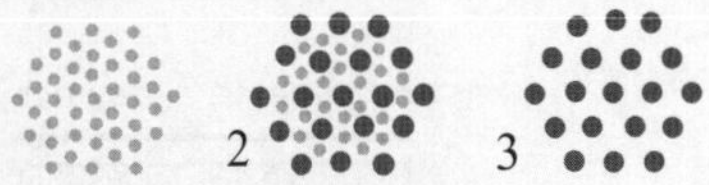

Sarcomere — a French mother with a dry sense of humour...

Blimey, there are an awful lot of similar-sounding names to learn on these pages. And then you've got your A-band, I-band, what-band, who-band to memorise too. But once you've learnt them, these are things you'll never forget — that's right, they'll take up vital brain space forever. And they'll also get you vital marks in your exam.

Effectors — Muscle Contraction

Myofilaments sliding over one another takes a lot of energy — probably why exercise is such hard work...

Myosin Filaments Have Globular Heads and Binding Sites

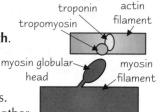

1) **Myosin filaments** have **globular heads** that are **hinged**, so they can move **back** and **forth**.
2) Each myosin head has a **binding site** for **actin** and a **binding site** for **ATP**.
3) **Actin filaments** have **binding sites** for **myosin heads**, called **actin-myosin** binding sites.
4) Two other **proteins** called **tropomyosin** and **troponin** are found between actin filaments. These proteins are **attached** to **each other** and they **help** myofilaments **move** past each other.

Binding Sites in Resting Muscles are Blocked by Tropomyosin

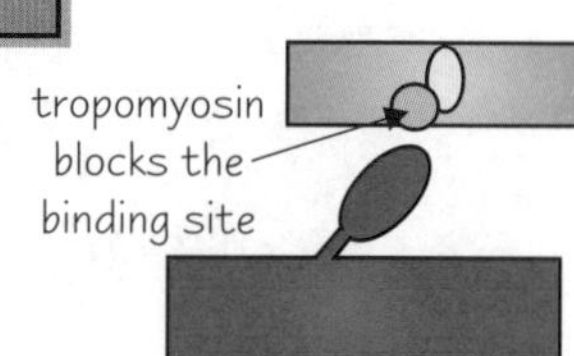

1) In a **resting** (unstimulated) muscle the **actin-myosin binding site** is **blocked** by **tropomyosin**, which is held in place by **troponin**.
2) So **myofilaments can't slide** past each other because the **myosin heads can't bind** to the actin-myosin binding site on the actin filaments.

Muscle Contraction is Triggered by an Influx of Calcium Ions

1) When an action potential from a motor neurone **stimulates** a muscle cell, it **depolarises** the **sarcolemma**. Depolarisation **spreads** down the **T-tubules** to the **sarcoplasmic reticulum** (see p. 150).
2) This causes the **sarcoplasmic reticulum** to **release** stored **calcium ions** (Ca^{2+}) into the **sarcoplasm**.

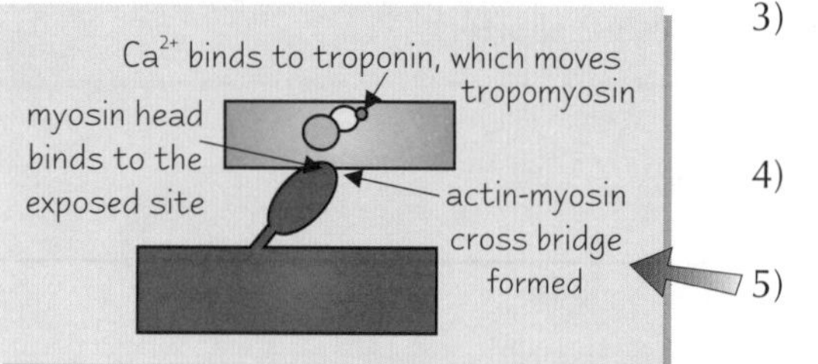

3) Calcium ions **bind** to **troponin**, causing it to **change shape**. This **pulls** the attached **tropomyosin out** of the **actin-myosin binding site** on the actin filament.
4) This **exposes** the **binding site**, which allows the **myosin head** to **bind**.
5) The bond formed when a **myosin head** binds to an **actin filament** is called an **actin-myosin cross bridge**.
6) **Calcium** ions also **activate** the enzyme **ATPase**, which **breaks down ATP** (into ADP + P_i) to **provide** the **energy** needed for muscle contraction.

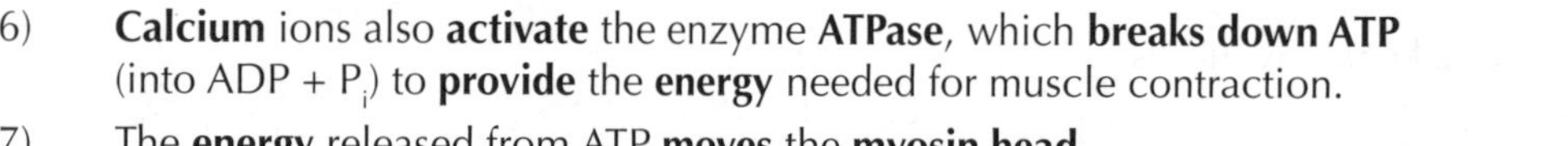

7) The **energy** released from ATP **moves** the **myosin head**, which **pulls** the **actin filament** along in a kind of **rowing action**.

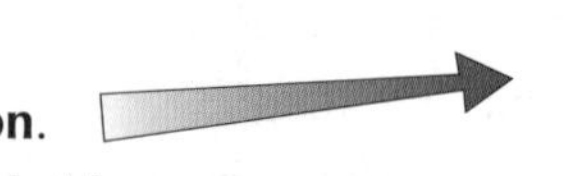

8) **ATP** also provides the **energy** to **break** the **actin-myosin cross bridge**, so the **myosin head detaches** from the actin filament **after** it's moved.
9) The **myosin head** then **reattaches** to a **different binding site** further along the actin filament. A **new actin-myosin cross bridge** is formed and the **cycle** is **repeated** (attach, move, detach, reattach to new binding site...).
10) **Many** actin-myosin cross bridges **form** and **break** very **rapidly**, pulling the actin filament along — which **shortens** the **sarcomere**, causing the **muscle** to **contract**.
11) The cycle will **continue** as long as **calcium ions** are **present** and **bound** to **troponin**.

When Excitation Stops, Calcium Ions Leave Troponin Molecules

1) When the muscle **stops** being **stimulated**, **calcium ions leave** their **binding sites** on the **troponin** molecules and are moved by **active transport** back into the **sarcoplasmic reticulum** (this needs **ATP** too).

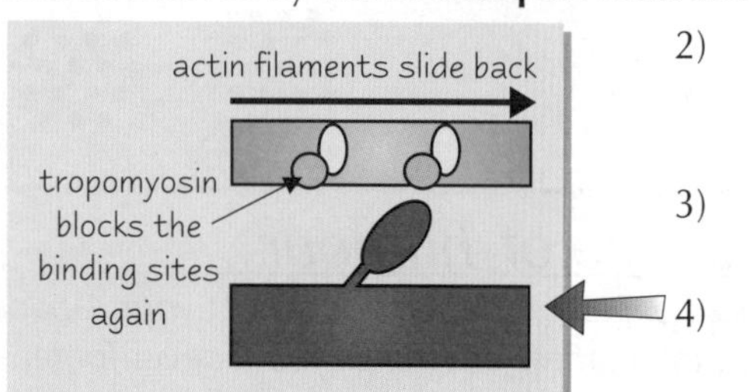

2) The **troponin** molecules return to their **original shape**, pulling the attached **tropomyosin** molecules with them. This means the **tropomyosin** molecules **block** the actin-myosin **binding sites** again.
3) Muscles **aren't contracted** because **no myosin heads** are **attached** to **actin** filaments (so there are no actin-myosin cross bridges).
4) The **actin** filaments **slide back** to their **relaxed** position, which **lengthens** the **sarcomere**.

Effectors — Muscle Contraction

ATP and Phosphocreatine Provide the Energy for Muscle Contraction

So much **energy** is **needed** when muscles contract that **ATP** gets **used up very quickly**.
ATP has to be **continually generated** so exercise can continue — this happens in **three main ways**:

1) Aerobic respiration
   - **Most ATP** is generated via **oxidative phosphorylation** in the cell's **mitochondria**.
   - **Aerobic** respiration only works when there's **oxygen** so it's good for **long periods** of **low-intensity exercise**.

See pages 112-115 for more on aerobic and anaerobic respiration.

2) Anaerobic respiration
   - ATP is made **rapidly** by **glycolysis**.
   - The **end product** of glycolysis is **pyruvate**, which is converted to **lactate** by **lactate fermentation**.
   - Lactate can **quickly build up** in the muscles and cause **muscle fatigue**.
   - Anaerobic respiration is good for **short periods** of **hard exercise**, e.g. a **400 m sprint**.

3) ATP-Phosphocreatine (PCr) System
   - **ATP** is made by **phosphorylating ADP** — adding a phosphate group taken from **PCr**.

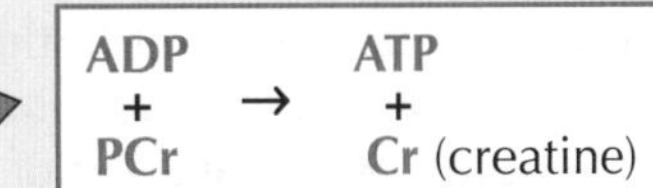

   - **PCr** is **stored** inside cells and the ATP-PCr system **generates ATP** very **quickly**.
   - **PCr runs out** after a few seconds so it's used during **short bursts** of **vigorous exercise**, e.g. a **tennis serve**.
   - The ATP-PCr system is **anaerobic** (it doesn't need oxygen) and it's **alactic** (it doesn't form any lactate).

Many activities use a combination of these systems.

Skeletal Muscles are Made of Slow Twitch and Fast Twitch Muscle Fibres

Skeletal muscles are made up of **two types** of **muscle fibres** — **slow twitch** and **fast twitch**.
Different muscles have **different proportions** of slow and fast twitch fibres. The two types have **different properties**:

SLOW TWITCH MUSCLE FIBRES	FAST TWITCH MUSCLE FIBRES
Muscle fibres that contract slowly.	Muscle fibres that contract very quickly.
Muscles you use for posture, e.g. those in the back, have a high proportion of them.	Muscles you use for fast movement, e.g. those in the eyes and legs, have a high proportion of them.
Good for endurance activities, e.g. maintaining posture, long-distance running.	Good for short bursts of speed and power, e.g. eye movement, sprinting.
Can work for a long time without getting tired.	Get tired very quickly.
Energy's released slowly through aerobic respiration. Lots of mitochondria and blood vessels supply the muscles with oxygen.	Energy's released quickly through anaerobic respiration using glycogen (stored glucose). There are few mitochondria or blood vessels.
Reddish in colour because they're rich in myoglobin — a red-coloured protein that stores oxygen.	Whitish in colour because they don't have much myoglobin (so can't store much oxygen).

Practice Questions

Q1 Describe one way that ATP can be generated in contracting muscles.

Q2 State three differences between slow and fast twitch skeletal muscle fibres.

Exam Questions

Q1 Rigor mortis is the stiffening of muscles in the body after death. It happens when ATP reserves are exhausted. Explain why a lack of ATP leads to muscles being unable to relax. [3 marks]

Q2 Bepridil is a drug that blocks calcium ion channels.
Describe and explain the effect this drug will have on muscle contraction. [3 marks]

What does muscle contraction cost? 80p...

Sorry, that's my favourite sciencey joke so I had to fit it in somewhere — a small distraction before you revisit this page. It's tough stuff but you know the best way to learn it. That's right, grab yourself a nice felt-tip pen and a pad of paper...

Responses in Animals

Right, that's enough pages on receptors, effectors and whatnot — here are some real-life examples to bring it all together.

Control of Heart Rate Involves the Brain and Autonomic Nervous System

1) The **sinoatrial node** (**SAN**) generates **electrical impulses** that cause the **cardiac muscles** to **contract**.
2) The **rate** at which the SAN fires (i.e. heart rate) is **unconsciously controlled** by a part of the **brain** called the **medulla**.
3) Animals need to **alter** their **heart rate** to **respond** to **internal stimuli**, e.g. to prevent fainting due to low blood pressure or to make sure the heart rate is high enough to supply the body with enough oxygen.
4) **Stimuli** are **detected** by **pressure receptors** and **chemical receptors**:
   - There are **pressure receptors** called **baroreceptors** in the **aorta** and the **vena cava**. They're stimulated by **high** and **low blood pressure**.
   - There are **chemical receptors** called **chemoreceptors** in the **aorta**, the **carotid artery** (a major artery in the neck) and in the **medulla**. They **monitor** the **oxygen** level in the **blood** and also **carbon dioxide** and **pH** (which are indicators of O_2 level).
5) Electrical impulses from receptors are sent **to the medulla** along **sensory** neurones. The medulla processes the information and sends impulses to the SAN along **sympathetic** or **parasympathetic** neurones (which are part of the **autonomic nervous system**). Here's how it all works:

There's more about the autonomic nervous system on page 140.

Stimulus	Receptor	Neurone and neurotransmitter	Effector	Response
High blood pressure.	Baroreceptors detect high blood pressure.	Impulses are sent to the medulla, which sends impulses along parasympathetic neurones. These secrete acetylcholine, which binds to receptors on the SAN.	Cardiac muscles	Heart rate slows down to reduce blood pressure back to normal.
Low blood pressure.	Baroreceptors detect low blood pressure.	Impulses are sent to the medulla, which sends impulses along sympathetic neurones. These secrete noradrenaline (a neurotransmitter), which binds to receptors on the SAN.	Cardiac muscles	Heart rate speeds up to increase blood pressure back to normal.
High blood O_2, low CO_2 or high pH levels.	Chemoreceptors detect chemical changes in the blood.	Impulses are sent to the medulla, which sends impulses along parasympathetic neurones. These secrete acetylcholine, which binds to receptors on the SAN.	Cardiac muscles	Heart rate decreases to return O_2, CO_2 and pH levels back to normal.
Low blood O_2, high CO_2 or low pH levels.	Chemoreceptors detect chemical changes in the blood.	Impulses are sent to the medulla, which sends impulses along sympathetic neurones. These secrete noradrenaline, which binds to receptors on the SAN.	Cardiac muscles	Heart rate increases to return O_2, CO_2 and pH levels back to normal.

Reflexes are Rapid, Automatic Responses to Stimuli

1) A **reflex** is where the body **responds** to a stimulus **without** making a **conscious decision** to respond.
2) Because you don't have to **spend time deciding** how to respond, information travels **really fast** from **receptors** to **effectors**.
3) So simple reflexes help organisms to **avoid damage** to the body because they're **rapid**.
4) The **pathway** of neurones linking receptors to effectors in a reflex is called a **reflex arc**. You need to **learn** a **simple reflex arc** involving three neurones — a **sensory**, a **relay** and a **motor** neurone.

E.g. the hand-withdrawal response to heat

- **Thermoreceptors** in the skin **detect** the **heat** stimulus.
- The **sensory neurone** carries impulses to the **relay neurone**.
- The **relay neurone** connects to the **motor neurone**.
- The **motor neurone** sends **impulses** to the **effector** (your biceps muscle).
- Your **muscle contracts** to **stop** your hand being **damaged**.

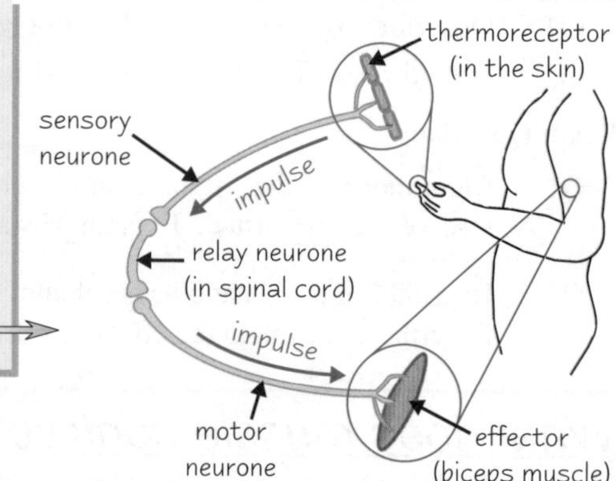

5) If there's a **relay neurone** involved in the simple reflex arc then it's possible to **override** the reflex, e.g. in the example above your **brain** could tell your hand to **withstand** the **heat**.

Responses in Animals

Simple Responses Keep *Simple Organisms* in a *Favourable Environment*

Simple organisms, e.g. woodlice and earthworms, have **simple responses** to keep them in a **favourable environment**. Their **response** can either be **tactic** or **kinetic**:

- **Tactic responses** (**taxes**) — the organisms move towards or away from a **directional stimulus**.

  For example, **woodlice** show a **tactic** response to light (**phototaxis**) — they move **away from** a **light source**. This helps them **survive** as it keeps them **concealed** under stones during the day (where they're **safe** from predators) and keeps them in **damp conditions** (which reduces water loss).

- **Kinetic responses** (**kineses**) — the organisms' movement is affected by a **non-directional** stimulus, e.g. **intensity**.

  For example, **woodlice** show a **kinetic** response to **humidity**. In **high humidity** they move **slowly** and **turn** more often, so that they **stay where they are**. As the air gets **drier**, they move **faster** and **turn less** often, so that they move into a **new area**. This response helps woodlice **move** from **drier air** to more **humid air**, and then **stay put**. This **improves** the **survival** chances of the organism — it **reduces** their **water loss** and it helps to keep them **concealed**.

Some *Cells Communicate* with Other Cells by *Secreting Chemical Mediators*

1) A **chemical mediator** is a **chemical messenger** that acts **locally** (i.e. on **nearby cells**).
2) Communication using chemical mediators is **similar** to communication using **hormones** (see p. 141) — cells release chemicals that bind to **specific receptors** on **target cells** to cause a **response**. But there are a few **differences**:
   - Chemical mediators are secreted from **cells** that are **all over** the **body** (not just from glands).
   - Their **target cells** are right **next to** where the chemical mediator's produced. This means they stimulate a **local response** (not a widespread one).
   - They only have to travel a **short distance** to their target cells, so produce a **quicker response** than hormones (which are transported in the blood).
3) You need to **know** about **two** types of **chemical mediator** — **histamine** and **prostaglandins**:

HISTAMINE

Histamine is a chemical mediator that's stored in **mast cells** and **basophils** (types of immune system cell). It's **released** in response to the body being **injured** or **infected**. It **increases** the **permeability** of the **capillaries nearby** to allow more immune system cells to move out of the blood to the infected or injured area.

PROSTAGLANDINS

Prostaglandins are a **group** of chemical mediators that are produced by **most cells** of the **body**. They're involved in loads of things like **inflammation**, **fever**, **blood pressure regulation** and **blood clotting**. E.g. one type of prostaglandin is released from blood vessel epithelium cells and causes the muscles around them to relax.

Practice Questions

Q1 Describe a simple reflex arc.

Q2 What's the difference between taxes and kineses?

Q3 What is a chemical mediator?

Exam Question

Q1 a) Explain how high blood pressure in the aorta causes the heart rate to slow down. [5 marks]

b) What would be the effect of severing the nerves from the medulla to the sinoatrial node (SAN)? [2 marks]

AAAAAAAAAAAAAAAAAAAAAAAAARGH — the reflex response to revision...

There's also the good old tactic response to revision — when you see your revision notes, you always move away from them. I suppose that wouldn't really help, unless you put your notes miles apart from each other and used it as a fitness exercise. Better still, get on and learn this stuff because you've nearly finished this section — bet that stimulates an excitable response.

Survival and Responses in Plants

Plants also have ways of responding to stimuli — OK so they're not as quick as animals, but they're important all the same.

Plants Need to Respond to Stimuli Too

Flowering plants, like animals, **increase** their chances of **survival** by **responding** to changes in their **environment**, e.g:

- They sense the direction of **light** and **grow** towards it to **maximise** light absorption for **photosynthesis**.
- They can sense **gravity**, so their roots and shoots **grow** in the **right direction**.
- **Climbing** plants have a sense of **touch**, so they can find things to climb and **reach** the **sunlight**.

A Tropism is a Plant's Growth Response to an External Stimulus

1) A **tropism** is the **response** of a plant to a **directional stimulus** (a stimulus coming from a particular direction).
2) Plants respond to stimuli by **regulating** their **growth**.
3) A **positive tropism** is growth **towards** the stimulus.
4) A **negative tropism** is growth **away** from the stimulus.

- **Phototropism** is the growth of a plant in response to **light**.
- **Shoots** are **positively phototropic** and grow **towards** light.
- **Roots** are **negatively phototropic** and grow **away** from light.

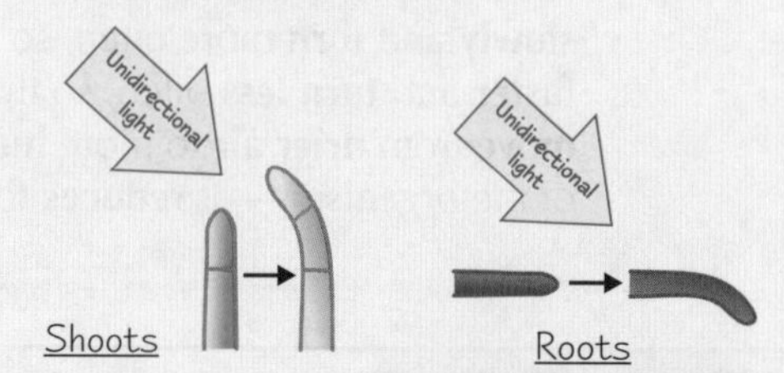
Unidirectional light
Shoots
Unidirectional light
Roots

- **Geotropism** is the growth of a plant in response to **gravity**.
- **Shoots** are **negatively geotropic** and grow **upwards**.
- **Roots** are **positively geotropic** and grow **downwards**.

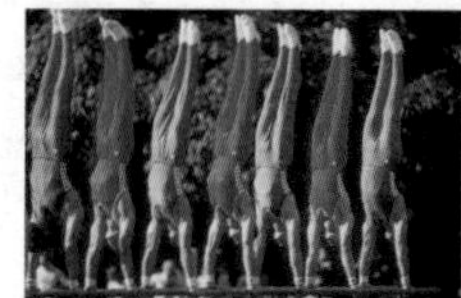

The men's gymnastics team were negatively geotropic.

Responses are Brought About by Growth Factors

1) Plants **don't** have a **nervous system** so they can't respond using neurones, and they **don't** have a **circulatory system** so they can't respond using hormones either.
2) Plants **respond** to stimuli using **growth factors** — these are chemicals that **speed up** or **slow down** plant **growth**.
3) Growth factors are **produced** in the **growing regions** of the plant (e.g. shoot tips, leaves) and they **move** to where they're needed in the **other parts** of the plant.
4) A growth factor called **gibberellin** stimulates **flowering** and **seed germination**.
5) Growth factors called **auxins** stimulate the **growth** of shoots by **cell elongation** — this is where **cell walls** become **loose** and **stretchy**, so the cells get **longer**.
6) **High** concentrations of auxins **inhibit growth** in **roots** though.

Indoleacetic Acid (IAA) is an Important Auxin

1) **Indoleacetic acid** (**IAA**) is an important **auxin** that's produced in the **tips** of **shoots** in flowering plants.
2) IAA is **moved** around the plant to **control tropisms** — it moves by **diffusion** and **active transport** over short distances, and via the **phloem** over long distances.
3) This results in **different parts** of the plants having **different amounts** of IAA. The **uneven distribution** of IAA means there's **uneven growth** of the plant, e.g:

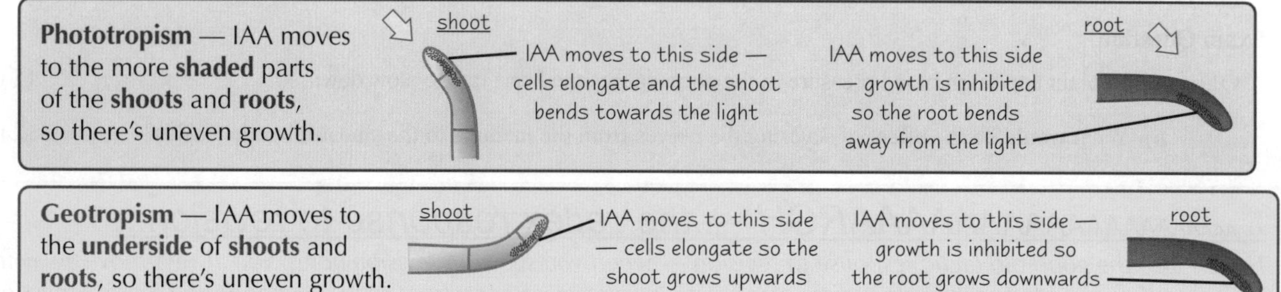
Phototropism — IAA moves to the more **shaded** parts of the **shoots** and **roots**, so there's uneven growth.
shoot
IAA moves to this side — cells elongate and the shoot bends towards the light
IAA moves to this side — growth is inhibited so the root bends away from the light
root
Geotropism — IAA moves to the **underside** of **shoots** and **roots**, so there's uneven growth.
shoot
IAA moves to this side — cells elongate so the shoot grows upwards
IAA moves to this side — growth is inhibited so the root grows downwards
root

Survival and Responses in Plants

You May Have to *Interpret Experimental Data* About *Auxins*

Here's some **data** similar to what you might get in your **exam**:

1) An experiment was carried out to **investigate** the role of **auxin** in **shoot growth.**
2) Eight shoots, **equal in height and mass**, had their **tips removed.**
3) **Sponges** soaked in **glucose and either auxin** or **water** were then ⟹ placed where the tip should be.
4) **Four shoots** were then placed in the **dark** (**experiment A**) and the **other** four shoots were exposed to a **light** source, directed at them from the **right** (**experiment B**).
5) After **two days** the **amount** of growth (in mm) and **direction** of growth was **recorded.** The results are shown in the table. ⟹

Sponge soaked in auxin and glucose
Sponge soaked in water and glucose
A B C D
Shoot minus the tip

	Growth			
	Shoot A	Shoot B	Shoot C	Shoot D
Experiment A (dark)	6 mm, right	6 mm, left	6 mm, straight	1 mm, straight
Experiment B (light)	8 mm, right	8 mm, right	8 mm, right	3 mm, straight

<u>You could be asked to **explain the data**:</u>

The results show how the **movement** of **auxin controls phototropism** in plant shoots.

In **experiment A shoot A**, the auxin **diffused** straight down from the sponge into the **left-hand side** of the shoot. This stimulated the cells on this side to **elongate**, so the shoot **grew towards the right.** In shoot B, the opposite occurred, making the shoot **grow towards the left.** In shoot C, **equal amounts** of auxin diffused down **both sides**, making **all** the cells elongate at the same rate.

In **experiment B**, the shoots were exposed to a **light source.** The auxin diffused into the shoot and **accumulated on the shaded side** (left-hand side) **regardless** of **where** the sponge was placed. Shoots A, B and C all **grew towards the right**, because most auxin **accumulated** on the left, **stimulating** cell elongation there.

<u>You could be asked to comment on the **experimental design**:</u>

A **control** (sponge soaked in water) was included to show that it was the auxin having an **effect** and nothing else. **Glucose** was included so that the shoots would have **energy to grow in the dark** (no photosynthesis can take place).

There's more on pages 90-92 and 196-198 about interpreting experimental data.

Practice Questions

Q1 What is a tropism?

Q2 What are plant growth factors?

Q3 How does the movement of IAA control geotropism in roots?

Exam Question

Q1 The table shows the results some students obtained when they investigated the effect of providing plants with auxins.

Week	Height of plant not given auxins / cm	Height of plant provided with auxins / cm
1	1	2
2	2	5
3	4	8
4	6	9
5	9	13

a) Describe and explain what the data shows. [2 marks]

b) Explain the role of auxins in the control of phototropism. [5 marks]

c) Suggest what the students should do to increase the reliability of their results. [2 marks]

d) Why might this data be useful to a commercial tomato producer? [1 mark]

IAA Productions — do you have the growth factor — with Simon Trowel...

Hoorah, you've finally reached the end of this mega-section. Unless you're casually skipping through the book, in which case you've got the treat of all treats waiting for you. There are just a few more gibberish names (sorry, I mean gibberellin) for you to learn, and some rather bendy responses to understand, then you've got responses all sorted. Aux-ilarating.

Homeostasis Basics

Ah, there's nothing like learning a nice long word to start you off on a new section — welcome to homeostasis.

Homeostasis is the Maintenance of a Constant Internal Environment

1) **Changes** in your **external environment** can affect your **internal environment** — the blood and tissue fluid that surrounds your cells.
2) **Homeostasis** involves **control systems** that keep your **internal environment** roughly **constant** (within **certain limits**).
3) **Keeping** your internal environment **constant** is vital for cells to **function normally** and to **stop** them being **damaged**.
4) It's particularly important to **maintain** the right **core body temperature** and **blood pH**. This is because temperature and pH affect **enzyme activity**, and enzymes **control** the **rate** of **metabolic reactions**:

Temperature

- If body temperature is **too high** (e.g. 40 °C) **enzymes** may become **denatured**. The enzyme's molecules **vibrate too much**, which **breaks** the **hydrogen bonds** that hold them in their **3D shape**. The **shape** of the enzyme's **active site** is **changed** and it **no longer works** as a **catalyst**. This means **metabolic reactions** are **less efficient**.
- If body temperature is **too low enzyme activity** is **reduced**, **slowing** the rate of **metabolic reactions**.
- The **highest rate** of **enzyme activity** happens at their **optimum temperature** (about **37 °C** in humans).

pH

- If blood pH is **too high** or **too low** (highly alkaline or acidic) **enzymes** become **denatured**. The **hydrogen bonds** that hold them in their 3D shape are affected so the **shape** of the enzyme's **active site is changed** and it **no longer works** as a **catalyst**. This means **metabolic reactions** are **less efficient**.
- The **highest rate** of **enzyme activity** happens at their **optimum pH** — usually **around pH 7** (neutral), but some enzymes work best at other pHs, e.g. enzymes found in the stomach work best at a low pH.

5) It's important to **maintain** the right **concentration** of **glucose** in the **blood** because cells need glucose for **energy**. Blood glucose concentration also affects the **water potential** of blood — this is the potential (likelihood) of water molecules to **diffuse** out of or into a solution.

Glucose

- If blood glucose concentration is **too high** the **water potential** of blood is **reduced** to a point where **water** molecules **diffuse out** of cells into the blood by osmosis. This can cause the cells to **shrivel up** and **die**.
- If blood glucose concentration is **too low**, cells are **unable** to carry out **normal activities** because there **isn't enough glucose** for respiration to provide **energy**.

Homeostatic Systems Detect a Change and Respond by Negative Feedback

1) Homeostatic systems involve **receptors**, a **communication system** and **effectors** (see page 140).
2) Receptors detect when a level is **too high** or **too low**, and the information's communicated via the **nervous** system or the **hormonal** system to **effectors**.
3) The effectors respond to **counteract** the change — bringing the level **back** to **normal**.
4) The mechanism that **restores** the level to **normal** is called a **negative feedback** mechanism.
5) Negative feedback **keeps** things around the **normal** level, e.g. body temperature is usually kept **within 0.5 °C** above or below **37 °C**.
6) Negative feedback only works within **certain limits** though — if the change is **too big** then the **effectors** may **not** be able to **counteract** it, e.g. a huge drop in body temperature caused by prolonged exposure to cold weather may be too large to counteract.

Control of **body temperature** by **negative feedback**:

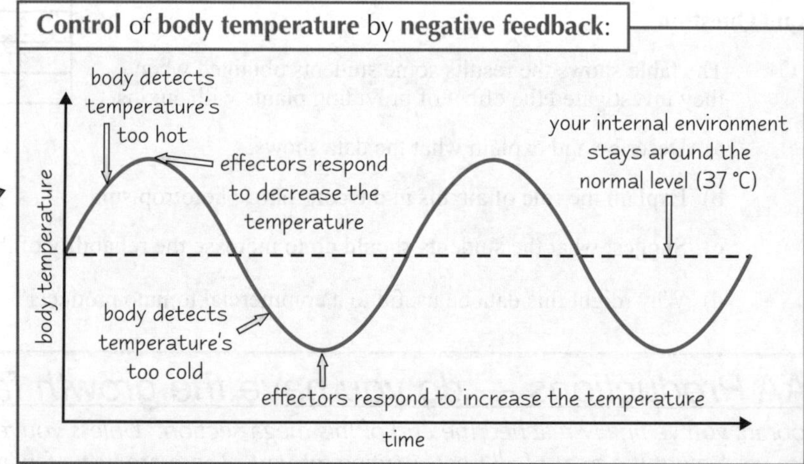

body detects temperature's too hot
effectors respond to decrease the temperature
your internal environment stays around the normal level (37 °C)
body temperature
body detects temperature's too cold
effectors respond to increase the temperature
time

Homeostasis Basics

Multiple Negative Feedback Mechanisms Give More Control

1) Homeostasis involves **multiple negative feedback mechanisms** for each thing being controlled. This is because having more than one mechanism gives **more control** over changes in your internal environment than just having one negative feedback mechanism.
2) Having multiple negative feedback mechanisms means you can **actively increase** or **decrease a level** so it returns to **normal**, e.g. you have feedback mechanisms to reduce your body temperature and you also have mechanisms to increase it.
3) If you only had **one negative feedback mechanism**, all you could do would be **turn it on** or **turn it off**. You'd only be able to actively change a level in **one direction** so it returns to normal, e.g. it's a bit like trying to slow down a car with only an accelerator — all you can do is take your foot off the accelerator (you'd have more control with a brake too).
4) Only **one** negative feedback mechanism means a **slower response** and **less control**.

There was plenty of negative feedback when Carl wore his new vest-pants combo out for dinner.

Positive Feedback Mechanisms Amplify a Change from the Normal Level

1) Some changes trigger a **positive feedback** mechanism, which **amplifies** the change.
2) The effectors respond to **further increase** the level **away** from the **normal** level.
3) Positive feedback is useful to **rapidly activate** something, e.g. a **blood clot** after an injury.

- **Platelets** become **activated** and release a **chemical** — this triggers **more platelets** to be activated, and so on.
- Platelets **very quickly** form a **blood clot** at the injury site.
- The process **ends with negative feedback**, when the body detects the **blood clot** has been **formed**.

4) Positive feedback can also happen when a **homeostatic system breaks down**, e.g. if you're too cold for too long:

Hypothermia involves **positive feedback**:

- **Hypothermia** is **low body temperature** (below 35 °C).
- It happens when **heat's lost** from the body **quicker** than it can be **produced**.
- As body temperature **falls** the **brain doesn't work** properly and **shivering stops** — this makes body temperature **fall even more**.
- **Positive feedback** takes body temperature **further away** from the normal level, and it continues to decrease unless action is taken.

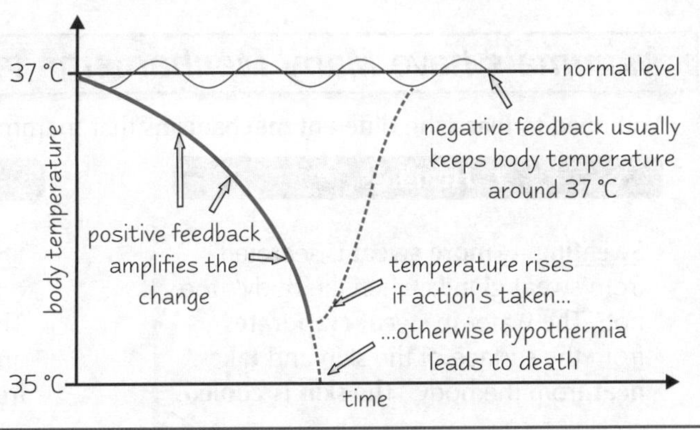

5) Positive feedback **isn't** involved in **homeostasis** because it **doesn't** keep your internal environment **constant**.

Practice Questions

Q1 What is homeostasis and why is it necessary?

Q2 Why is it important to control blood pH?

Q3 Why is it important to control blood glucose concentration?

Exam Questions

Statement A: "Hyperthermia happens when the brain can't work properly and body temperature continues to increase."
Statement B: "When body temperature is low, mechanisms return the temperature to normal."

Q1 Look at statements A and B in the box.

a) Which statement is describing a positive feedback mechanism? Give a reason for your answer. [2 marks]

b) Describe and explain what effect a very high body temperature has on metabolic reactions. [2 marks]

Q2 Describe the importance of multiple negative feedback mechanisms in homeostasis. [2 marks]

Homeostasis works like a teacher — everything always gets corrected...

The key to understanding homeostasis is to get your head around negative feedback. Basically, if one thing goes up, the body responds to bring it down — and vice versa. And when you're ready, turn over the page for some exciting examples.

Control of Body Temperature

So, negative feedback is a good thing — well, in biological terms anyway. Being able to control your temperature is really important and there are some pretty nifty mechanisms that help you do this. Read on, oh chosen one, read on...

Temperature is Controlled Differently in Ectotherms and Endotherms

Animals are classed as either **ectotherms** or **endotherms**, depending on how they **control** their body temperature:

Ectotherms — e.g. **reptiles, fish**	Endotherms — e.g. **mammals, birds**
Ectotherms **can't control** their body temperature **internally** — they **control** their temperature by **changing** their **behaviour** (e.g. reptiles gain heat by basking in the sun).	Endotherms **control** their body temperature **internally** by **homeostasis**. They can also control their temperature by **behaviour** (e.g. by finding shade).
Their **internal** temperature **depends** on the **external temperature** (their surroundings).	Their internal temperature is **less affected** by the **external temperature** (within certain limits).
Their **activity** level **depends** on the external temperature — they're **more** active at **higher** temperatures and **less** active at **lower** temperatures.	Their **activity** level is largely **independent** of the **external temperature** — they can be active at any temperature (within certain limits).
They have a **variable metabolic rate** and they **generate** very **little heat** themselves.	They have a constantly **high metabolic rate** and they **generate** a **lot of heat** from metabolic reactions.

Mammals have Many Mechanisms to Change Body Temperature

You need to **learn** the different **mechanisms** that mammals use to **change body temperature**:

Heat loss

__Sweating__ — **more sweat** is secreted from **sweat glands** when the body's too hot. The water in sweat **evaporates** from the surface of the skin and **takes heat** from the body. The **skin** is **cooled**.

__Hairs lie flat__ — mammals have a layer of **hair** that provides **insulation** by **trapping air** (air is a poor conductor of heat). When it's hot, **erector pili muscles relax** so the hairs lie flat. **Less air** is trapped, so the skin is **less insulated** and **heat** can be **lost** more easily.

__Vasodilation__ — when it's hot, **arterioles** near the surface of the skin **dilate** (this is called **vasodilation**). **More blood** flows through the **capillaries** in the surface layers of the dermis. This means **more heat** is **lost** from the skin by **radiation** and the **temperature** is **lowered**.

Heat production

__Shivering__ — when it's cold, **muscles contract** in **spasms**. This makes the body **shiver** and **more heat** is **produced** from **increased respiration**.

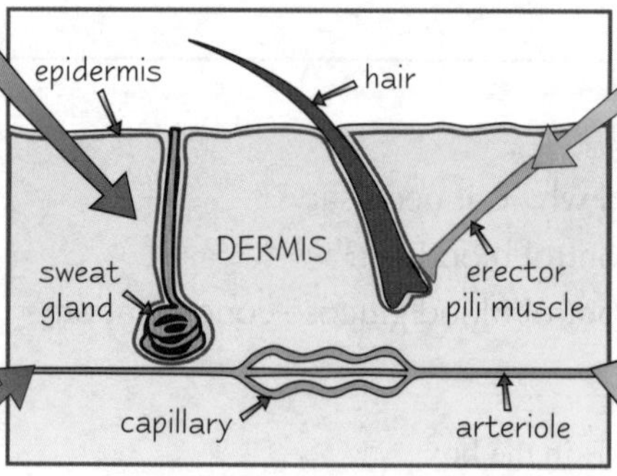

__Hormones__ — the body releases **adrenaline** and **thyroxine**. These **increase metabolism** and so **more heat** is **produced**.

Heat conservation

__Much less sweat__ — **less sweat** is secreted from sweat glands when it's cold, **reducing** the amount of **heat loss**.

__Hairs stand up__ — **erector pili muscles contract** when it's cold, which makes the **hairs stand up**. This **traps more air** and so **prevents heat loss**.

__Vasoconstriction__ — when it's cold, **arterioles** near the surface of the skin **constrict** (this is called **vasoconstriction**) so **less blood** flows through the **capillaries** in the surface layers of the dermis. This **reduces heat loss**.

Control of Body Temperature

The **Hypothalamus Controls** Body Temperature in **Mammals**

1) **Body temperature** in mammals is **maintained** at a **constant level** by a part of the **brain** called the **hypothalamus**.
2) The hypothalamus **receives information** about both **internal** and **external temperature** from **thermoreceptors** (temperature receptors):

Information about **internal temperature** comes from **thermoreceptors** in the **hypothalamus** that detect **blood temperature**.	Information about **external temperature** comes from **thermoreceptors** in the **skin** that detect **skin temperature**.

3) Thermoreceptors send **impulses** along **sensory neurones** to the **hypothalamus**, which sends **impulses** along **motor neurones** to **effectors** (muscles and glands).
4) The neurones are part of the **autonomic nervous system**, so it's all done **unconsciously**.
5) The effectors respond to **restore** the body temperature **back** to **normal**. Here's how it all works:

See p. 140 for more about the autonomic nervous system.

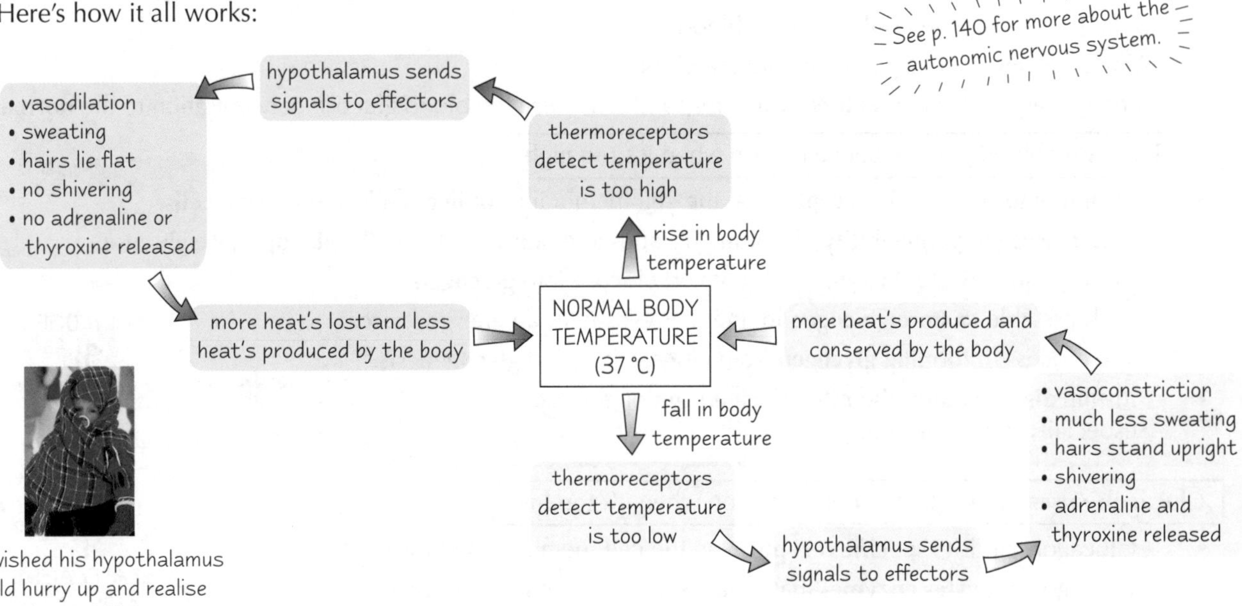

NORMAL BODY TEMPERATURE (37 °C)
rise in body temperature
thermoreceptors detect temperature is too high
hypothalamus sends signals to effectors
• vasodilation
• sweating
• hairs lie flat
• no shivering
• no adrenaline or thyroxine released
more heat's lost and less heat's produced by the body
fall in body temperature
thermoreceptors detect temperature is too low
hypothalamus sends signals to effectors
• vasoconstriction
• much less sweating
• hairs stand upright
• shivering
• adrenaline and thyroxine released
more heat's produced and conserved by the body

Billy wished his hypothalamus would hurry up and realise just how cold he was.

Practice Questions

Q1 Give four differences between ectotherms and endotherms.

Q2 Which type of animal has more control over their body temperature, ectotherms or endotherms?

Q3 How does sweating reduce body temperature?

Q4 What role do erector pili muscles play in controlling body temperature in mammals?

Q5 Which part of the brain is responsible for maintaining a constant body temperature in mammals?

Exam Questions

Q1 Mammals that live in cold climates have thick fur and layers of fat beneath their skin to keep them warm. Describe and explain two other ways they maintain a constant body temperature in cold conditions. [4 marks]

Q2 Describe and explain how the body detects a high external temperature. [2 marks]

Q3 Snakes are usually found in warm climates. Suggest why they are not usually found in cold climates. Explain your answer. [4 marks]

Sweat, hormones and erector muscles — ooooh errrrrrr...

Blimey, I'm glad this is all done unconsciously — you'd waste tons of time if you had to think about every single response. Mind you, I reckon I could think up some slightly less embarrassing ways of controlling temperature, rather than getting all red-faced and stinky. Ectotherms have got it sussed with their whole sunbathing thing — now that's the life...

Control of Blood Glucose Concentration

These pages are all about how negative feedback helps you to not go totally hyper when you stuff your face with sweets.

Eating *and* ***Exercise*** *Change the* ***Concentration*** *of* ***Glucose*** *in your* ***Blood***

1) **All cells** need a constant **energy supply** to work — so **blood glucose concentration** must be carefully **controlled**.
2) The **concentration** of **glucose** in the blood is **normally** around **90 mg per 100 cm³** of blood. It's **monitored** by cells in the **pancreas**.
3) Blood glucose concentration **rises** after **eating food** containing **carbohydrate**.
4) Blood glucose concentration **falls** after **exercise**, as **more glucose** is used in **respiration** to **release energy**.

Insulin *and* ***Glucagon Control*** *Blood Glucose Concentration*

The hormonal system (see p. 141) **controls** blood glucose concentration using **two hormones** called **insulin** and **glucagon**. They're both **secreted** by clusters of cells in the **pancreas** called the **islets of Langerhans**:

- **Beta (β) cells** secrete **insulin** into the blood.
- **Alpha (α) cells** secrete **glucagon** into the blood.

Insulin and glucagon act on **effectors**, which respond to **restore** the blood glucose concentration to the **normal level**:

Insulin lowers blood glucose concentration when it's **too high**

- Insulin binds to **specific receptors** on the cell membranes of **liver cells** and **muscle cells**.
- It **increases** the **permeability** of cell membranes to glucose, so the cells **take up more glucose**.
- Insulin also **activates enzymes** that convert **glucose** into **glycogen**.
- Cells are able to **store glycogen** in their cytoplasm, as an **energy source**.
- The process of **forming glycogen** from glucose is called **glycogenesis**.
- Insulin also **increases** the **rate** of **respiration** of glucose, especially in muscle cells.

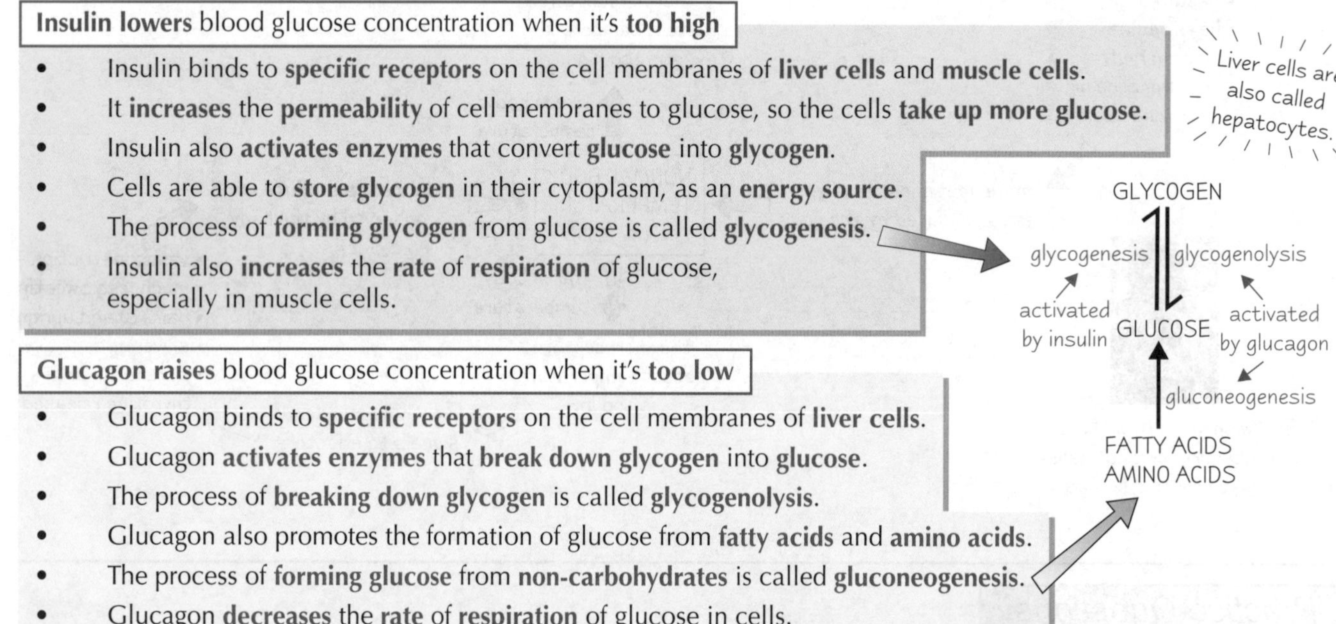

Liver cells are also called hepatocytes.
GLYCOGEN
glycogenesis
glycogenolysis
activated by insulin
GLUCOSE
activated by glucagon
gluconeogenesis
FATTY ACIDS
AMINO ACIDS

Glucagon raises blood glucose concentration when it's **too low**

- Glucagon binds to **specific receptors** on the cell membranes of **liver cells**.
- Glucagon **activates enzymes** that **break down glycogen** into **glucose**.
- The process of **breaking down glycogen** is called **glycogenolysis**.
- Glucagon also promotes the formation of glucose from **fatty acids** and **amino acids**.
- The process of **forming glucose** from **non-carbohydrates** is called **gluconeogenesis**.
- Glucagon **decreases** the **rate** of **respiration** of glucose in cells.

Negative Feedback Mechanisms *Keep Blood Glucose Concentration* ***Normal***

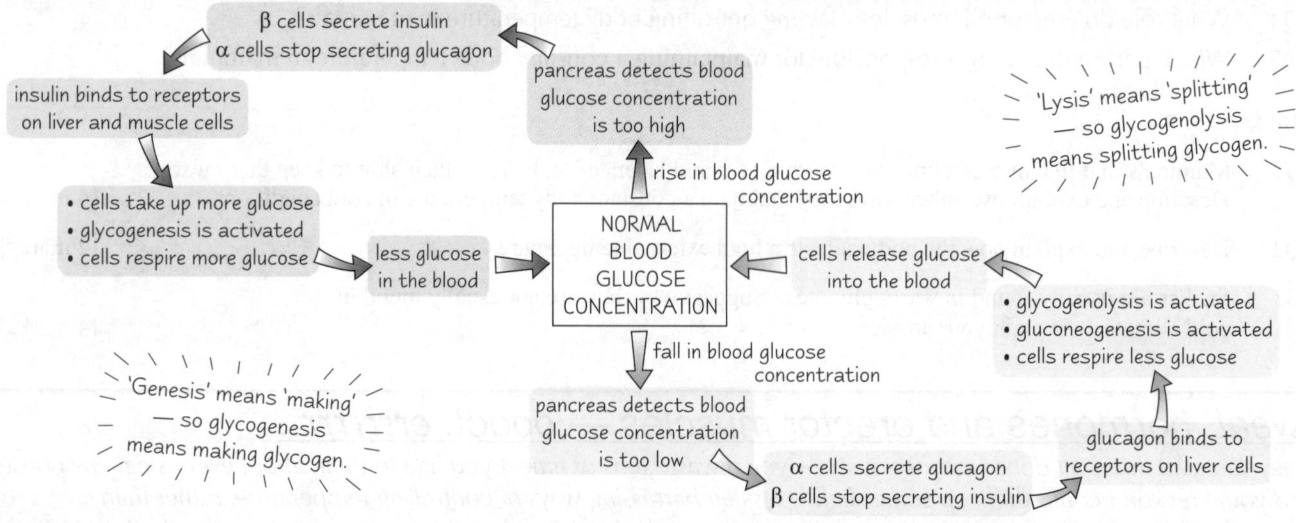

β cells secrete insulin
α cells stop secreting glucagon
insulin binds to receptors on liver and muscle cells
pancreas detects blood glucose concentration is too high
'Lysis' means 'splitting' — so glycogenolysis means splitting glycogen.
• cells take up more glucose
• glycogenesis is activated
• cells respire more glucose
rise in blood glucose concentration
less glucose in the blood
NORMAL BLOOD GLUCOSE CONCENTRATION
cells release glucose into the blood
• glycogenolysis is activated
• gluconeogenesis is activated
• cells respire less glucose
fall in blood glucose concentration
'Genesis' means 'making' — so glycogenesis means making glycogen.
pancreas detects blood glucose concentration is too low
α cells secrete glucagon
β cells stop secreting insulin
glucagon binds to receptors on liver cells

Control of Blood Glucose Concentration

Adrenaline Increases your Blood Glucose Concentration Too

1) **Adrenaline** is a **hormone** that's secreted from your **adrenal glands** (found just above your kidneys).
2) It's secreted when there's a **low concentration** of **glucose** in your blood, when you're **stressed** and when you're **exercising**.
3) Adrenaline binds to **receptors** in the cell membrane of **liver cells**:
   - It **activates glycogenolysis** (the breakdown of glycogen to glucose).
   - It **inhibits glycogenesis** (the synthesis of glycogen from glucose).

GLYCOGEN
activated by adrenaline
inhibited by adrenaline
glycogenesis
glycogenolysis
GLUCOSE

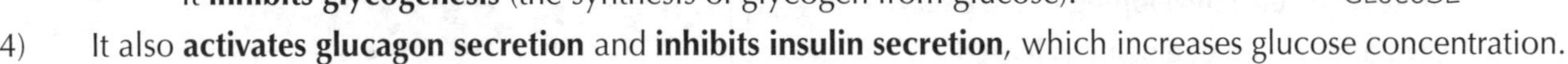

4) It also **activates glucagon secretion** and **inhibits insulin secretion**, which increases glucose concentration.
5) Adrenaline gets the **body ready** for **action** by making **more glucose** available for **muscles** to respire.
6) Both **adrenaline** and **glucagon** can activate glycogenolysis **inside** a cell even though they bind to **receptors** on the **outside** of the cell. Here's **how** they do it:

- Adrenaline and glucagon **bind** to their specific receptors and **activate** an **enzyme** called **adenylate cyclase**.
- Activated adenylate cyclase converts **ATP** into a **chemical signal** called a '**second messenger**'.
- The second messenger is called **cyclic AMP** (**cAMP**).
- cAMP **activates** a **cascade** (a chain of reactions) that break down glycogen into glucose (**glycogenolysis**).

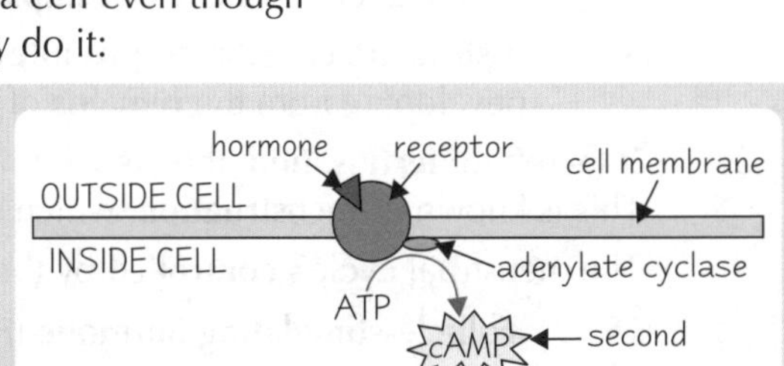

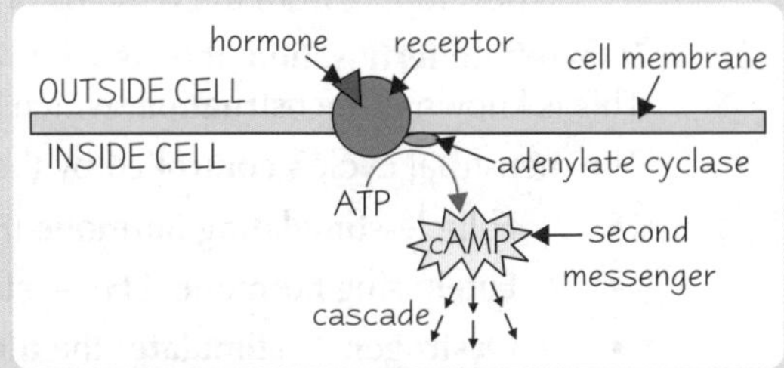

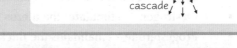

Diabetes Occurs when Blood Glucose Concentration is Not Controlled

Diabetes mellitus is a condition where **blood glucose** concentration **can't** be **controlled** properly. There are **two types**:

Type I

1) In **Type I** diabetes, the **β cells** in the islets of Langerhans **don't produce** any **insulin**.
2) After **eating**, the blood glucose level **rises** and **stays high** — this is called **hyperglycaemia** and can result in **death** if left untreated. The kidneys **can't reabsorb** all this glucose, so some of it's **excreted** in the urine.
3) It can be treated by regular **injections** of **insulin**. But this has to be **carefully controlled** because too much can produce a **dangerous drop** in blood glucose levels — this is called **hypoglycaemia**.
4) **Eating regularly** and **controlling simple carbohydrate intake** (sugars) helps to **avoid** a **sudden rise** in glucose.

Type II

1) **Type II** diabetes is usually acquired **later** in **life** than Type I, and it's often linked with **obesity**.
2) It occurs when the β cells **don't produce enough insulin** or when the body's **cells don't respond** properly to **insulin**. Cells don't respond properly because the insulin **receptors** on their membranes **don't work** properly, so the cells **don't** take up enough glucose. This means the **blood glucose concentration** is **higher** than normal.
3) It can be treated by **controlling simple carbohydrate intake** and **losing weight**. **Glucose-lowering tablets** can be taken if diet and weight loss can't control it.

Practice Questions

Q1 Give three functions of glucagon.

Q2 What is a second messenger?

Exam Questions

Q1 Describe and explain how hormones return blood glucose concentration to normal after a meal. [5 marks]

Q2 Explain why someone with diabetes can produce insulin but can't control their blood glucose concentration. [3 marks]

My α cells detect low glucose — urgent tea and biscuit break needed...

Aaaaargh there are so many stupidly complex names to learn and they all look and sound exactly the same to me. You can't even get away with sneakily misspelling them all in your exam — like writing 'glycusogen' or 'glucogenesisolysis'. Nope, examiners have been around for centuries, so I'm afraid old tricks like that just won't work on them. Grrrrrrr.

Control of the Menstrual Cycle

Sorry lads — these two pages are pretty much devoted to the inner workings of the ladies. Now they're bound to start going on about how easy you have it in comparison. Got a point though...

The Human ***Menstrual Cycle*** *is* ***Controlled*** *by* ***Hormones***

You don't need to learn this diagram (phew), but it shows you what's going on.

1) The human **menstrual cycle** (also called the **oestrous cycle**) lasts about **28 days**.
2) The menstrual cycle involves:
    - A **follicle** (an egg and its surrounding protective cells) **developing** in the **ovary**.
    - Ovulation — an **egg** being **released**.
    - The **uterus lining** becoming **thicker** so that a fertilised egg can **implant**.
    - A structure called a **corpus luteum** developing from the **remains** of the **follicle**.

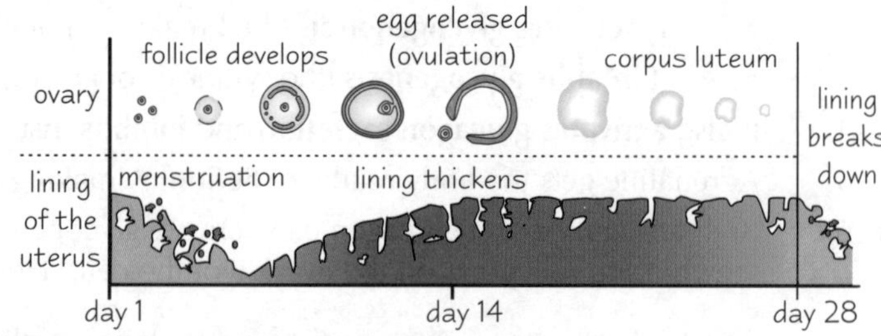

3) If there's **no fertilisation**, the uterus lining **breaks down** and leaves the body through the **vagina**. This is known as **menstruation**, which marks the **end** of one cycle and the **start** of another.
4) The menstrual cycle's **controlled** by the action of **four hormones**:
    - **Follicle-stimulating hormone** (**FSH**) — does just what it says, it **stimulates** the **follicle** to develop.
    - **Luteinising hormone** (**LH**) — **stimulates ovulation** and **stimulates** the **corpus luteum** to develop.
    - **Oestrogen** — **stimulates** the **uterus lining** to **thicken**.
    - **Progesterone** — **maintains** the **thick uterus lining**, ready for implantation of an embryo.
5) **FSH** and **LH** are secreted by the **anterior pituitary gland**. **Oestrogen** and **progesterone** are secreted by the **ovaries**.

Hormone Concentrations Change *During* ***Different Stages*** *of the* ***Cycle***

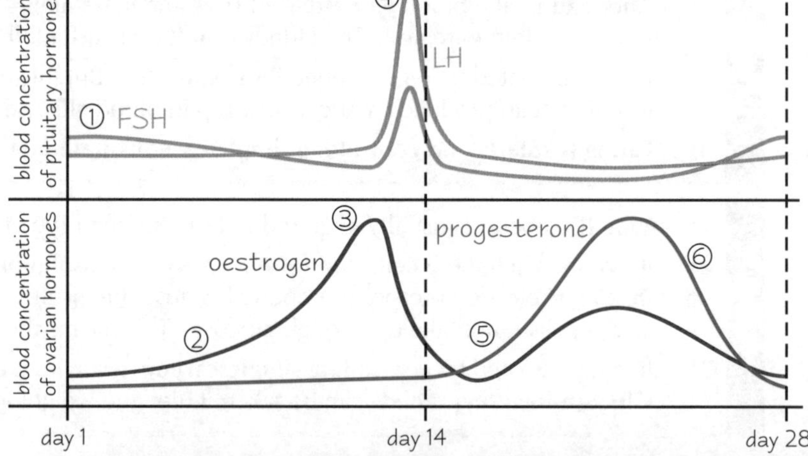

① High FSH concentration in the blood

- FSH stimulates **follicle development**.
- The **follicle** releases **oestrogen**.
- FSH **stimulates** the **ovaries** to release **oestrogen**.

② Rising concentration of **oestrogen**

- Oestrogen **stimulates** the **uterus lining** to **thicken**.
- Oestrogen **inhibits FSH** being released from the **pituitary** gland.

③ Oestrogen concentration **peaks**

High oestrogen concentration **stimulates** the pituitary gland to release **LH** and **FSH**.

④ LH surge (a rapid increase)

- **Ovulation** is stimulated by LH — the **follicle ruptures** and the **egg** is **released**.
- **LH** stimulates the **ruptured follicle** to turn into a **corpus luteum**.
- The **corpus luteum** releases **progesterone**.

⑤ Rising concentrations of **progesterone**

- Progesterone **inhibits FSH** and **LH** release from the **pituitary**.
- The **uterus lining** is **maintained** by progesterone.
- If **no embryo** implants, the **corpus luteum breaks down** and **stops releasing** progesterone.

⑥ Falling concentration of **progesterone**

- **FSH** and **LH** concentrations **increase** because they're **no longer inhibited** by progesterone.
- The uterus lining **isn't maintained** so it **breaks down** — **menstruation** happens and the **cycle starts again**.

Control of the Menstrual Cycle

Negative and *Positive Feedback* Mechanisms *Control* the Level of *Hormones*

The **different concentrations** of **hormones** in the blood during the menstrual cycle are **controlled** by **feedback loops**.

Negative feedback — example one

1) **FSH** stimulates the **ovary** to release **oestrogen.**
2) Oestrogen **inhibits** further release of **FSH.**

After FSH has stimulated follicle development, **negative feedback** keeps the **FSH** concentration **low.** This makes sure that **no more follicles develop.**

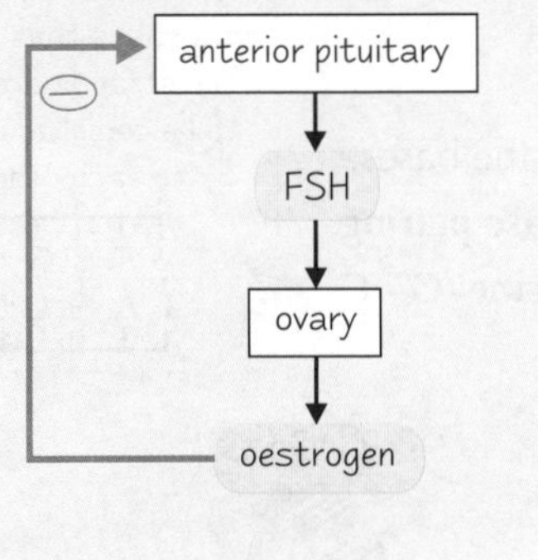

Negative feedback — example two

1) **LH** stimulates the **corpus luteum** to develop, which produces **progesterone.**
2) Progesterone **inhibits** further release of **LH.**

Negative feedback makes sure that **no more follicles develop** when the corpus luteum is developing.
It also makes sure the uterus lining **isn't** maintained if **no embryo** implants.

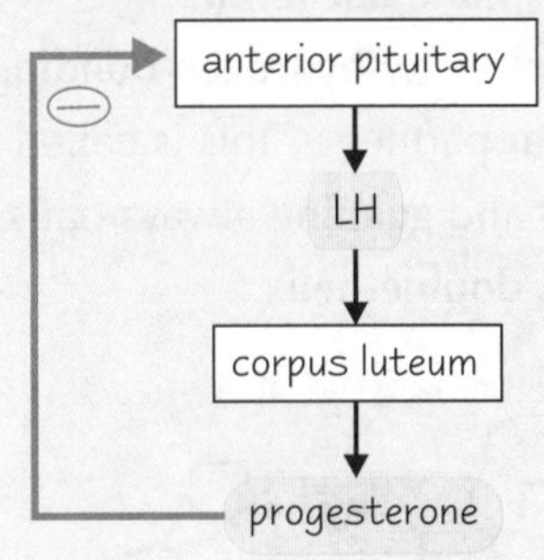

Positive feedback — example

1) **Oestrogen** stimulates the **anterior pituitary** to release **LH.**
2) **LH** stimulates the **ovary** to release **more oestrogen.**
3) Oestrogen **further stimulates** the **anterior pituitary** to release **LH,** and **so on.**

High oestrogen concentration triggers **positive feedback** to make **ovulation** happen.

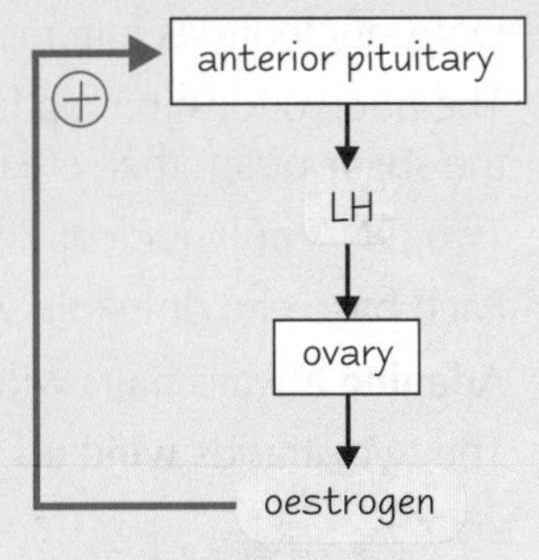

It's a bit weird — oestrogen usually inhibits FSH and LH release, but at really high levels it switches to stimulating their release.

Practice Questions

Q1 Name the hormones released by the anterior pituitary gland.

Q2 Which hormone stimulates the corpus luteum to develop?

Q3 What's the main role of oestrogen in the uterus?

Q4 What's the main role of progesterone in the uterus?

Q5 What happens to the uterus lining if no embryo implants?

Exam Questions

Q1 The human menstrual cycle is controlled by pituitary and ovarian hormones, which are present at different concentrations during the cycle.

a) Explain how negative feedback ensures only one main follicle develops. [5 marks]

b) Explain how positive feedback is involved in ovulation. [3 marks]

Q2 The contraceptive pill contains synthetic equivalents of the hormones oestrogen and progesterone. Suggest how taking the pill can prevent pregnancy. [3 marks]

Sometimes it's hard to be a woman...

What on earth... talk about women being hard to understand. Let's treat it like a good pair of shoes and take it in steps. Start by learning the four main hormones and what they do, then learn when concentrations are highest and why. Finally, get scribbling down those feedback loops. And when you know it all, it must be time for an end-of-section break.

DNA and RNA

__D__eoxyribo__n__ucleic __a__cid (a.k.a. __DNA__) is remarkably clever stuff — as you might remember from AS. Not only do you revisit it for A2, you get to go a step further and head into the mysterious world of RNA. Let the good times roll.

DNA is Made of Nucleotides that Contain a Sugar, a Phosphate and a Base

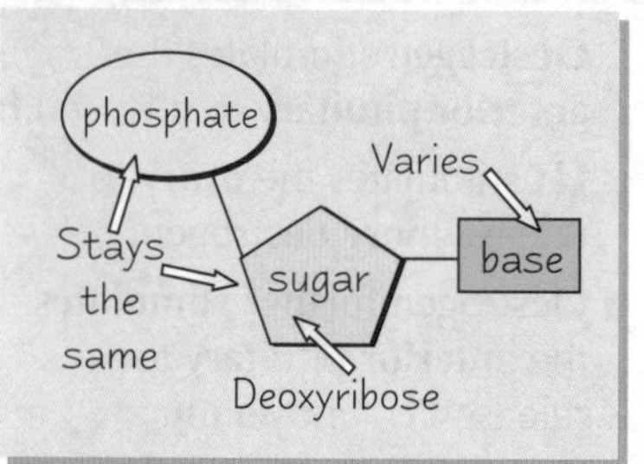

1) DNA is a **polynucleotide** — it's made up of lots of **nucleotides** joined together.
2) Each nucleotide is made from a **pentose sugar** (with 5 carbon atoms), a **phosphate** group and a **nitrogenous base**.
3) The **sugar** in DNA nucleotides is a **deoxyribose** sugar.
4) Each nucleotide has the **same sugar and phosphate**. The **base** on each nucleotide can **vary** though.
5) There are **four** possible bases — adenine (**A**), thymine (**T**), cytosine (**C**) and guanine (**G**).

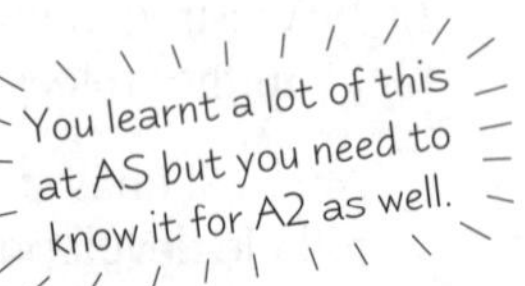

You learnt a lot of this at AS but you need to know it for A2 as well.

Two Polynucleotide Strands Join Together to Form a Double-Helix

1) DNA nucleotides join together to form **polynucleotide strands**.
2) The nucleotides join up between the **phosphate** group of one nucleotide and the **sugar** of another, creating a **sugar-phosphate backbone**.
3) **Two** DNA polynucleotide strands join together by **hydrogen bonding** between the bases.
4) Each base can only join with one particular partner — this is called **specific base pairing**.
5) **Adenine** always pairs with **thymine** (**A - T**) and **guanine** always pairs with **cytosine** (**G - C**).
6) The two strands **wind up** to form the **DNA double-helix**.

When two strands have bases that pair up the strands are said to be complementary to each other:

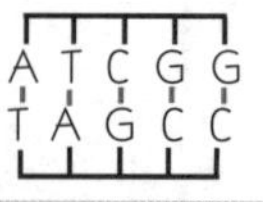

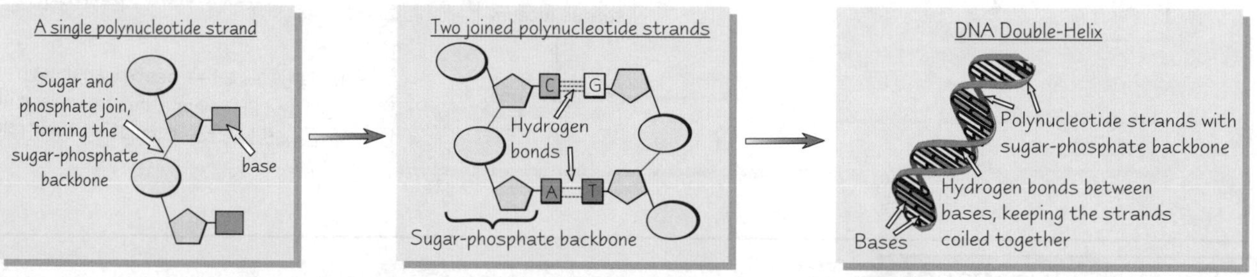

DNA Contains Genes Which are Instructions for Proteins

1) Genes are **sections of DNA**. They're found on **chromosomes**.
2) Genes **code** for **proteins** (polypeptides) — they contain the **instructions** to make them.
3) Proteins are made from **amino acids**. Different proteins have a **different number** and **order** of amino acids.
4) It's the **order** of **nucleotide bases** in a gene that determines the **order of amino acids** in a particular **protein**.
5) Each amino acid is coded for by a sequence of **three bases** (called a **triplet**) in a gene. A DNA triplet is also called a **base triplet** or a **codon**.
6) **Different sequences** of bases code for different amino acids. This is the **genetic code** — see page 170 for more.
7) So the **sequence of bases** in a section of DNA is a **template** that's used to make a **protein** during **protein synthesis**.

Polypeptide is just another word for a protein.

Bases on DNA
G T C T G A
DNA triplet = one amino acid

DNA is Copied into RNA for Protein Synthesis

1) DNA molecules are found in the **nucleus** of the cell, but the organelles for protein synthesis (**ribosomes**) are found in the **cytoplasm**.
2) DNA is too large to move out of the nucleus, so a section is **copied** into **RNA**. This process is called **transcription** (see page 168).
3) The RNA **leaves** the nucleus and joins with a **ribosome** in the cytoplasm, where it can be used to synthesise a **protein**. This process is called **translation** (see page 169).

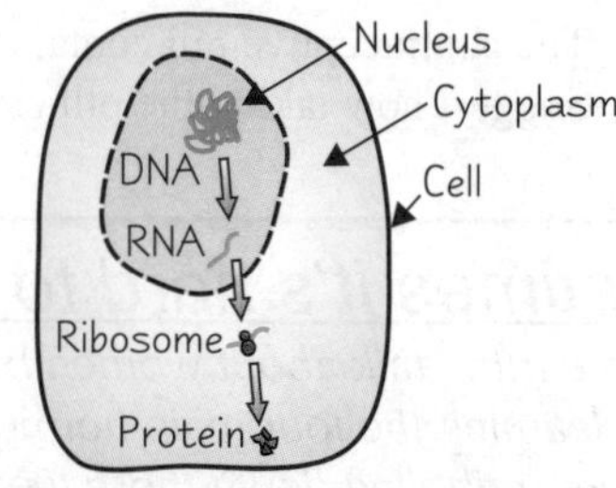

DNA and RNA

RNA is Very Similar to DNA

Like DNA, **RNA** (**ribo**nucleic **a**cid) is made of **nucleotides** that contain one of **four different bases**. The nucleotides also form a **polynucleotide strand** with a sugar-phosphate backbone. But RNA **differs** from DNA in **three** main ways:

1) The **sugar** in RNA nucleotides is a **ribose sugar** (not deoxyribose).
2) The nucleotides form a **single polynucleotide strand** (not a double one).
3) **Uracil** (**U**) replaces thymine as a base. Uracil **always pairs** with **adenine** during protein synthesis.

You need to know about **two different types** of RNA — **messenger RNA** (**mRNA**) and **transfer RNA** (**tRNA**).

Messenger RNA (mRNA)

mRNA is a **single polynucleotide strand**. In mRNA, groups of three adjacent bases are usually called **codons** (they're sometimes called **triplets** or **base triplets**). mRNA is made in the **nucleus** during **transcription**. It **carries the genetic code** from the DNA in the **nucleus** to the **cytoplasm**, where it's used to make a **protein** during **translation**.

Base
A
C
codon
Phosphate
U
Ribose sugar

Transfer RNA (tRNA)

tRNA is a **single polynucleotide strand** that's folded into a **clover shape**. **Hydrogen bonds** between **specific base pairs** hold the molecule in this shape. Every tRNA molecule has a **specific sequence** of **three bases** at one end called an **anticodon**. They also have an **amino acid binding site** at the other end. tRNA is found in the **cytoplasm** where it's involved in **translation**. It **carries** the amino acids that are used to make **proteins** to the **ribosomes**.

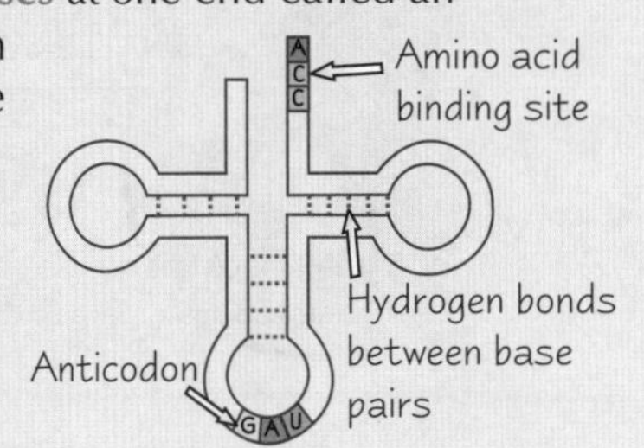

You need to be able to Compare DNA, mRNA and tRNA

You need to know the **structure** and **composition** of **DNA**, **mRNA** and **tRNA** really well — you could be asked to **compare** them in your exam. The table below outlines the **main differences** between them:

	DNA	mRNA	tRNA
Shape	Double-stranded — twisted into a double-helix and held together by hydrogen bonds	Single-stranded	Single-stranded — folded into a clover shape and held together by hydrogen bonds
Sugar	Deoxyribose sugar	Ribose sugar	Ribose sugar
Bases	A, T, C, G	A, U, C, G	A, U, C, G
Other features	Three adjacent bases are called a triplet (sometimes a base triplet or codon)	Three adjacent bases are called a codon (sometimes a triplet or base triplet)	Each tRNA molecule has a specific sequence of three bases called an anticodon and an amino acid binding site

tRNA growing in its natural environment.

Practice Questions

Q1 What are the nucleotides in DNA made of?

Q2 Name the bases found in DNA.

Q3 Describe the shape of a tRNA molecule.

Exam Question

Q1 a) Describe the differences in the composition of DNA and RNA molecules. [2 marks]

b) Name each of the following molecules:

i) A single-stranded molecule that contains ribose sugar and has an amino acid binding site. [1 mark]

ii) A double-stranded molecule that contains deoxyribose sugars and the base thymine. [1 mark]

iii) A single-stranded molecule that contains the base uracil and has an anticodon. [1 mark]

Genes, genes are good for your heart, the more you eat, the more you...

An easy way to remember where mRNA and tRNA come into the whole protein synthesis game is to look at the first letters. mRNA is a messenger — it carries the code from DNA to a ribosome. tRNA transfers amino acids. Easy as that.

Protein Synthesis

OK, so you know all about DNA and RNA — what they're made of, what they look like and the different types. Now you find out what they actually do. It gets kind of complicated but bear with it — it's impressive stuff.

First Stage of Protein Synthesis — Transcription

During transcription an **mRNA copy** of a gene (a section of DNA) is made in the **nucleus**:

1) Transcription starts when **RNA polymerase** (an **enzyme**) **attaches** to the **DNA** double-helix at the **beginning** of a **gene**.
2) The **hydrogen bonds** between the two DNA strands in the gene **break**, **separating** the strands, and the DNA molecule **uncoils** at that point.
3) One of the strands is then used as a **template** to make an **mRNA copy**.

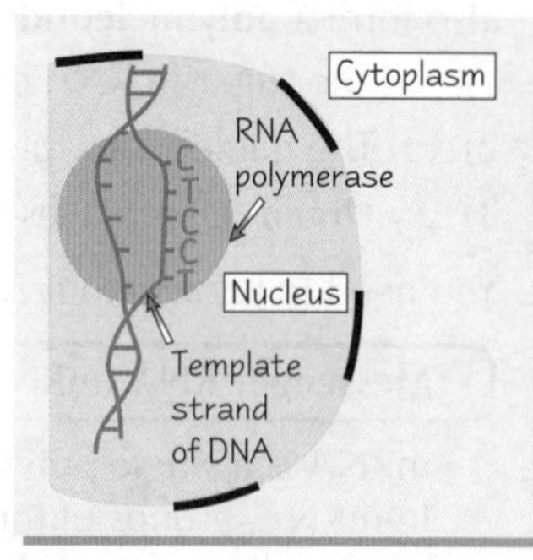

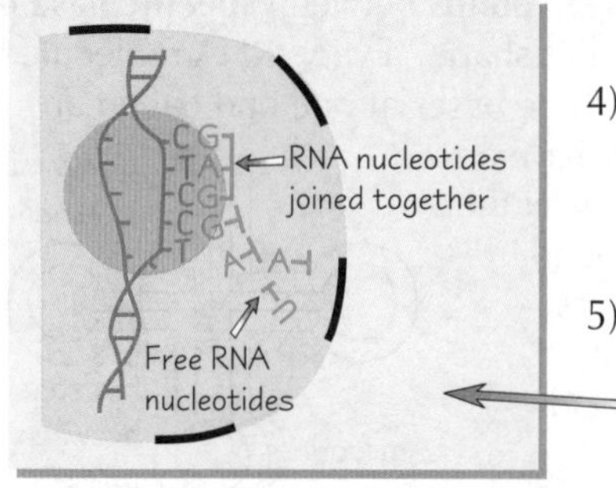

4) The RNA polymerase lines up free **RNA nucleotides** alongside the template strand. **Specific base pairing** means that the mRNA strand ends up being a **complementary copy** of the DNA template strand (except the base **T** is replaced by **U** in **RNA**).
5) Once the RNA nucleotides have **paired up** with their **specific bases** on the DNA strand they're **joined together**, forming an **mRNA** molecule.

6) The RNA polymerase moves **along** the DNA, separating the strands and **assembling** the mRNA strand.
7) The **hydrogen bonds** between the uncoiled strands of DNA **re-form** once the RNA polymerase has passed by and the strands **coil back into a double-helix**.

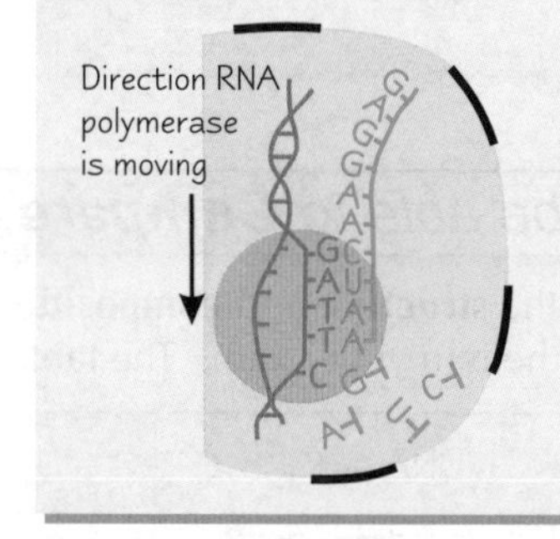

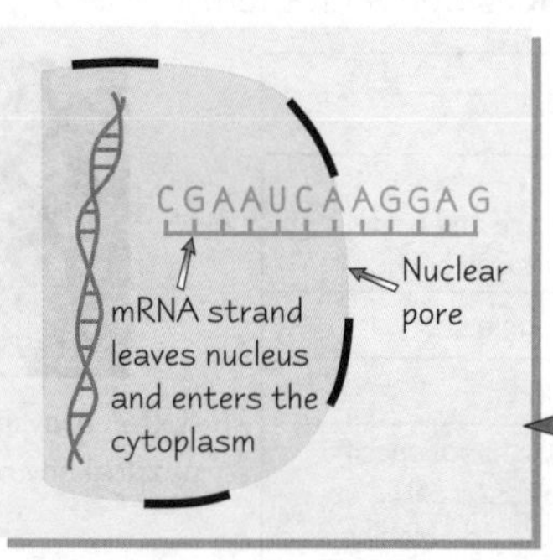

8) When RNA polymerase reaches a particular sequence of DNA called a **stop signal**, it stops making mRNA and **detaches** from the DNA.
9) The **mRNA** moves **out** of the **nucleus** through a nuclear pore and attaches to a **ribosome** in the cytoplasm, where the next stage of protein synthesis takes place (see next page).

mRNA is Edited in Eukaryotic Cells

1) Genes in **eukaryotic DNA** contain sections that **don't code** for amino acids.
2) These sections of DNA are called **introns**. All the bits that **do** code for amino acids are called **exons**.
3) During transcription the introns and exons are both **copied** into mRNA. mRNA strands containing introns and exons are called **pre-mRNA**.
4) Introns are **removed** from pre-mRNA strands by a process called **splicing** — introns are removed and exons joined forming **mRNA** strands. This takes place in the **nucleus**.
5) The mRNA then **leaves** the nucleus for the next stage of protein synthesis (**translation**).

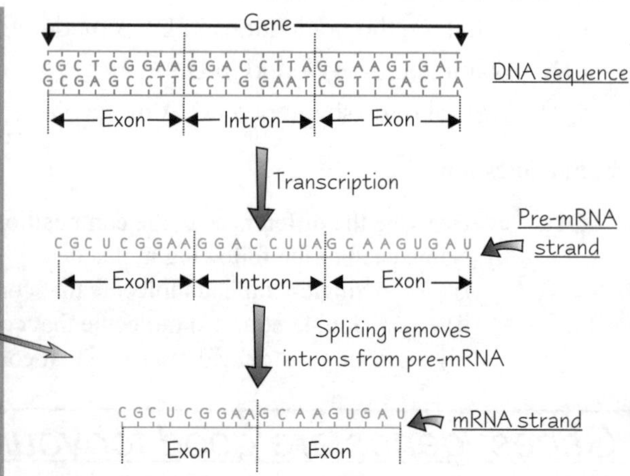

Protein Synthesis

Second Stage of Protein Synthesis — Translation

Translation occurs at the **ribosomes** in the **cytoplasm**. During **translation**, **amino acids** are **joined together** to make a **polypeptide chain** (protein), following the sequence of **codons** (triplets) carried by the mRNA.

1) The **mRNA attaches** itself to a **ribosome** and **transfer RNA** (**tRNA**) molecules **carry amino acids** to the ribosome.
2) A tRNA molecule, with an **anticodon** that's **complementary** to the **first codon** on the mRNA, attaches itself to the mRNA by **specific base pairing**.

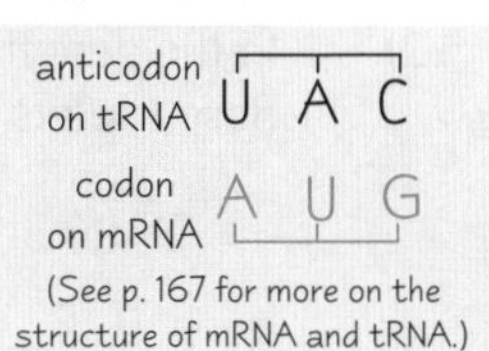

3) A second tRNA molecule attaches itself to the **next codon** on the mRNA in the **same way**.
4) The two amino acids attached to the tRNA molecules are **joined** by a **peptide bond**. The first tRNA molecule **moves away**, leaving its amino acid behind.
5) A third tRNA molecule binds to the **next codon** on the mRNA. Its amino acid **binds** to the first two and the second tRNA molecule **moves away**.
6) This process continues, producing a chain of linked amino acids (a **polypeptide chain**), until there's a **stop signal** on the mRNA molecule.
7) The polypeptide chain (**protein**) **moves away** from the ribosome and translation is complete.

Protein synthesis is also called polypeptide synthesis as it makes a polypeptide (protein).

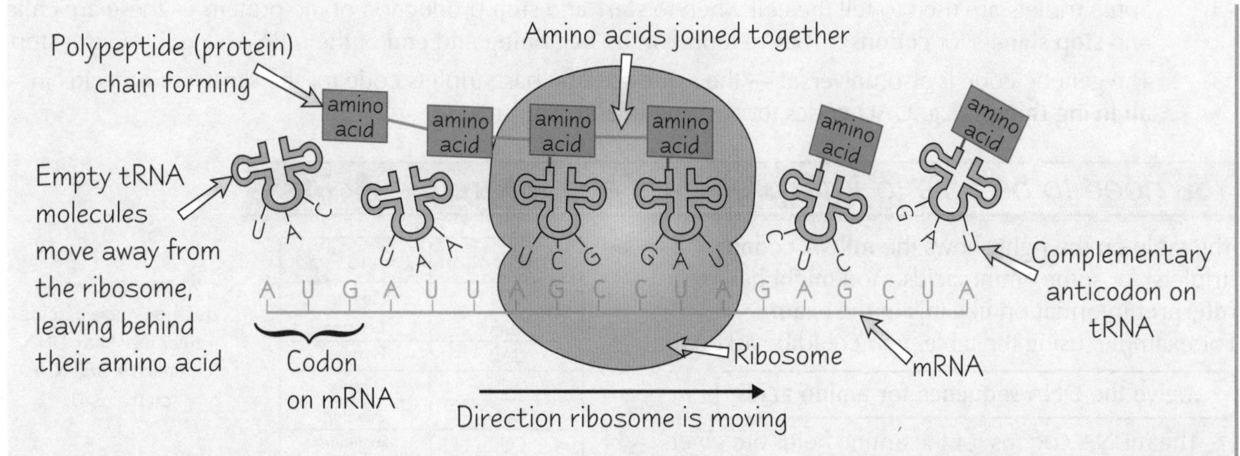

Practice Questions

Q1 What are the two stages of protein synthesis called?

Q2 Where does the first stage of protein synthesis take place?

Q3 When does RNA polymerase stop making mRNA?

Q4 What is an exon?

Q5 Where does the second stage of protein synthesis take place?

Q6 What is a polypeptide chain (protein) made up of?

Exam Questions

Q1 A drug that inhibits cell growth is found to be able to bind to DNA, preventing RNA polymerase from binding. Explain how this drug will affect protein synthesis. [2 marks]

Q2 A polypeptide chain (protein) from a eukaryotic cell is 10 amino acids long.
a) Predict how long the mRNA for this protein would be in nucleotides. Explain your answer. [2 marks]
b) Would you expect the number of nucleotides in the gene (DNA sequence) for this protein to be greater or fewer than your answer for part a)? Explain your answer. [3 marks]

The only translation I'm interested in is a translation of this page into English...

So you start off with DNA, lots of cleverness happens and bingo... you've got a protein. Only problem is you need to know the cleverness bit in quite a lot of detail. So scribble it down, recite it to yourself, explain it to your best mate or do whatever else helps you remember the joys of protein synthesis. And then think how clever you must be to know it all.

The Genetic Code and Nucleic Acids

The genetic code is exactly as it sounds — a code found in your genes that tells your body how to make proteins. It can be interpreted, just like any other code, which is exactly what you might have to do in your exam. So get cracking...

*The Genetic Code is **Non-Overlapping**, **Degenerate** and **Universal***

1) The genetic code is the **sequence of base triplets** (codons) in **mRNA** which **code** for specific **amino acids**.
2) In the genetic code, each base triplet is **read** in sequence, **separate** from the triplet **before** it and **after** it. Base triplets **don't share** their **bases** — the code is **non-overlapping**.

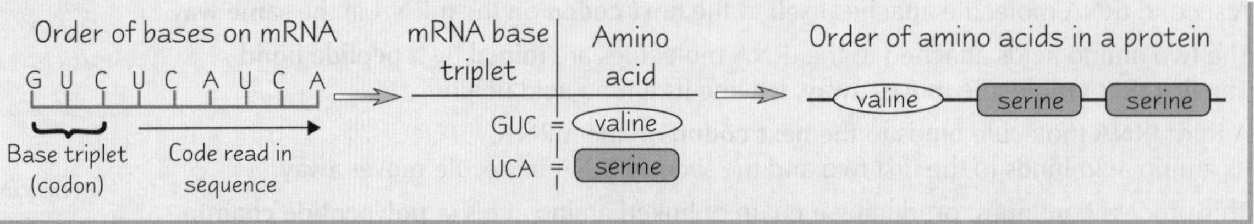

3) The genetic code is also **degenerate** — there are **more** possible combinations of **triplets** than there are amino acids (20 amino acids but 64 possible triplets). This means that some **amino acids** are coded for by **more than one** base triplet, e.g. tyrosine can be coded for by UAU or UAC.
4) Some triplets are used to tell the cell when to **start** and **stop** production of the protein — these are called **start** and **stop** signals (or **codons**). They're found at the **beginning** and **end** of the mRNA. E.g. UAG is a stop signal.
5) The genetic code is also **universal** — the **same** specific base triplets code for the **same** amino acids in **all living things**. E.g. UAU codes for tyrosine in all organisms.

*You need to be able to **Interpret Data** about **Nucleic Acids***

The table on the right shows the **mRNA codons** (triplets) for some amino acids. You might have to **interpret** information like this in the exam. For example, using the table, you could be asked to...

mRNA codon	Amino Acid
UCU	Serine
CUA	Leucine
UAU	Tyrosine
GUG	Valine
GCA	Alanine
CGC	Arginine

When interpreting data on nucleic acids remember that DNA contains T and RNA contains U.

...give the DNA sequence for amino acids

The mRNA codons for the amino acids are given in the table. Because **mRNA** is a **complementary copy** of the **DNA** template, the DNA sequence for each amino acid is made up of bases that would **pair** with the mRNA sequence:

mRNA codon	Amino Acid	DNA sequence (of template strand)
UCU	Serine	AGA
CUA	Leucine	GAT
UAU	Tyrosine	ATA
GUG	Valine	CAC
GCA	Alanine	CGT
CGC	Arginine	GCG

You could also be asked to work out the amino acids from a given DNA sequence and a table.

...give the tRNA anticodons from mRNA codons

tRNA anticodons are **complementary copies** of **mRNA codons**, so you can work out the tRNA anticodon from the mRNA codon:

mRNA codon	tRNA anticodon
UCU	AGA
CUA	GAU
UAU	AUA
GUG	CAC
GCA	CGU
CGC	GCG

You might be asked to name the amino acid coded for by a tRNA anticodon using a table like the one above.

...write the amino acid sequence for a section of mRNA

You might have to work out the sequence of some mRNA from a sequence of amino acids and a table.

To **work out** the sequence of **amino acids** from some mRNA, you need to break the genetic code into **codons** and then use the information in the table to work out what **amino acid** they code for.

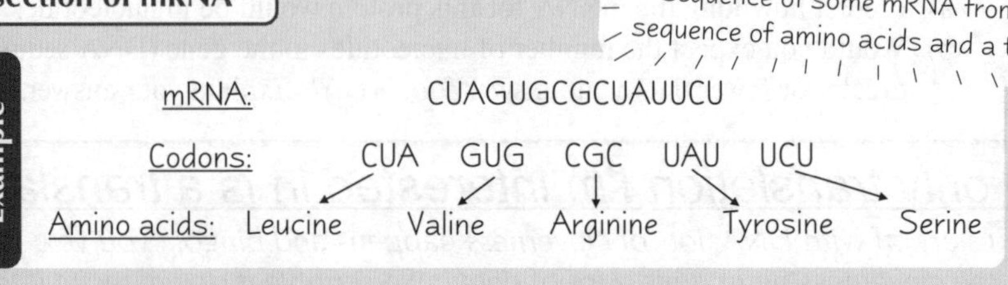

The Genetic Code and Nucleic Acids

You Might Have to **Interpret Data** About The **Role** of **Nucleic Acids**

In the exam you might have to **interpret data** from experiments done to **investigate nucleic acids** and their **role** in **protein synthesis**. Here's an example (you **don't** need to **learn** it):

Investigating the effect of new drugs on nucleic acids

1) To investigate **how** two new drugs affect **nucleic acids** and their **role** in protein synthesis, **bacteria** were **grown in normal conditions** for a few generations, then moved to media containing the drugs.
2) After a short period of time, the **concentration** of **protein** and **complete strands** of **mRNA** in the bacteria were analysed. The results are shown in the **bar graph**.
3) Both mRNA **and** protein concentration were **lower** in the presence of **drug 1 compared** to the **no-drug control**. This suggests that drug 1 **affects the production** of **full length mRNA**, so there's no mRNA for protein synthesis during **translation**.
4) **mRNA production** in the presence of **drug 2** was **unaffected**, but **less protein** was produced — **3 mg cm^{-3}** compared to **8 mg cm^{-3}**. This suggests that drug 2 **interferes** with **translation**. **mRNA was produced**, but **less protein** was **translated** from it.
5) **Further tests** to establish the **nature** of the two drugs were carried out.
6) **Drug 1** was found to be a **ribonuclease** (an enzyme that **digests RNA**). This could **explain** the results of the first experiment — **any strands** of **mRNA** produced by the cell would be **digested** by drug 1, so **couldn't be used** in **translation** to make proteins.
7) **Drug 2** was found to be a **single-stranded**, **clover-shaped** molecule capable of binding to the **ribosome**. Again, this helps to **explain** the **results** from the first experiment — drug 2 could work by **binding** to the ribosome, **blocking tRNAs** from binding to it and so **preventing translation**.

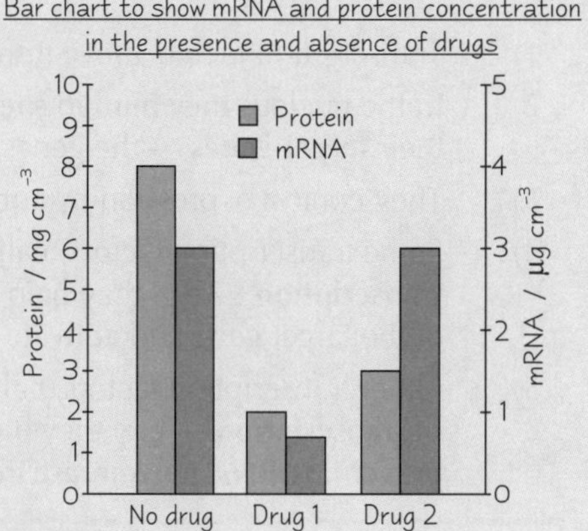

Transcription and translation are on pages 168-169.

Practice Questions

Q1 What is the genetic code?

Q2 Why is the genetic code described as degenerate?

Q3 Why is the genetic code described as universal?

mRNA codon	amino acid
UGU	Cysteine
CGC	Arginine
GGG	Glycine
GUG	Valine
GCA	Alanine
UUG	Leucine
UUU	Phenylalanine

Exam Questions

Q1 An artificial mRNA was synthesized and used in an experiment to form a protein (polypeptide). The mRNA sequence was: UUGUGUGGGUUUGCAGCA and the protein produced was: Leucine–Cysteine–Glycine–Phenylalanine–Alanine–Alanine. Use the table above to help you answer the following questions.

a) Explain how the result suggests that the genetic code is based on triplets of nucleotides in mRNA. [2 marks]

b) Explain how the result suggests that the genetic code is non-overlapping. [2 marks]

Q2 The table shows the mRNA codons for some amino acids. Show your working for the following questions.

a) Give the amino acid sequence for the mRNA sequence GUGUGUCGCGCA. [2 marks]

b) Give the mRNA sequence for the amino acid sequence arginine, alanine, leucine, phenylalanine. [2 marks]

c) Give the DNA template strand sequence that codes for the amino acid sequence valine, arginine, alanine. [3 marks]

—•• ——— ••• ——— —— • •—— ——— •—• —•—

Hurrah — a page with slightly fewer confusing terms and a lot less to remember. The key to the genetic code is to be able to interpret it, so if you know how DNA, mRNA and tRNA work together to make a protein you should be able to handle any data they can throw at you. Now repeat after me, C pairs with G, A pairs with T. Unless it's RNA — then it's U.

Regulation of Transcription and Translation

Oh yes, you read that right — it's back to the incredibly important and immensely clever transcription and translation.

Transcription Factors Control the Transcription of Target Genes

All the **cells** in an organism carry the **same genes** (DNA) but the **structure** and **function** of different cells **varies**. This is because **not all** the **genes** in a cell are **expressed** (transcribed and used to make a protein). Because **different genes** are expressed, **different proteins** are made and these proteins modify the cell — they determine the **cell structure** and control **cell processes** (including the expression of more genes, which produce more proteins).

The **transcription** of genes is **controlled** by protein molecules called **transcription factors**:

1) Transcription factors **move** from the **cytoplasm** to the **nucleus**.
2) In the nucleus they **bind** to **specific DNA sites** near the start of their **target genes** — the genes they **control** the expression of.
3) They control expression by controlling the **rate** of transcription.
4) Some transcription factors, called **activators**, **increase** the **rate of transcription** — e.g. they help **RNA polymerase bind** to the start of the target gene and **activate** transcription.
5) Other transcription factors, called **repressors**, **decrease** the **rate of transcription** — e.g. they **bind** to the start of the target gene, **preventing RNA polymerase** from **binding**, **stopping** transcription.

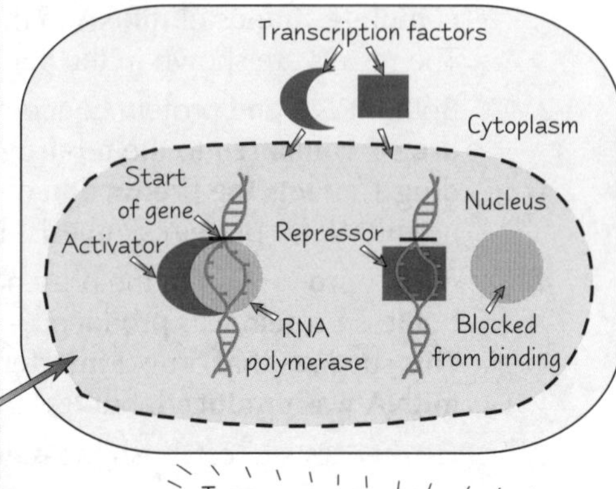

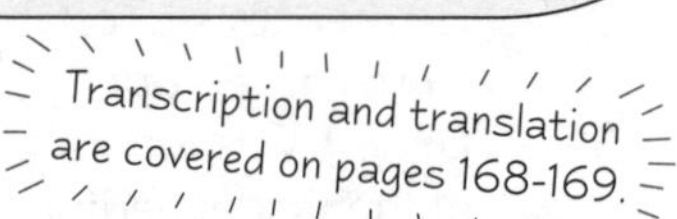
Transcription and translation are covered on pages 168-169.

Oestrogen Affects the Transcription of Target Genes

The **expression** of **genes** can also be **affected** by **other molecules** in the cell, e.g. **oestrogen**:

1) Oestrogen is a **hormone** that can affect transcription by **binding** to a **transcription factor** called an **oestrogen receptor**, forming an **oestrogen-oestrogen receptor complex**.
2) The complex moves from the **cytoplasm** into the **nucleus** where it **binds** to **specific DNA sites** near the **start** of the **target gene**.
3) The complex can **either** act as an **activator**, e.g. **helping** RNA polymerase, or as a **repressor**, e.g. **blocking** RNA polymerase.
4) Whether the complex **acts** as a repressor **or** activator **depends** on the **type of cell** and the target gene.
5) So, the **level of oestrogen** in a particular cell affects the **rate of transcription** of target genes.

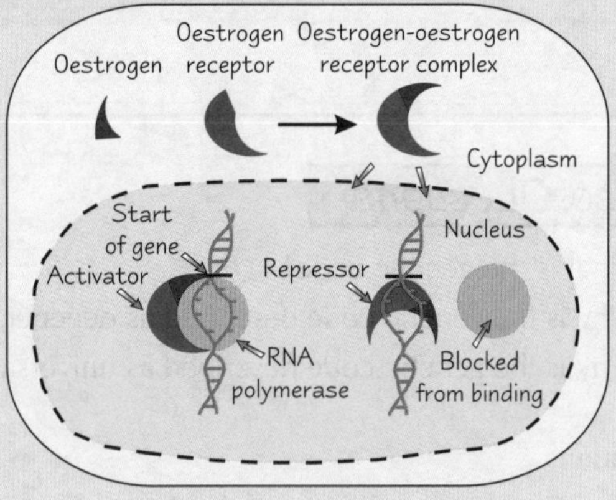

siRNA Interferes with Gene Expression

Gene expression is also affected by a **type of RNA** called **small interfering RNA** (**siRNA**):

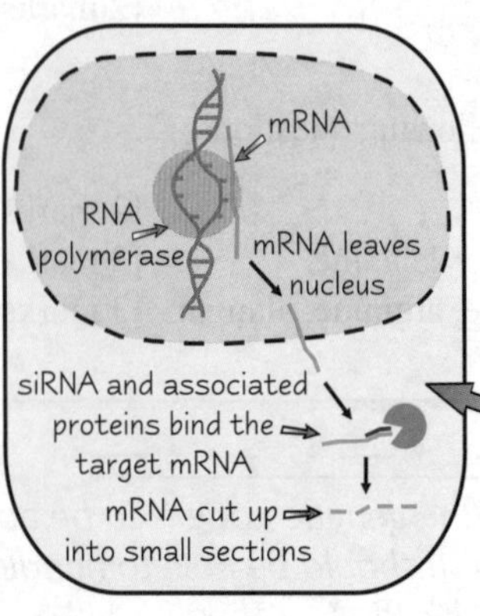

1) siRNA molecules are **short**, **double-stranded RNA** molecules that can **interfere** with the **expression** of specific genes.
2) Their bases are **complimentary** to **specific sections** of a **target gene** and the mRNA that's formed from it.
3) siRNA can **interfere** with both the **transcription** and **translation** of genes.
4) siRNA affects **translation** through a mechanism called **RNA interference**:
   - In the **cytoplasm**, siRNA and associated proteins **bind** to the **target mRNA**.
   - The proteins **cut up** the mRNA into sections so it can no longer be translated.
   - So, the siRNA **prevents the expression** of the specific gene as its protein can no longer be made during **translation**.

Regulation of Transcription and Translation

You need to be able to Interpret Experimental Data on Gene Expression

You could get a question in the exam where you have to **interpret data** about gene expression. It could be on **anything** you've learnt on the **previous page**, e.g. transcription factors, oestrogen or siRNAs. Below is an example of a **gene expression system** in bacteria and an experiment that **investigates** how it works. You **don't** need to **learn** the information, just **understand** what the results of the experiment tell you about how the expression of the gene is **controlled**.

Transcribing — I can do it with my eyes closed.

The *lac* repressor:

1) ***E. coli*** is a **bacterium** that respires **glucose**, but it can use **lactose** if glucose **isn't available**.
2) If lactose is present, *E. coli* makes an **enzyme** (**β-galactosidase**) to **digest** it. But if there's **no** lactose, it doesn't **waste energy** making an enzyme it **doesn't need**. The enzyme's **gene** is **only expressed** when lactose is **present**.
3) The production of the enzyme is **controlled** by a **transcription factor** — the ***lac* repressor**.
4) When there's **no** lactose, the *lac* repressor **binds** to the **DNA** at the start of the gene, **stopping transcription**.
5) When lactose **is present** it **binds** to the *lac* repressor, **stopping** it binding to the DNA, so the gene **is transcribed**.

Experiment:

Medium	Mutant	mRNA	Colour
Glucose	Normal	No	No yellow
Lactose	Normal	Yes	Yellow
Glucose	Mutant 1	Yes	Yellow
Lactose	Mutant 1	Yes	Yellow
Glucose	Mutant 2	No	No yellow
Lactose	Mutant 2	Yes	No yellow

1) Different ***E. coli*** mutants were isolated and grown in **different media**, e.g. with lactose or glucose.
2) The mutants have **mutations** (**changes** in their **DNA bases**, see next page) that mean they **act differently** from normal *E. coli*, e.g. they **produce β-galactosidase** when grown with glucose.
3) To **detect** whether active (working) β-galactosidase was produced, a **chemical** that turns **yellow** in the presence of active β-galactosidase was **added** to the medium.
4) The production of **mRNA** that **codes for β-galactosidase** was also measured. The results are shown in the **table**.
5) In **mutant 1**, mRNA and active β-galactosidase **were produced** even when they were grown with **only glucose** — the gene is **always** being expressed.
6) This suggests that mutant 1 has a **faulty *lac* repressor**, e.g. in the **absence** of lactose the repressor **isn't able** to bind DNA, so transcription **can** occur and mRNA and active β-galactosidase **are produced**.
7) In **mutant 2**, **mRNA** is produced but **active β-galactosidase isn't** when **lactose** is present — the **gene** is being **transcribed** but it **isn't** producing **active** β-galactosidase.
8) This suggests mutant 2 is producing **faulty β-galactosidase**, e.g. because a **mutation** has affected its active site.

Practice Questions

Q1 Name two types of transcription factor.

Q2 Name the transcription factor that oestrogen can bind to.

Q3 What does siRNA stand for?

Exam Question

Tube	Medium	Bacteria	Full length mRNA	Protein
1	+ Oestrogen	Normal	Yes	Active
2	– Oestrogen	Normal	No	No
3	+ Oestrogen	Mutant	No	No
4	– Oestrogen	Mutant	No	No

Q1 An experiment was carried out to investigate gene expression of the Chi protein in genetically engineered bacteria. A mutant bacterium was isolated and analysed to look for mRNA coding for Chi, and active Chi protein production. The results are shown in the table above.

a) What do the results of tubes 1 and 2 suggest about the control of gene expression? Explain your answer. [4 marks]

b) What do the results of tubes 3 and 4 suggest could be wrong with the mutant? Explain your answer. [3 marks]

c) If an siRNA complimentary to the Chi gene was added to tube 1, what would you expect the results to be? Explain your answer. [3 marks]

Transcription Factor — not quite as eXciting as that other factor programme...

If it was a competition, oestrogen would totally win — it's very jazzy and awfully controlling. Flexible too — sometimes it helps to activate and other times it helps to repress. Although I'm not sure it can hold a note or wiggle in time to music.

Mutations, Genetic Disorders and Cancer

Mutations — as featured in numerous superhero movies. Well, I'm sorry to be the one to break it to you, but they don't usually give you special powers or superhuman strength — in fact, they can cause a lot of problems.

Mutations are Changes to the Base Sequence of DNA

Any change to the **base sequence** of DNA is called a **mutation**:

1) They can be caused by **errors** during **DNA replication**.
2) They can also be caused by **mutagenic agents** (see below).
3) The **types** of errors that can occur include:

Errors can also be caused by insertion, duplication and inversion of bases.

- **Substitution** — one base is substituted with another, e.g. ATGCCT becomes ATTCCT (G is **swapped** for T).
- **Deletion** — one base is deleted, e.g. ATGCCT becomes ATCCT (G is **deleted**).

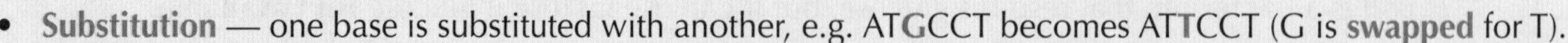

4) The **order** of **DNA bases** in a gene determines the **order of amino acids** in a particular **protein** (see p. 166). If a mutation occurs in a gene, the **sequence** of **amino acids** that it codes for (and the protein formed) could be **altered**:

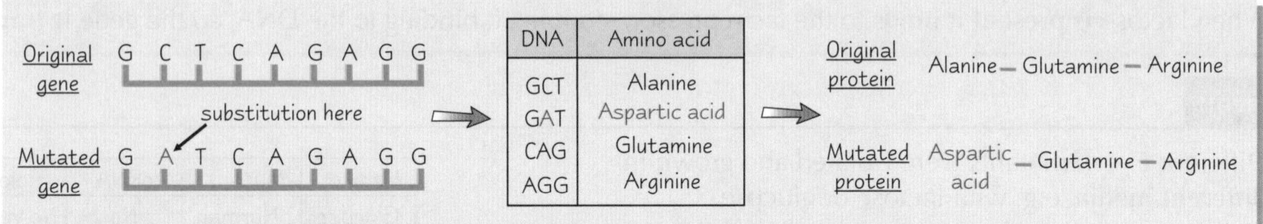

Not All Mutations Affect the Order of Amino Acids

The **degenerate nature** of the genetic code (see page 170) means that some amino acids are coded for by **more than one DNA triplet** (e.g. tyrosine can be coded for by TAT or TAC in DNA). This means that **not all** substitution mutations will result in a change to the amino acid sequence of the protein — some substitutions will still **code** for the **same amino acid**. For example:

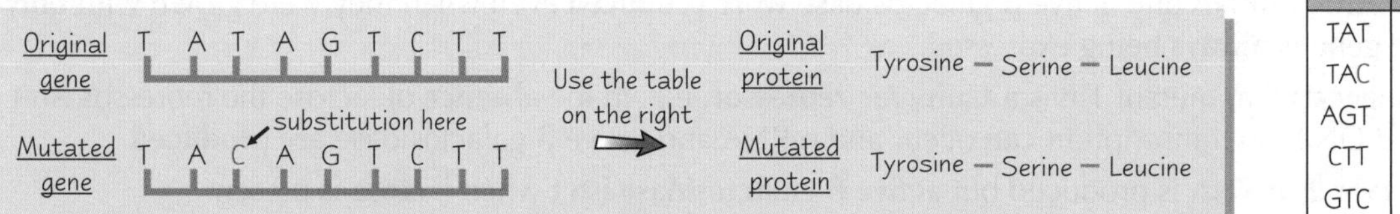

DNA	Amino acid
TAT	Tyrosine
TAC	Tyrosine
AGT	Serine
CTT	Leucine
GTC	Valine

Substitution mutations **won't always** lead to changes in the amino acid sequence, but **deletions will** — the deletion of a base will change the **number** of bases present, which will cause a **shift** in all the base triplets after it:

Mutagenic Agents Increase the Rate of Mutation

Mutations occur **spontaneously**, e.g. when DNA is **misread** during **replication**. But some things can cause an **increase** in the **rate** of **mutations** — these are called **mutagenic agents**. **Ultraviolet radiation**, **ionising radiation**, some **chemicals** and some **viruses** are examples of mutagenic agents. They can increase the rate of mutations by:

1) **Acting as a base** — chemicals called **base analogs** can **substitute** for a base during DNA replication, **changing** the **base sequence** in the new DNA. E.g. **5-bromouracil** is a base analog that can substitute for **thymine**. It can pair with **guanine** (**instead of adenine**), causing a **substitution mutation** in the new DNA.
2) **Altering bases** — some chemicals can **delete** or **alter bases**. E.g. **alkylating agents** can add an alkyl group to **guanine**, which **changes** the **structure** so that it pairs with **thymine** (**instead of cytosine**).
3) **Changing** the **structure of DNA** — some types of **radiation** can change the structure of DNA, which causes **problems** during DNA replication. E.g. **UV radiation** can cause adjacent **thymine** bases to **pair up** together.

Mutations, Genetic Disorders and Cancer

Genetic Disorders and Cancer are Caused By Mutations

Hereditary Mutations Cause Genetic Disorders and Some Cancers

Some mutations can cause **genetic disorders** — inherited disorders caused by **abnormal genes** or **chromosomes**, e.g. cystic fibrosis. Some mutations can **increase** the **likelihood** of developing certain **cancers**, e.g. mutations of the gene **BRCA1** can increase the chances of developing **breast cancer**. If a **gamete** (sex cell) containing a mutation for a genetic disorder or certain cancer is **fertilised**, the mutation will be present in the new **fetus** formed — these are called **hereditary mutations** because they are passed on to the offspring.

Acquired Mutations Can Cause Cancer

1) Mutations that occur in individual cells **after** fertilisation (e.g. in adulthood) are called **acquired mutations**.
2) If these mutations occur in the **genes** that **control** the rate of **cell division**, it can cause **uncontrolled cell division**.
3) If a cell divides uncontrollably the result is a **tumour** — a mass of abnormal cells. Tumours that **invade** and **destroy surrounding tissue** are called **cancers**.
4) There are **two types** of **gene** that control cell division — **tumour suppressor genes** and **proto-oncogenes**. Mutations in these genes can cause cancer:

Tumour suppressor genes can be **inactivated** if a **mutation** occurs in the DNA sequence.

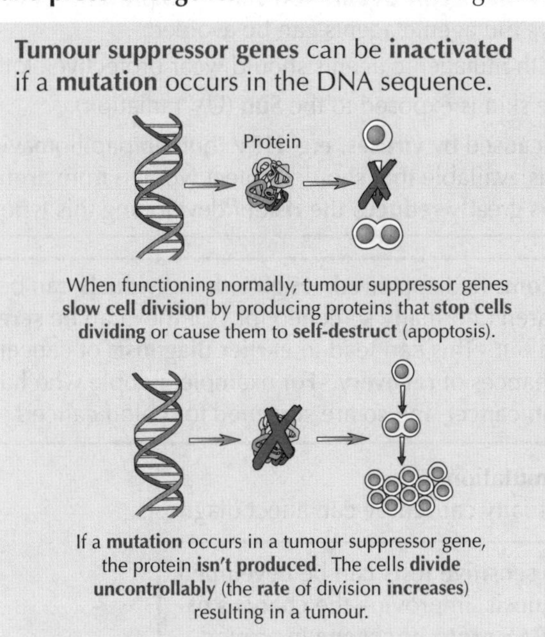

When functioning normally, tumour suppressor genes **slow cell division** by producing proteins that **stop cells dividing** or cause them to **self-destruct** (apoptosis).

If a **mutation** occurs in a tumour suppressor gene, the protein **isn't produced**. The cells **divide uncontrollably** (the **rate** of division **increases**) resulting in a tumour.

The **effect** of a **proto-oncogene** can be **increased** if a **mutation** occurs in the DNA sequence. A mutated proto-oncogene is called an **oncogene**.

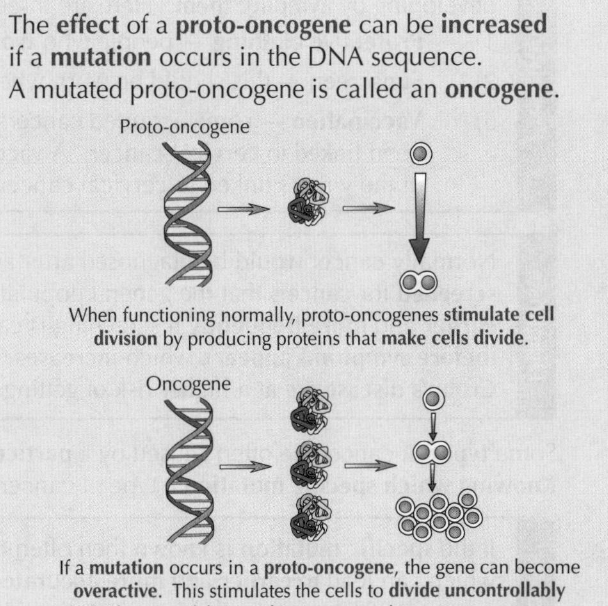

When functioning normally, proto-oncogenes **stimulate cell division** by producing proteins that **make cells divide**.

If a **mutation** occurs in a **proto-oncogene**, the gene can become **overactive**. This stimulates the cells to **divide uncontrollably** (the **rate** of division **increases**) resulting in a **tumour**.

Practice Questions

Q1 What is a substitution mutation?

Q2 What are mutagenic agents?

Q3 What is a genetic disorder?

Exam Question

Before exposure: A G T T A T C A G G C T

After exposure: A G G T A T G A G G C C

DNA	Amino acids	DNA	Amino acids
AGT	Serine	GAG	Glutamic acid
AGG	Arginine	GCT	Alanine
TAT	Tyrosine	GCC	Alanine
CAG	Glutamine		

Q1 The order of bases in a liver cell's proto-oncogene before and after exposure to a mutagenic agent is shown above.
a) Underline any mutation(s) that have occurred. [1 mark]
b) Use the table to explain the changes that the mutations would cause to the sequence of amino acids. [5 marks]
c) Would you describe these mutations as acquired or inherited? Explain your answer. [2 marks]
d) Explain how the mutation(s) may lead to cancer. [3 marks]

Just hope your brain doesn't have a deletion mutation during the exam...

Right, there's plenty to learn on these pages and some of it's a bit complicated, so you know the drill. Don't read it all through at once — take the sections one by one and get all the facts straight. There could be nothing more fun...

Diagnosing and Treating Cancer and Genetic Disorders

Before you start this page, make sure you've got one thing straight in your head from the previous page — make sure you understand the difference between acquired and hereditary disorders.

*Knowing the **Mutation** is Useful for the **Diagnosis** and **Treatment** of **Disorders***

1) **Cancer** and most **genetic disorders** are caused by **mutations** (see previous page).
2) Knowing whether a disorder is caused by an **acquired** or **inherited mutation** affects the **prevention** and **diagnosis** of the disorder.
3) **Identifying** the **specific mutation** that causes a disorder in an individual affects the prevention, diagnosis and treatment.
4) Here are some **examples** for each type of **disorder**:

See page 192 for how to screen for mutations.

*1 **Cancer** — Caused by **Acquired** Mutations*

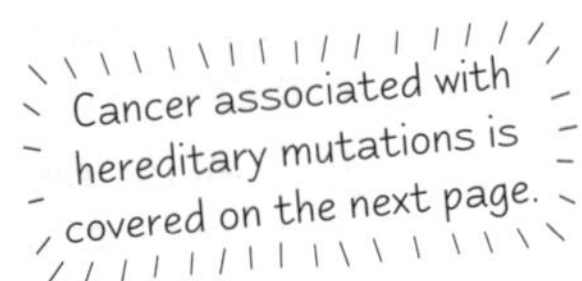

Cancer associated with
hereditary mutations is
covered on the next page.

Acquired mutations can **occur spontaneously** or be **caused by** exposure to **mutagenic agents** (see page 174). Knowing that cancer can be caused by **acquired mutations** affects the prevention and diagnosis:

Prevention

If you know that **acquired mutations** are caused by **mutagenic agents** you can try to prevent cancer developing by **avoiding them**. Here are three ways mutagenic agents can be avoided:
1) **Protective clothing** — people who **work** with mutagenic agents should wear protective clothing.
2) **Sunscreen** — this should be worn when the skin is exposed to the **Sun** (UV radiation).
3) **Vaccination** — some acquired cancers are caused by **viruses**, e.g. **HPV** (human papillomavirus) has been linked to **cervical cancer**. A vaccine is available that should protect women from **around 80%** of the viruses linked to cervical cancer. This greatly **reduces the risk** of developing this type of cancer.

Diagnosis

Normally cancer would be diagnosed **after symptoms** had **appeared**. **High-risk individuals** can be **screened** for cancers that the general population aren't normally screened for. Or they can be **screened earlier** and **more frequently** if screening is carried out. This can lead to **earlier diagnosis** of cancer (**before symptoms appear**), which **increases** the chances of **recovery**. For example, people who have **Crohn's disease** are at a higher risk of getting **colon cancer** and so are **screened** for colon cancer.

Some **types** of cancer are often caused by a **particular mutation**.
Knowing which specific mutation a type of cancer is usually caused by can affect diagnosis:

Diagnosis

If the **specific mutation** is known then often **more sensitive tests** can be **developed**, which can lead to **earlier** and **more accurate diagnosis**, improving the **chances of recovery**. For example, there's a **mutation** in the **RAS proto-oncogene** in around **half** of all **bowel cancers**. Bowel cancer can be **detected early** by looking for RAS mutations in the DNA of **bowel cells**.

Individuals diagnosed with **cancer** can also have the **DNA** from the **cancerous cells analysed** to see which mutation has caused it. Knowing **which specific mutation** the cancer is caused by affects treatment:

Treatment

1) The **treatment** can be **different** for different mutations. For example, **breast cancer** caused by mutation of the **HER2 proto-oncogene** can be treated with a drug called **Herceptin®**. This drug binds **specifically** to the altered HER2 protein receptor and **suppresses cell division and tumour growth**. Breast cancer caused by other mutations is not treated with this drug as it doesn't work.
2) The **aggressiveness** of the **treatment** can **differ** depending on the mutation. Different mutations produce **different types** of cancer, which affects the treatment. For example, if the mutation is known to cause an **aggressive** (**fast-growing**) cancer it may be treated with **higher doses** of **radiotherapy** or by **removing larger areas** of the tumour and surrounding tissue during **surgery**.
3) If the specific mutation is known, **gene therapy** (see page 194) may be able to treat it. For example, if you know it's caused by **inactivated tumour suppressor genes** (see previous page), gene therapy could be used to provide **working versions** of the genes.

Diagnosing and Treating Cancer and Genetic Disorders

2 Cancer — Caused by Hereditary Mutations

Cancer caused by **hereditary mutations** usually results in a **family history** of a certain type of cancer. If an individual has a family history of cancer, things can be done to **prevent** it **developing** and **diagnose it earlier** if it does:

Prevention

Most cancers are caused by mutations in **multiple genes**. So people with a family history should **avoid gaining extra acquired mutations** by **avoiding mutagenic agents**, e.g. those with a family history of **lung cancer shouldn't smoke**.

Diagnosis

Screening, or **increased** and **earlier screening**, if there's a **family history** can lead to **early detection** (i.e. before symptoms appear) and **increased** chances of recovery. E.g. **more frequent breast examinations** if there's a family history of **breast cancer**.

Individuals with a **family history** of cancer can have their **DNA analysed** to see if they carry the **specific mutation**. Knowing **which specific mutation** the cancer is caused by affects prevention, diagnosis and treatment:

Prevention

If the mutation causes a very high risk of cancer **preventative surgery** may be carried out — removing the **organ** the cancer is likely to affect **before cancer develops**. E.g. women with a mutation in **BRCA1** sometimes choose to have a **mastectomy** (removal of one or both breasts) to **prevent breast cancer** from developing.

Treatment

Treatment is **similar** to treating cancer caused by acquired mutations (see previous page). E.g. the treatment depends on the **particular mutation**. But cancer caused by hereditary mutations is often **diagnosed earlier**, which can **change** the treatment used.

Diagnosis

Screening, or **increased** and **early screening** of those with a **hereditary mutation** can lead to **early detection** and **increased** chances of recovery. E.g. **frequent colonoscopies** for those with a mutated **APC gene** to diagnose hereditary **colon cancer earlier**.

3 Genetic Disorders — Caused by Hereditary Mutations

Diagnosis

If a person has a **family history** of a genetic disorder they can have their **DNA analysed** to see if they have the **mutation** that causes it or if they are a **carrier** (see p. 134). If they're **tested** and **diagnosed before symptoms develop**, any **treatment** available can **begin earlier**. Also, knowing if they have the disorder or if they're a carrier can help to figure out if any **children** they have (or might have) are at **risk**.

Treatment

1) **Gene therapy** (see page 194) — this may be able to treat **some genetic disorders**. E.g. scientists have shown it's possible to **treat** symptoms of **cystic fibrosis** by inserting a **normal copy** of the **mutated gene**.
2) The **treatment** can be **different** for different **mutations** — for example, the **exact gene mutation** for Huntington's disease affects **symptom treatment options** as it affects the **time of onset** of symptoms.
3) **Early diagnosis** can affect treatment options — for example, if **sickle cell anaemia** is diagnosed at **birth**, treatments that **relieve symptoms** and work to **avoid complications** can be given straight away.

Prevention

Carriers or **sufferers** of genetic disorders can undergo **preimplantation genetic diagnosis** during ***in vitro* fertilisation** (IVF) to **prevent** any **offspring** having the disease. **Embryos** are produced by IVF and **screened** for the mutation. Only embryos **without the mutation** are **implanted** in the womb.

Practice Questions

Q1 Describe how knowing which specific mutation a cancer is caused by affects diagnosis.

Q2 Describe how knowing which specific mutation a genetic disorder is caused by affects treatment.

Exam Question

Q1 Discuss how knowing if a disorder is caused by a hereditary or acquired mutation affects prevention and diagnosis of the disorder. [25 marks]

My genes need therapy — they've all got holes in the knees...

So whether a disorder is caused by an acquired or hereditary gene mutation has a pretty big effect on things like prevention, diagnosis and treatment. Make sure you understand the differences so you're not caught out in the exam.

Stem Cells

Stem cells — they're the daddy of all cells, the big cheese, the top dog, and the head honcho. And here's why...

Stem Cells are Able to **Mature** into **Any Type** of **Body Cell**

1) **Multicellular organisms** are made up from many **different cell types** that are **specialised** for their function, e.g. liver cells, muscle cells, white blood cells.
2) **All** these specialised cell types originally came from **stem cells**.
3) Stem cells are **unspecialised** cells that can develop into **other types** of cell.
4) Stem cells divide to become **new** cells, which then become **specialised**.
5) All multicellular organisms have some form of stem cell.
6) Stem cells are found in the **embryo** (where they become all the **specialised cells** needed to form a **fetus**) and in **some adult tissues** (where they become **specialised** cells that need to be **replaced**, e.g. stem cells in the **bone marrow** can become **red blood cells**).
7) Stem cells that can mature (develop) into **any type** of **body cell** in an organism are called **totipotent cells**.
8) Totipotent stem cells in humans are **only present** in the **early life** of an **embryo**.
9) After this point the embryonic stem cells **lose** their ability to **specialise** into **all** types of cells.
10) The few **stem cells** that remain in mature animals are called **multipotent stem cells**. They can only develop into a **few types** of cells.

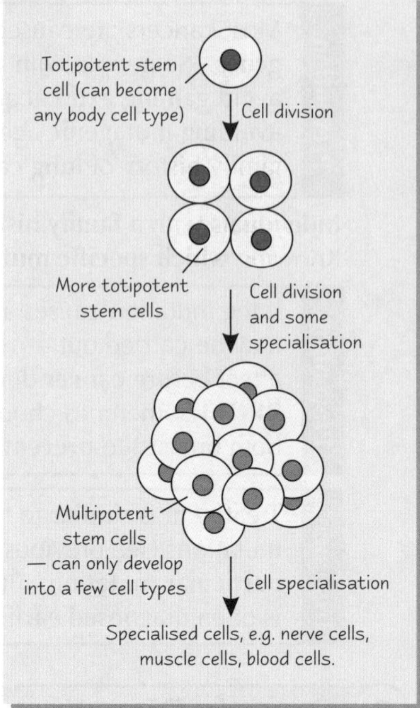

Plants Contain **Totipotent Stem Cells**

1) Mature **plants** also have **stem cells** — they're found in areas where the plant is **growing**, e.g. in roots and shoots.
2) All stem cells in plants are **totipotent** — they can mature into any **cell type**.
3) This means they can be used to grow **plant organs** (e.g. roots) or **whole new plants *in vitro*** (artificially). Growing plant tissue artificially is called **tissue culture** (see next page).

Stem Cells Become **Specialised** Because **Different Genes** are **Expressed**

Stem cells become **specialised** because during their development they only **transcribe** and **translate part** of their **DNA**:

1) **Stem cells** all contain the **same genes** — but during **development not all** of them are **transcribed** and **translated** (expressed).
2) Under the **right conditions**, some **genes** are **expressed** and others are switched off.
3) **mRNA** is only **transcribed** from **specific genes**.
4) The mRNA from these genes is then **translated** into **proteins**.
5) These proteins **modify** the cell — they determine the cell **structure** and **control cell processes** (including the expression of **more genes**, which produces more proteins).
6) **Changes** to the cell produced by these proteins cause the cell to become **specialised**. These changes are **difficult** to **reverse**, so once a cell has specialised it **stays** specialised.

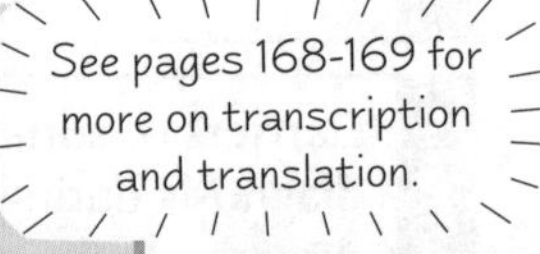
See pages 168-169 for more on transcription and translation.

All of the girls expressed different jeans.

EXAMPLE: RED BLOOD CELLS

1) **Red blood cells** are produced from a type of **stem cell** in the **bone marrow**. They contain lots of **haemoglobin** and have **no nucleus** (to make room for more haemoglobin).
2) The stem cell produces a new cell in which the genes for **haemoglobin production** are **expressed**. Other genes, such as those involved in **removing the nucleus**, are **expressed** too. Many other genes are expressed or switched off, resulting in a specialised red blood cell.

Stem Cells

Tissue Culture Can be Used to Grow Plants from a Single Totipotent Cell

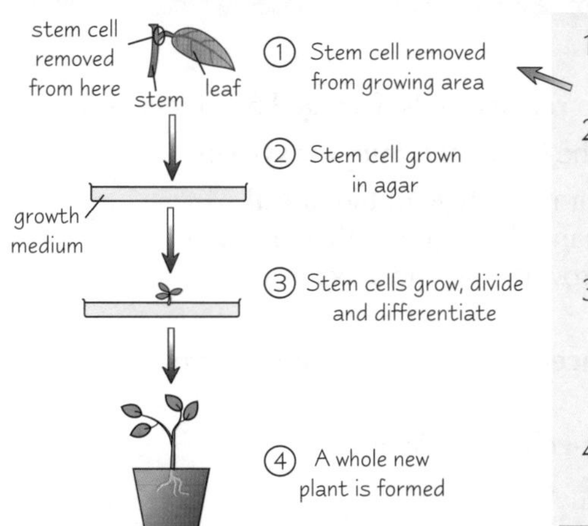

1) A **single totipotent stem cell** is taken from a **growing area** on a plant (e.g. a **root** or **shoot**).
2) The cell is placed in some **growth medium** (e.g. agar) that contains **nutrients** and **growth factors**. The growth medium is **sterile**, so microorganisms can't grow and compete with the plant cells.
3) The plant stem cell will **grow** and **divide** into a **mass** of **unspecialised** cells. If the **conditions** are **suitable** (e.g. the plant cells are given the **right growth factors**), the cells will **mature** (develop) into **specialised** cells.
4) The cells grow and specialise to form a **plant organ** or an **entire plant** depending on the **growth factors** used.

See p. 156 for more on growth factors.

You Need to be Able to Interpret Data About Tissue Culture

Here's an example of the kind of data that might crop up in the exam:

The **table** on the right shows the results of a **tissue culture experiment** — samples of **plant tissue** were taken from a **shoot** and grown on media with **varying ratios** of the growth factors **auxin** and **cytokinin**.

Growth medium	Ratio of auxin : cytokinin	Growth after 2 months
1	1 : 1	Growth but no specialised cells
2	1 : 25	Shoot formation
3	25 : 1	Root formation

Jacob had probably overdone the growth factors for this one.

1) Growth medium **1** showed **no specialised cell growth** — so an **equal ratio** (1 : 1) of auxin : cytokinin promotes **unspecialised cell growth**.
2) Growth medium **2** showed **shoot formation** — so a **high cytokinin : auxin ratio** (25 : 1) promotes the growth of **specialised shoot cells**.
3) Growth medium **3** showed **root formation** — so a **high auxin : cytokinin ratio** (25 : 1) promotes the growth of **specialised root cells**.

The results of this experiment show that the **ratio** of these **growth factors** helps to control the **specialisation** of different tissues in this plant.

Practice Questions

Q1 What is a stem cell?

Q2 Where are stem cells found in plants?

Exam Question

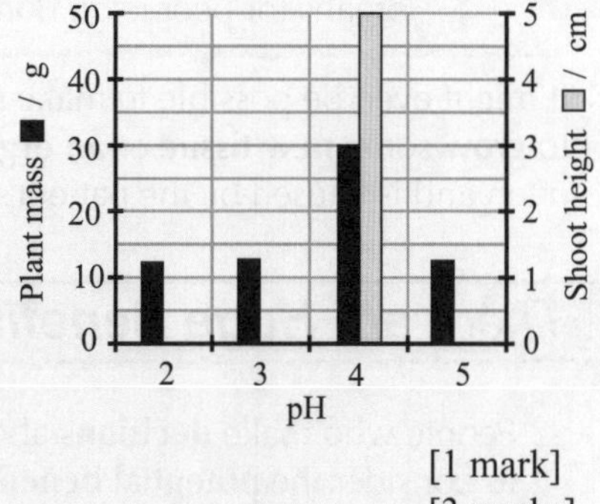

Q1 Samples of plant tissue containing totipotent cells were taken from a stem and placed on growth media of different pH values. Each tissue sample weighed less than 1 g at the start of the experiment. The graph shows the results.

a) What are totipotent cells? [1 mark]

b) Describe and explain the results. [3 marks]

A tissue culture — what you need when you have a cold...

Jokes aside, all your biology knowledge is going to stem from some good old revision... Get it...? Stem... Sorry, I couldn't resist that one. But I mean it now — you need to know all about stem cells and how they become specialised to carry out a particular function. When you've got that straight, take a look at the tissue culture stuff. Plants are très exciting.

Stem Cells in Medicine

Like I said before, stem cells really are the daddy of all cells because they can divide and turn into other cell types. And it's the ability of stem cells to produce other cell types that's got scientists and doctors fairly excited...

Some **Stem Cell Therapies Already Exist**

1) Stem cells can divide into **other cell types**, so they could be used to **replace** cells **damaged** by illness or injury.
2) Some stem cell therapies **already exist** for some diseases affecting the **blood** and **immune system**.
3) **Bone marrow** contains **stem cells** that can become specialised to form **any type** of **blood cell**. **Bone marrow transplants** can be used to replace the **faulty** bone marrow in patients that produce **abnormal blood cells**. The stem cells in the transplanted bone marrow **divide** and **specialise** to produce healthy blood cells.
4) This technique has been used successfully to treat **leukaemia** (a **cancer** of the blood or bone marrow) and **lymphoma** (a cancer of the **lymphatic system**).
5) It has also been used to treat some **genetic disorders**, such as **sickle-cell anaemia** and **severe combined immunodeficiency** (**SCID**):

Example

Severe combined immunodeficiency (**SCID**) is a genetic disorder that affects the immune system. People with SCID have a **poorly functioning immune system** as their **white blood cells** (made in the bone marrow from stem cells) are **defective**. This means they **can't defend** the body against infections by identifying and destroying microorganisms. So SCID sufferers are **extremely susceptible** to **infections**. Treatment with a **bone marrow transplant** replaces the faulty bone marrow with donor bone marrow that contains **stem cells without** the **faulty genes** that cause SCID. These then **differentiate** to produce **functional** white blood cells. These cells can identify and destroy invading pathogens, so the **immune system functions properly**.

Stem Cells Could be Used to **Treat Other Diseases**

As **stem cells** can divide into specialised cell types, scientists think they could be used to **replace damaged tissues** in a **range** of **diseases**. Scientists are **researching** the use of stem cells as **treatment** for lots of conditions, including:

- **Spinal cord injuries** — stem cells could be used to replace damaged **nerve tissue**.
- **Heart disease** and **damage caused by heart attacks** — stem cells could be used to replace damaged **heart tissue**.
- **Bladder conditions** — stem cells could be used to grow **whole bladders**, which are then **implanted** in patients to replace diseased ones.
- **Respiratory diseases** — **donated windpipes** can be stripped down to their simple collagen structure and then covered with **tissue** generated by stem cells. This can then be **transplanted** into patients.
- **Organ transplants** — organs could be **grown** from stem cells to provide new organs for people on **donor waiting lists**.

These treatments aren't available yet but some are in the early stages of clinical trials.

It might even be possible to make **stem cells genetically identical** to a **patient's own cells**. These could then be used to **grow** some **new tissue** or **an organ** that the patient's body **wouldn't reject** (rejection of transplants occurs quite often and is caused by the patient's immune system recognising the tissue as **foreign** and **attacking it**).

There are **Huge Benefits** to Using **Stem Cells** in **Medicine**

People who make **decisions** about the **use** of stem cells to treat human disorders have to consider the **potential benefits** of stem cell therapies:

- They could **save** many **lives** — e.g. many people waiting for organ transplants **die** before a **donor organ** becomes available. Stem cells could be used to **grow organs** for those people awaiting transplants.
- They could **improve** the **quality of life** for many people — e.g. stem cells could be used to replace damaged cells in the eyes of people who are **blind**.

Stem Cells in Medicine

Human **Stem Cells** Can Come from **Adult Tissue** or **Embryos**

To **use stem cells** scientists have to get them from somewhere. There are **two** potential **sources** of human stem cells:

1 Adult stem cells

1) These are obtained from the **body tissues** of an **adult**. For example, adult stem cells are found in **bone marrow**.
2) They can be obtained in a relatively **simple operation** — with very **little risk** involved, but quite **a lot of discomfort**.
3) Adult stem cells **aren't** as **flexible** as embryonic stem cells — they can only specialise into a **limited** range of cells, not all body cell types (they're **multipotent**). Although scientists are **trying** to find ways to make adult stem cells **specialise** into **any cell type**.

2 Embryonic stem cells

1) These are obtained from **embryos** at an **early stage of development**.
2) Embryos are created in a **laboratory** using ***in vitro*** **fertilisation** (IVF) — **egg cells** are **fertilised** by sperm **outside the womb**.
3) Once the embryos are approximately **4 to 5 days old**, **stem cells** are **removed** from them and the rest of the embryo is **destroyed**.
4) Embryonic stem cells can develop into **all types** of specialised cells (they're **totipotent**).

There are **Ethical Issues Surrounding Stem Cell Use**

1) Obtaining stem cells from **embryos** created by IVF raises **ethical issues** because the procedure results in the **destruction** of an embryo that **could** become a fetus if placed in a **womb**.
2) Some people believe that at the moment of **fertilisation** an **individual** is formed that has the **right** to **life** — so they believe that it's **wrong** to **destroy** embryos.
3) Some people have **fewer objections** to stem cells being **obtained** from **unfertilised embryos** — embryos made from **egg cells** that **haven't** been fertilised by sperm. This is because the embryos **couldn't survive** past a few days and **wouldn't** produce a fetus if placed in a womb.
4) Some people think that **scientists** should **only use** adult stem cells because their production **doesn't** destroy an embryo. But adult stem cells **can't** develop into all the specialised cell types that embryonic stem cells can.
5) The decision makers in **society** have to take into account **everyone's views** when making decisions about **important scientific work** like stem cell research and its use to treat human disorders.

Practice Questions

Q1 What types of cells can bone marrow stem cells produce?

Q2 Name two conditions that stem cells could potentially be used to treat.

Q3 Describe one difference between embryonic and adult stem cells.

Exam Questions

Q1 Explain one way in which stem cell therapy is currently being used. [4 marks]

Q2 Explain why some people object to the use of embryonic stem cells in treating human disorders. [2 marks]

It's OK — you can grow yourself a new brain especially for this revision...

And that's the end of this section — whoopdidoo. It was a whopper — in size, I mean, not as in the famous 1980s chewy bars. Before you zoom off to something else (because I know you can't wait), take the time to learn all of the pages, including these last two. It might take a little while, but you'll be glad of it when it comes to the exam. Promise.

Making DNA Fragments

You might have just done a section about genetics, but the good stuff's all in here. Three, two, one... go.

Gene Technology — Techniques Used to Study Genes

Gene technology is basically all the **techniques** that can be used to **study genes** and their **function**. Examples include:

- The **polymerase chain reaction** (**PCR**) — produces **lots** of **identical copies** of a specific **gene** (see next page).
- ***In vivo* gene cloning** — also produces **lots** of **identical copies** of a specific **gene** (see page 184).
- **DNA probes** — used to **identify** specific genes (see page 190).

Scientists use these techniques to do many things (as well as study genes) — e.g. they use them for **genetic engineering** (see p. 186), **DNA fingerprinting** (see p. 188), **diagnosing diseases** (see p. 189) and **treating genetic disorders** (see p. 193).

Gene Technology Uses DNA Fragments

As gene technology is all about **studying genes**, a good place to start is learning how to **get a copy** of the **DNA fragment** containing the gene you're **interested** in (the **target gene**). There are **three ways** that DNA fragments can be produced:

1 Using Reverse Transcriptase

1) Many **cells** only contain **two copies** of each gene, making it **difficult** to obtain a DNA fragment containing the target gene. But they can contain **many mRNA** molecules (see p. 168) complementary to the gene, so mRNA is often **easier** to obtain.
2) The mRNA molecules can be used as **templates** to **make lots of DNA.** The **enzyme reverse transcriptase makes DNA** from an RNA template. The DNA produced is called **complementary DNA (cDNA)**.
3) For example, **pancreatic cells** produce the protein **insulin**. They have loads of mRNA molecules complementary to the **insulin gene**, but only **two copies** of the gene **itself**. So reverse transcriptase could be used to **make cDNA** from the **insulin mRNA**.
4) To do this, **mRNA** is first isolated from cells. Then it's **mixed** with **free DNA nucleotides** and **reverse transcriptase**. The reverse transcriptase uses the mRNA as a **template** to synthesise a **new strand** of cDNA.

2 Using Restriction Endonuclease Enzymes

1) Some sections of DNA have **palindromic** sequences of **nucleotides**. These sequences consist of **antiparallel base pairs** (base pairs that read the **same** in **opposite directions**).

palindromic sequence
Reads: GAATTC→
C G A A T T C G A
G C T T A A G C T
double-stranded DNA
← Reads: GAATTC

2) **Restriction endonucleases** are enzymes that **recognise specific** palindromic sequences (known as **recognition sequences**) and **cut** (**digest**) the DNA at these places.
3) Different restriction endonucleases cut at **different specific** recognition sequences, because the **shape** of the recognition sequence is **complementary** to an enzyme's **active site**. E.g. the restriction endonuclease *Eco*RI cuts at GAATTC, but *Hind*III cuts at AAGCTT.
4) If recognition sequences are present at **either side** of the DNA fragment you want, you can use restriction endonucleases to **separate** it from the rest of the DNA.
5) The DNA sample is **incubated** with the specific restriction endonuclease, which **cuts** the DNA fragment out via a **hydrolysis reaction**.
6) Sometimes the cut leaves **sticky ends** — **small tails** of **unpaired bases** at **each end** of the fragment. Sticky ends can be used to **bind** (**anneal**) the DNA fragment to another piece of DNA that has sticky ends with **complementary sequences** (there's more about this on p. 184).

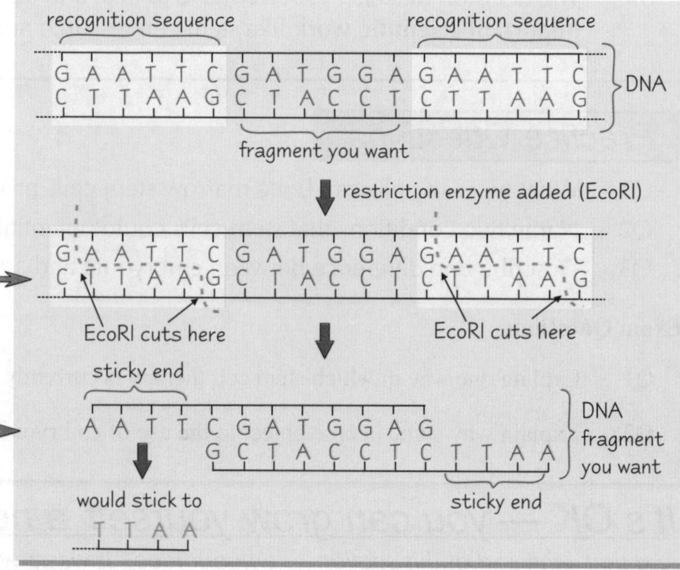

Making DNA Fragments

3 Using the **Polymerase Chain Reaction (PCR)**

The **polymerase chain reaction** (PCR) can be used to make **millions of copies** of a fragment of DNA in just a few hours. PCR has **several stages** and is **repeated** over and over to make lots of copies:

1) A reaction mixture is set up that contains the **DNA sample, free nucleotides, primers** and **DNA polymerase.**
   - **Primers** are short pieces of DNA that are **complementary** to the bases at the **start** of the fragment you want.
   - **DNA polymerase** is an **enzyme** that creates new DNA strands.
2) The DNA mixture is **heated** to **95 °C** to break the **hydrogen bonds** between the two strands of DNA.
3) The mixture is then **cooled** to between **50** and **65 °C** so that the primers can **bind** (**anneal**) to the strands.

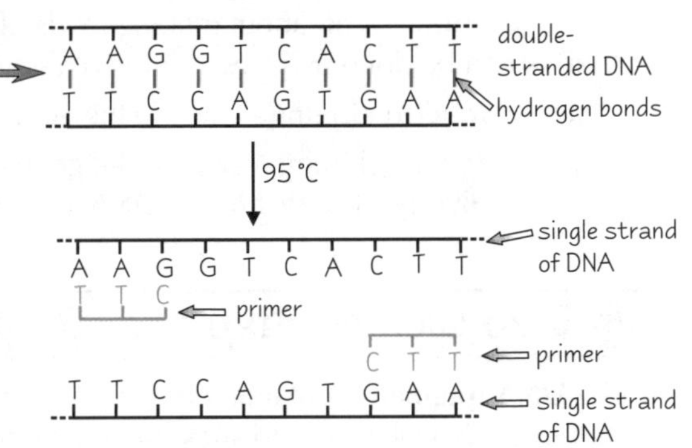

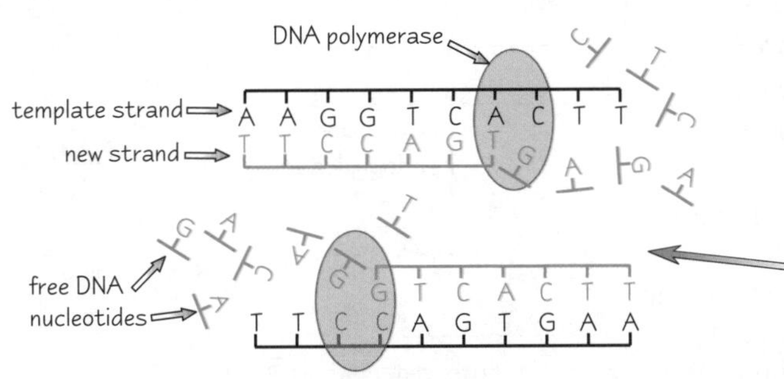

4) The reaction mixture is heated to **72 °C**, so **DNA polymerase** can **work**.
5) The DNA polymerase **lines up** free DNA nucleotides **alongside each template strand.** Specific **base pairing** means **new complementary strands** are formed.

6) **Two new copies** of the fragment of DNA are formed and **one cycle** of PCR is **complete.**
7) The cycle starts again, with the mixture being heated to 95 °C and this time **all four strands** (two original and two new) are used as **templates.**
8) Each PCR cycle **doubles** the amount of DNA, e.g. **1st cycle = 2 × 2 = 4 DNA fragments, 2nd cycle = 4 × 2 = 8 DNA fragments, 3rd cycle = 8 × 2 = 16 DNA fragments**, and so on.

PCR produces loads of identical copies of DNA, so it can be used to clone genes — this is called *in vitro* cloning.

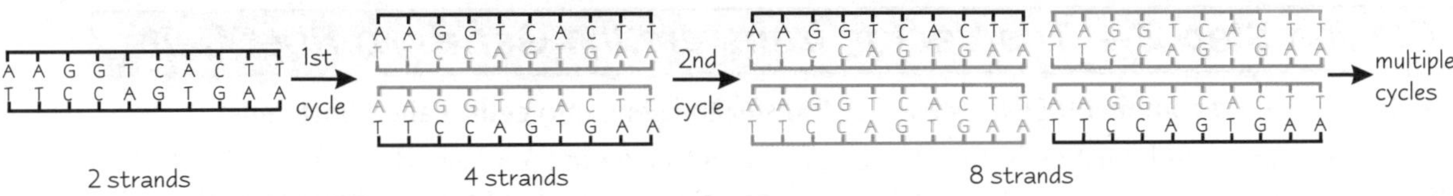

Practice Questions

Q1 Name three ways a DNA fragment can be produced.

Q2 What is reverse transcriptase?

Q3 What are sticky ends?

Q4 What does PCR stand for?

Q5 What is a primer?

G G A T C C G T T T C A G G A T C C
C C T A G G C A A A G T C C T A G G

DNA fragment wanted

Exam Question

Q1 A fragment of DNA (shown above) needs to be isolated from some bacterial DNA. The restriction endonuclease *Bam*HI recognises the sequence GGATCC and cuts between G and G.

a) Explain how *Bam*HI could be used to isolate the DNA fragment. [2 marks]

b) Describe and explain how to produce multiple copies of this DNA fragment using PCR. [5 marks]

Sticky ends — for once a name that actually makes sense.

Okay, your eyes might have gone funny from seeing so many nucleotides on these two pages. But once you've recovered, it's really important to go over these pages as many times as you need to, 'cause examiners love throwing in a few questions on restriction enzymes or PCR. Bless 'em — examiners get excited about the strangest things.

Gene Cloning

Hmmmmm... nope, can't think of any exciting or funny ways to start this double page. Sorry.

Gene Cloning can be done *In Vitro* or *In Vivo*

In vitro is Latin for within glass. *In vivo* is Latin for within the living.

Gene cloning is all about making loads of **identical copies** of a gene.
This can be done using **two** different techniques:

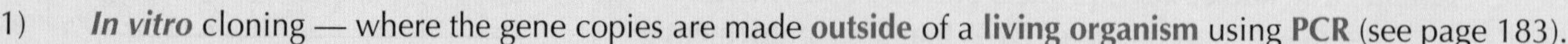

1) ***In vitro*** cloning — where the gene copies are made **outside** of a **living organism** using **PCR** (see page 183).
2) ***In vivo*** cloning — where the gene copies are made **within** a **living organism**. As the organism **grows** and **divides**, it **replicates** its **DNA**, creating multiple copies of the gene (see below).

In Vivo Cloning *Step 1* — *The Gene* is *Inserted* into a *Vector*

The **DNA fragment** containing the **target gene** has been isolated using one of the techniques on pages 182-183. The first step in *in vivo* cloning is to **stick** the fragment **into** a **vector** using **restriction endonuclease** and **ligase** (an enzyme):

1) The DNA fragment is inserted into vector DNA — a **vector** is something that's used to **transfer DNA** into a **cell**. They can be **plasmids (small, circular molecules** of DNA in **bacteria**) or **bacteriophages** (**viruses** that **infect** bacteria).
2) The vector DNA is **cut open** using the **same** restriction endonuclease that was used to **isolate** the DNA fragment containing the target gene (see p. 182). So the **sticky ends** of the vector are **complementary** to the sticky ends of the DNA fragment containing the gene.
3) The vector DNA and DNA fragment are **mixed together** with **DNA ligase**. DNA ligase **joins** the sticky ends of the DNA fragment to the sticky ends of the vector DNA. This process is called **ligation**.

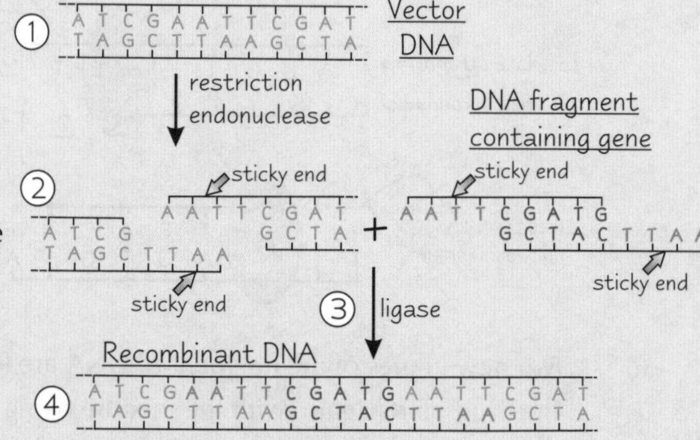

4) The new combination of bases in the DNA (vector DNA + DNA fragment) is called **recombinant DNA**.

In Vivo Cloning *Step 2* — The *Vector Transfers* the *Gene* into *Host Cells*

1) The **vector** with the **recombinant DNA** is used to **transfer** the gene into **cells** (called **host** cells).
2) If a **plasmid vector** is used, **host cells** have to be **persuaded** to **take in** the plasmid vector and its DNA. E.g. host bacterial cells are placed into ice-cold **calcium chloride** solution to make their cell walls more **permeable**. The **plasmids** are **added** and the mixture is **heat-shocked** (heated to around **42 °C** for **1-2 minutes**), which encourages the cells to take in the plasmids.
3) With a **bacteriophage** vector, the bacteriophage will **infect** the host bacterium by **injecting** its **DNA** into it. The phage DNA (with the target gene in it) then **integrates** into the bacterial DNA.
4) **Host cells** that **take up** the vectors containing the gene of interest are said to be **transformed**.

In Vivo Cloning *Step 3* — *Identifying Transformed* Host Cells

Not all host cells will have **taken up** the vector and its DNA. **Marker genes** can be used to **identify** the **transformed** cells:

1) **Marker genes** can be inserted into vectors at the **same time** as the gene to be cloned. This means any **transformed host cells** will contain the gene to be cloned **and** the marker gene.
2) Host cells are **grown** on **agar plates** and each cell **divides** and **replicates** its DNA, creating a **colony** of **cloned cells**.
3) Transformed cells will produce colonies where **all the cells** contain the cloned gene and the marker gene.
4) The marker gene can code for **antibiotic resistance** — host cells are grown on agar plates **containing** the specific **antibiotic**, so **only** transformed cells that have the **marker gene** will **survive** and **grow**.
5) The marker gene can code for **fluorescence** — when the agar plate is placed under a **UV light** **only** transformed cells will **fluoresce**.
6) **Identified** transformed cells are allowed to **grow more**, producing **lots** and **lots** of **copies** of the **cloned gene**.

Gene Cloning

There are Advantages and Disadvantages to Both In Vivo and In Vitro Cloning

Depending on the **reason** why you want to clone a gene, you can choose to do it either by ***in vivo* cloning** or ***in vitro* cloning**. Both techniques have **advantages** and **disadvantages**:

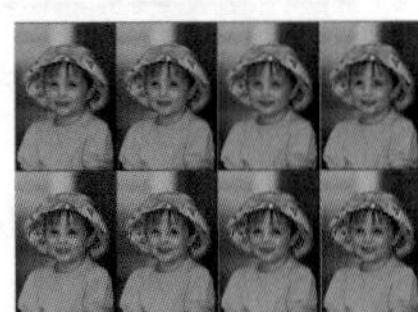

Either the cloning had worked, or Professor Dim's eyesight had gone.

In Vivo Cloning

1) Cloning *in vivo* can produce **mRNA** and **protein** as well as DNA because it's done in a **living cell** (which has the ribosomes and all the enzymes needed to produce them).
2) Cloning *in vivo* can also produce **modified DNA**, **modified mRNA** or **modified protein** — they have **modifications added** to them, e.g. **sugar** or **methyl** ($-CH_3$) groups.
3) **Large fragments** of DNA can be cloned using *in vivo* cloning, e.g. between **20 to 45 kilobases** of DNA can be inserted into some **plasmids** and **bacteriophages**.
4) *In vivo* cloning can be a **relatively cheap method**, depending on **how much** DNA you want to produce.

In vivo cloning also has **disadvantages** — the DNA fragment has to be **isolated** from other cell components, you may **not want modified** DNA, and it can be quite a **slow** process (because some types of bacteria **grow quite slowly**).

A kilobase is 1000 nucleotide bases. It's often shortened to kb.

In Vitro Cloning (PCR)

1) *In vitro* cloning can be used to produce **lots** of **DNA** (but **not** mRNA or protein).
2) The DNA produced **isn't modified** (see above) — an advantage if you **don't want** it to be modified.
3) This technique **only** replicates the **DNA fragment** of **interest** (e.g. the target gene). This means that you **don't** have to **isolate** the DNA fragment from **host DNA** or **cell components**.
4) *In vitro* cloning is a **fast** process — PCR can clone **millions of copies** of DNA in just a **few hours**.

In vitro cloning also has **disadvantages** — it can only replicate a **small DNA fragment** (compared to *in vitro*), you may want a **modified** product, **mRNA** and **protein** aren't made as well, and it can be **expensive** if you want to produce a lot of DNA.

Practice Questions

Q1 What is *in vitro* gene cloning?

Q2 What is a vector?

Q3 Other than a plasmid, give an example of a vector.

Q4 Name the type of enzyme that can be used to cut DNA.

Q5 What is the name of the type of DNA formed from vector DNA and an inserted DNA fragment?

Q6 What is a marker gene?

Q7 Give two advantages of *in vivo* cloning.

Q8 *In vitro* cloning is a slow process — is this statement correct?

Agar plate under UV light

agar plate

colony A

colony B

colony not visible

colony visible

Exam Question

Q1 A scientist has cloned a gene by transferring a plasmid containing the target gene and a fluorescent marker gene into some bacterial cells. The cells were grown on an agar plate. The plate was then placed under UV light (see above).

a) Explain why the scientist thinks colony A contains transformed host cells, but colony B doesn't. [2 marks]

b) Explain how the scientist might have inserted the target gene into the plasmid. [3 marks]

Transformed boyfriends — made to listen, tidy up and agree with you...

This page is quite scary I know. But don't worry, it's not as difficult as photosynthesis or respiration — you just need to keep going over the steps of all these different techniques until they make sense. And they do make sense really, promise. I know I've said it before, but drawing out the diagram will help — then you'll know inserting DNA into vectors like a pro.

Genetic Engineering

Now that you know how to make a DNA fragment and clone a gene, it's probably a good time to tell you why you might want to. Don't worry — it's not evil stuff, but I promise to do my evil laugh. Mwah ha hah.

Genetic Engineering is the Manipulation of an Organism's DNA

1) **Genetic engineering** is also known as **recombinant DNA technology**.
2) Organisms that have had their **DNA altered** by genetic engineering are called **transformed organisms**.
3) These organisms have **recombinant DNA** — **DNA** formed by **joining together** DNA from **different sources**.
4) **Microorganisms**, **plants** and **animals** can all be **genetically engineered** to **benefit humans**.
5) **Transformed microorganisms** can be made using the same technology as ***in vivo*** cloning (see page 184). For example, **foreign DNA** can be **inserted** into **microorganisms** to produce **lots** of **useful protein**, e.g. insulin:

Transformed organisms are also known as genetically engineered or genetically modified organisms.

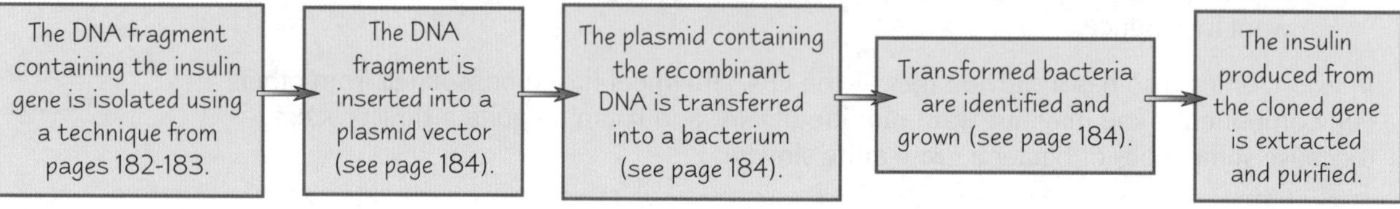
The DNA fragment containing the insulin gene is isolated using a technique from pages 182-183.
The DNA fragment is inserted into a plasmid vector (see page 184).
The plasmid containing the recombinant DNA is transferred into a bacterium (see page 184).
Transformed bacteria are identified and grown (see page 184).
The insulin produced from the cloned gene is extracted and purified.

6) **Transformed plants** can also be produced — a gene that codes for a **desirable characteristic** is inserted into a **plasmid**. The plasmid is added to a **bacterium** and the bacterium is used as a **vector** to get the gene into the **plant cells**. The transformed plant will have the desirable characteristic coded for by that gene.
7) **Transformed animals** can be produced too — a gene that codes for a **desirable characteristic** is inserted into an **animal embryo**. The transformed animal will have the desirable characteristic coded for by that gene.

Transformed (Genetically Engineered) Organisms can Benefit Humans

Producing **transformed organisms** (microorganisms, plants and animals) can benefit **humans** in lots of ways:

1 Agriculture

- **Agricultural crops** can be **transformed** so that they give **higher yields** or are **more nutritious**. This means these plants can be used to reduce the risk of **famine** and **malnutrition**. Crops can also be transformed to have **pest resistance**, so that **fewer pesticides** are needed. This **reduces costs** and reduces any **environmental problems** associated with using pesticides.
- For example, ***Golden Rice*** is a variety of **transformed rice**. It contains **one gene** from a **maize plant** and **one gene** from a **soil bacterium**, which together enable the rice to produce **beta-carotene**. The beta-carotene is used by our bodies to produce **vitamin A**. *Golden Rice* is being developed to **reduce vitamin A deficiency** in areas where there's a **shortage** of **dietary vitamin A**, e.g. **south Asia**, **Africa**. Vitamin A deficiency is a big problem in these areas, e.g. up to **500 000 children per year** worldwide go **blind** due to vitamin A deficiency.

2 Industry

- **Industrial processes** often use **biological catalysts (enzymes)**. These enzymes can be produced from **transformed organisms**, so they can be produced in **large quantities** for **less money, reducing costs.**
- For example, **chymosin** (or **rennin**) is an enzyme used in **cheese-making**. It used to be made from **rennet** (a substance produced in the **stomach of cows**), but it can now be produced by **transformed organisms**. This means it can be made in **large quantities**, relatively **cheaply** and **without killing** any **cows**, making some cheese suitable for **vegetarians**.

3 Medicine

- Many **drugs** and **vaccines** are produced by transformed organisms, using recombinant DNA technology. They can be made **quickly**, **cheaply** and in **large quantities** using this method.
- For example, **insulin** is used to treat **Type 1 diabetes** and used to come from **animals** (cow, horse or pig pancreases). This insulin **wasn't** human insulin though, so it **didn't work quite as well**. Human insulin is now made from **transformed microorganisms**, using a **cloned human insulin gene** (see above).

Genetic Engineering

Many People are **Concerned** About the Use of **Genetic Engineering**

There are **ethical**, **moral** and **social concerns** associated with the **use** of **genetic engineering**:

1 Agriculture

- **Farmers** might plant only **one type** of transformed crop (this is called **monoculture**). This could make the **whole crop vulnerable** to **disease** because the plants are **genetically identical**.
- Some people are concerned about the possibility of '**superweeds**' — weeds that are **resistant** to **herbicides**. These could occur if transformed crops **interbreed** with **wild plants**.

2 Industry

- **Without proper labelling**, some people think they **won't** have a **choice** about whether to consume food made using genetically engineered organisms.
- Some people are worried that the process used to **purify** proteins (from genetically engineered organisms) could lead to the introduction of **toxins** into the **food industry**.

3 Medicine

- Companies who **own** genetic engineering technologies may **limit** the **use** of technologies that could be **saving lives**.
- Some people worry this technology could be used **unethically**, e.g. to make **designer babies** (babies that have characteristics **chosen** by their parents). This is currently **illegal** though.

Humanitarians Think **Genetic Engineering** will **Benefit People**

Genetic engineering has **many** potential **humanitarian benefits**:

1) **Agricultural crops** could be produced that help **reduce** the risk of **famine** and **malnutrition**, e.g. **drought-resistant** crops for **drought-prone** areas.
2) **Transformed crops** could be used to produce **useful pharmaceutical products** (e.g. **vaccines**) which could make drugs **available** to **more people**, e.g. in areas where **refrigeration** (usually needed for **storing** vaccines) **isn't available**.
3) **Medicines** could be produced more **cheaply**, so more people can **afford** them.

You need to be able to balance the humanitarian benefits with opposing views from environmentalists and anti-globalisation activists.

But some **environmentalists** and **anti-globalisation activists** have concerns:

1) **Environmentalists** — Many **oppose** recombinant DNA technology because they think it could **potentially damage** the **environment**. E.g. transformed crops could encourage **farmers** to carry out monoculture (see above), which **decreases biodiversity**. There are also fears that if **transformed crops breed** with **wild plants** there'll be **uncontrolled spread** of **recombinant DNA**, with **unknown consequences**.
2) **Anti-globalisation activists** — These are people who **oppose globalisation** (e.g. the **growth** of **large multinational companies** at the **expense** of **smaller ones**). A **few**, **large** biotechnology companies **control** some forms of genetic engineering. As the **use** of this technology **increases**, these companies get **bigger** and **more powerful**. This may **force** smaller companies **out of business**, e.g. by making it **harder** for them to **compete**.

Practice Questions

Q1 What is recombinant DNA?

Q2 Give an example of a transformed agricultural crop.

Exam Question

Q1 A large agricultural company has isolated a gene from bacteria that may increase the drought resistance of wheat plants.
a) Briefly explain how this gene could be used to make a transformed wheat plant. [3 marks]
b) Suggest how the transformed wheat plants might be beneficial to humans. [2 marks]
c) Suggest why anti-globalisation activists may be against the use of this gene. [1 mark]

Neapolitan — recombinant ice cream...

Ahhh, sitting in the sun, licking an ice cream, exams all over. That's where you'll be in a few months' time. After revising all this horrible stuff that is. As genetic engineering advances, more questions will pop up about its implications. So it's a good idea to know all sides of the argument — you need to know them for the exam anyway.

Genetic Fingerprinting

100 years ago they were starting to identify people using their fingerprints, but now we can use their DNA instead.

Genomes Contain Repetitive, Non-Coding DNA Sequences

1) **Not all** of an organism's **genome** (all the genetic material in an organism) **codes** for **proteins**.
2) Some of the genome consists of **repetitive**, **non-coding base sequences** — base sequences that **don't** code for proteins and **repeat** next to each other over and over (sometimes thousands of times), e.g. CATGCATGCATGCATG is a repeat of the non-coding base sequence CATG.
3) The **number of times** these sequences are **repeated differs** from person to person, so the **length** of these sequences in nucleotides differs too. E.g. a **four** nucleotide sequence might be repeated **12 times** in one person = **48 nucleotides** (12 × 4), but repeated **16 times** in another person = **64 nucleotides** (16 × 4).
4) The repeated sequences occur in **lots of places** in the **genome**. The **number** of times a **sequence is repeated** (and so the number of nucleotides) at **different places** in their genome can be **compared** between **individuals** — this is called **genetic fingerprinting**.
5) The **probability** of **two individuals** having the **same** genetic fingerprint is **very low** because the **chance** of **two individuals** having the **same number** of sequence repeats at **each place** they're found in DNA is **very low**.

Electrophoresis Separates DNA Fragments to Make a Genetic Fingerprint

So **genetic fingerprints** can be **compared** between **different individuals**. Now you need to know how one is **made**:

1) A **sample** of **DNA** is obtained, e.g. from a person's **blood**, **saliva** etc.
2) **PCR** (see page 183) is used to make **many copies** of the **areas** of DNA that contain the repeated sequences — **primers** are used that bind to **either side** of these **repeats** and so the **whole** repeat is amplified.
3) You end up with **DNA fragments** where the **length** (in nucleotides) corresponds to the **number of repeats** the person has at each specific position, e.g. one person may have 80 nucleotides, another person 120.
4) A **fluorescent tag** is added to all the DNA fragments so they can be viewed under **UV light**.
5) The DNA fragments undergo **electrophoresis**:
   - The DNA mixture is placed into a **well** in a slab of **gel** and covered in a **buffer solution** that **conducts electricity**.
   - An **electrical current** is passed through the gel — DNA fragments are **negatively charged**, so they **move towards** the **positive electrode** at the far end of the gel.
   - **Small** DNA fragments move **faster** and **travel further** through the gel, so the DNA fragments **separate** according to **size**.
6) The DNA fragments are viewed as **bands** under **UV light** — this is the **genetic fingerprint**.
7) Two genetic fingerprints can be **compared** — e.g. if both fingerprints have a band at the **same location** on the **gel** it means they have the **same number** of **nucleotides** and so the same number of **sequence repeats** at that place — it's a **match**.

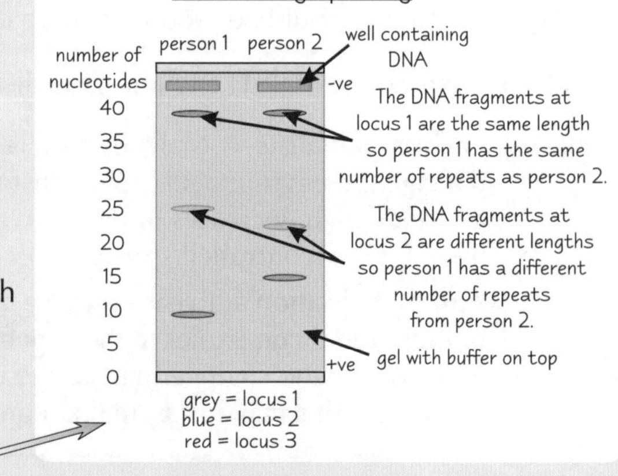

Genetic Fingerprinting is Used to Determine Relationships and Variability

Genetic fingerprinting has **many uses**, which include:

- **Determining genetic relationships** — We **inherit** the repetitive, non-coding base sequences from our **parents**. Roughly **half** of the sequences come from **each parent**. This means the **more bands** on a genetic fingerprint that match, the more **closely related** (**genetically similar**) two people are. E.g. **paternity tests** are used to determine the **biological father** of a child by comparing genetic fingerprints. If lots of bands on the fingerprint **match**, then that person is **most probably** the child's father. The **higher** the **number** of places in the genome compared, the more **accurate** the test result.
- **Determining genetic variability within a population** — The **greater** the **number of bands** that **don't** match on a genetic fingerprint, the more **genetically different** people are. This means you can **compare** the **number of repeats** at **several places** in the genome for a population to find out how **genetically varied** that population is. E.g. the **more** the **number of repeats** varies at **several places**, the **greater** the **genetic variability** within a population.

Genetic Fingerprinting

Genetic Fingerprinting can be Used in *Forensic Science...*

Forensic scientists use genetic fingerprinting to **compare** samples of **DNA** collected from **crime scenes** (e.g. DNA from **blood**, **semen**, **skin cells**, **saliva**, **hair** etc.) to samples of DNA from **possible suspects**, to **link them** to crime scenes.

1) The **DNA** is **isolated** from all the collected samples (from the crime scene and from the suspects).
2) Each sample is **replicated** using **PCR** (see p. 183).
3) The **PCR products** are run on an **electrophoresis gel** and the genetic fingerprints produced are **compared** to see if any match.
4) If the samples match, it **links** a **person** to the **crime scene**. E.g. this gel shows that the genetic fingerprint from **suspect C** **matches** that from the crime scene, **linking** them to the crime scene. All five bands match, so suspect C has the **same number** of repeats (nucleotides) at **five** different places.

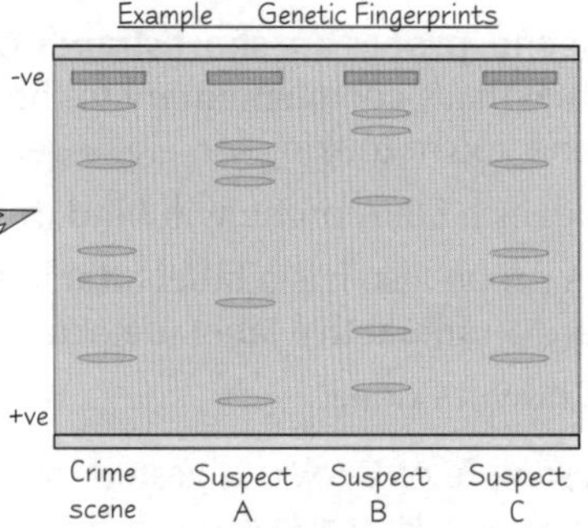

PCR amplifies the DNA, so enough is produced for it to be seen on the gel.

...Medical Diagnosis...

- In medical diagnosis, a genetic fingerprint can refer to a **unique pattern** of **several alleles**.
- It can be used to **diagnose genetic disorders** and **cancer**. It's useful when the **specific** mutation **isn't** known or where **several mutations** could have caused the disorder, because it identifies a **broader**, **altered** genetic pattern.

  EXAMPLE: **Preimplantation genetic haplotyping** (**PGH**) **screens embryos** created by **IVF** for genetic disorders **before** they're **implanted** into the uterus. The **faulty regions** of the **parents' DNA** are used to produce **genetic fingerprints**, which are **compared** to the genetic fingerprint of the **embryo**. If the fingerprints **match**, the embryo has **inherited** the **disorder**. It can be used to screen for **cystic fibrosis**, **Huntington's disease** etc.

  EXAMPLE: Genetic fingerprinting can be used to **diagnose sarcomas** (types of **tumour**). Conventional methods of identifying a tumour (e.g. biopsies) only show the **physical differences** between tumours. Now the **genetic fingerprint** of a known sarcoma (e.g. the **different mutated alleles**) can be **compared** to the genetic fingerprint of a **patient's tumour**. If there's a **match**, the sarcoma can be specifically **diagnosed** and the **treatment** can be targeted to that specific type (see page 193).

A specific mutation can be found using gene probes and sequencing (see p. 190-191).

...and Animal and Plant Breeding

Genetic fingerprinting can be used on **animals** and **plants** to **prevent inbreeding**, which causes **health**, **productivity** and **reproductive problems**. Inbreeding **decreases** the **gene pool** (the number of **different alleles** in a population, see p. 136), which can lead to an **increased risk** of **genetic disorders**, leading to **health problems** etc. Genetic fingerprinting can be used to **identify** how **closely-related** individuals are — the **more closely-related** two individuals are, the **more similar** their genetic fingerprint will be (e.g. **more bands** will **match**). The **least related** individuals will be **bred together**.

Practice Questions

Q1 Why are two people unlikely to have the same genetic fingerprint?

Q2 In gel electrophoresis, which electrode do DNA fragments move towards?

Q3 Why might genetic fingerprinting be used in forensic science?

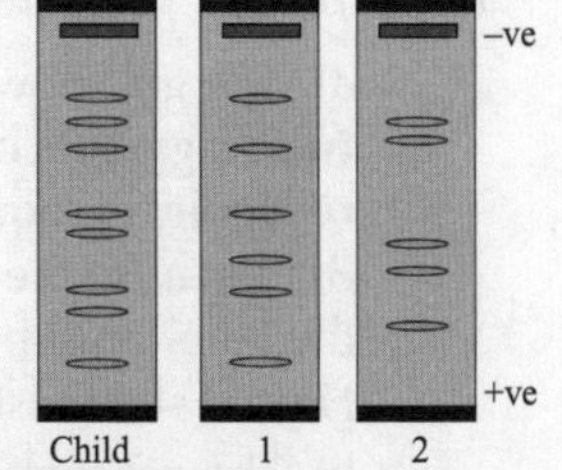

Exam Question

Q1 The diagram on the right shows three genetic fingerprints — one from a child and two from possible fathers.

a) Describe how the genetic fingerprint is made. [5 marks]

b) Which genetic fingerprint is most likely to be from the child's father? Explain your answer. [2 marks]

c) Give another use of genetic fingerprint technology. [1 mark]

Fingerprinting — in primary school it involved lots of paint and paper...

Who would have thought that tiny pieces of DNA on a gel would be that important? Well, they are and you need to know all about them. Make sure you know the theory behind fingerprinting as well as its applications. And remember, it's very unlikely that two people will have the same genetic fingerprint (except identical twins that is).

Locating and Sequencing Genes

We're gonna take gene technology to a whole new level now, with restriction mapping and DNA sequencing.

*You can **Look** for **Genes** Using **DNA Probes** and **Hybridisation***

1) DNA probes can be used to **locate genes** (e.g. on **chromosomes**) or see if a person's DNA **contains** a **mutated gene** (e.g. a gene that causes a **genetic disorder**).
2) **DNA** (gene) **probes** are **short strands** of **DNA**. They have a **specific base sequence** that's **complementary** to the base sequence of part of a **target gene** (the gene you're looking for, e.g. a gene that causes a genetic disorder).
3) This means a DNA probe will **bind** (**hybridise**) to the **target gene** if it's **present** in a **sample** of DNA.
4) A DNA probe also has a **label attached**, so that it can be **detected**. The two most common types of label are a **radioactive** label (detected using **X-ray film**) or a **fluorescent** label (detected using **UV light**).
5) Here's how it's done:

- A **sample** of **DNA** is **digested** into fragments using **restriction enzymes** (see page 182) and separated using electrophoresis (see page 188).
- The separated DNA fragments are then transferred to a **nylon membrane** and **incubated** with the **fluorescently labelled DNA probe**.
- If the gene is **present**, the **DNA probe** will **hybridise** (**bind**) to it.
- The **membrane** is then **exposed** to **UV light** and if the gene is present there will be a **fluorescent band**. E.g. **sample 3** has a visible band, so this patient has the **gene**.

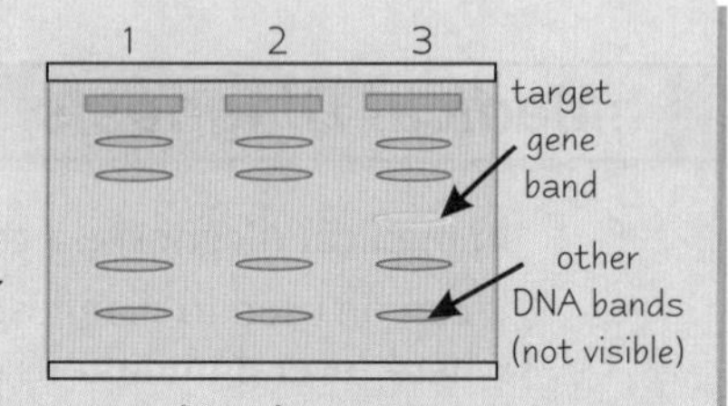

*The **Base Sequence** of a Gene can be **Determined** by **Restriction Mapping...***

As well as locating a **gene**, you might also want to know its **sequence** — this is done by **DNA sequencing** (see next page). But most genes are **too long** to be sequenced **all in one go**, so they're **cut** into **smaller sections** using **restriction enzymes**, then the smaller parts are sequenced. These smaller sections are then put back in the **correct order**, so the **entire gene sequence** can be **read** in the **right order** — **restriction mapping** can be used to do this:

1) **Different restriction enzymes** are used to **cut** labelled DNA into fragments (see page 182).
2) The DNA fragments are then **separated** by **electrophoresis** (see page 188).
3) The **size** of the **fragments** produced is used to **determine** the **relative locations** of **cut sites**.
4) A **restriction map** of the **original DNA** is made — a **diagram** of the piece of **DNA** showing the **different cut sites**, and so where the recognition sites of the restriction enzymes used are found.

Here's an example:

1) Some DNA, **10 kilobases** long (1 kb = **1000 nucleotides**) was **radioactively labelled**.
2) The DNA was **digested** using **two restriction enzymes**, ***Hind*III** and ***Eco*RI**, and the digested fragments were **separated** using **electrophoresis**.
3) The gel was used to build up a **restriction map** of the original DNA:

   a) The gel shows that the DNA was cut into **two fragments** by ***Hind*III**, so there's **one** *Hind*III **recognition sequence** in one of two places. But because the **2 kb piece** is **radioactive**, the label must be on the 2 kb piece. So the *Hind*III site must be 2 kb from the label.

   b) The gel shows that the DNA was cut into **two fragments** both **5 kb** long by ***Eco*RI**, so there's **one** *Eco*RI **recognition sequence** in the middle of the piece.

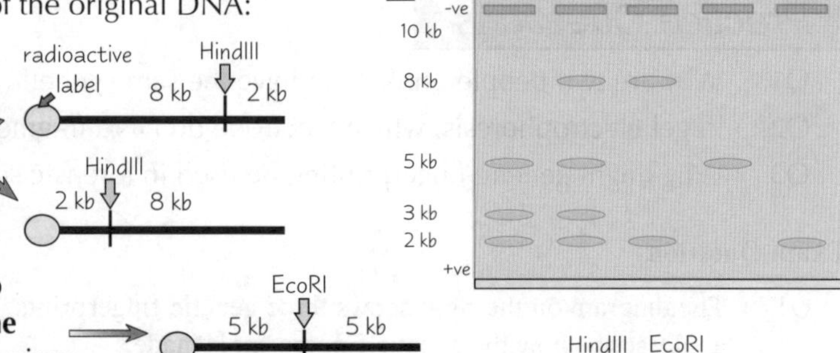

   c) Finally, putting both of these together, the **complete restriction map** must be:

4) The restriction map matches the fragments of the **total digest** (where both enzymes are present and the DNA is cut at **all** of the **recognition sequences** present) — the **radioactive label** is on the **2 kb *Hind*III** piece.
5) A **partial digest** is where the restriction enzymes **haven't** been **left long enough** to **cut** at all of their **recognition sequences**, producing **fragments** of **other lengths**, e.g. if *Eco*RI **doesn't** cut there'll be an **8 kb** fragment produced.

Locating and Sequencing Genes

... and Gene Sequencing

Gene sequencing is used to determine the **order** of **bases** in a section of **DNA** (gene). It can be carried out by the **chain termination method**, which lets you sequence **small fragments** of DNA, up to **750 base pairs**:

1) The following mixture is added to **four separate** tubes:
   - A **single-stranded DNA template** — the DNA to be sequenced.
   - **DNA polymerase** — the enzyme that joins DNA nucleotides together.
   - Lots of **DNA primer** — short pieces of DNA (see p. 183).
   - **Free nucleotides** — lots of free A, T, C and G nucleotides.
   - **Fluorescently-labelled modified nucleotide** — like a regular nucleotide, but once it's added to a DNA strand, **no more** bases can be added after it. A **different** modified nucleotide is added to **each** tube (**A*, T*, C*, G***).

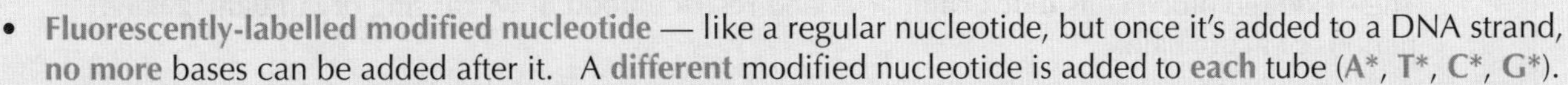

Jane and her friends had already made their costumes for the sequin-cing lesson.

2) The tubes undergo **PCR**, which produces many **strands of DNA**. The strands are **different lengths** because each one **terminates** at a **different point** depending on where the modified nucleotide was added.
3) For example, in tube A (with the **modified adenine** nucleotide **A***) sometimes A* is **added** to the DNA at point 4 **instead** of A, **stopping** the **addition** of any more bases (the strand is **terminated**). Sometimes A is added at point 4, then **A*** is added at **point 5**. Sometimes A is added at **point 4**, A again at point 5, G at point 6 and **A*** is added at **point 7**. So strands of **three different lengths** (4 bases, 5 bases and 7 bases) all ending in **A*** are produced.
4) The DNA fragments in each tube are separated by **electrophoresis** and **visualised** under **UV light** (because of the **fluorescent label**).
5) The **complementary base sequence** can be **read** from the gel. The **smallest** nucleotide (e.g. one base) is at the **bottom** of the gel. Each band after this represents **one more base** added. So by reading the bands **from the bottom** of the gel **to the top**, you can build up the **DNA sequence** one base at a time.

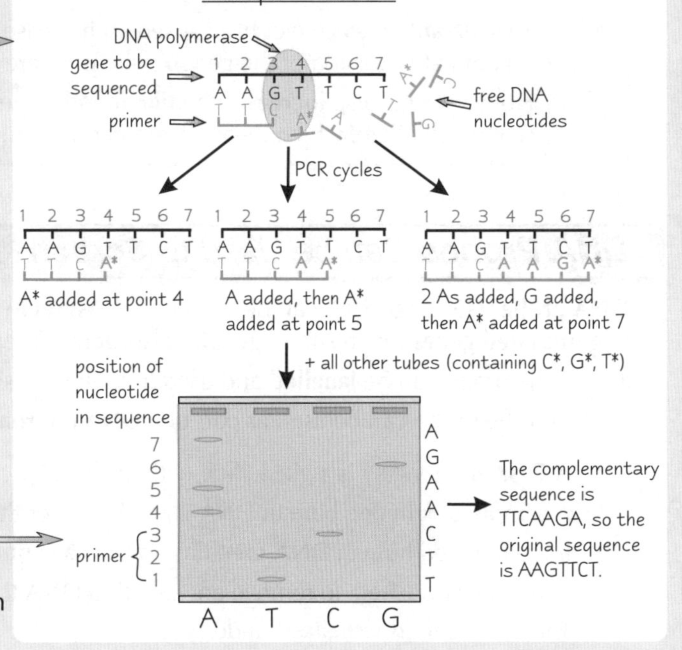

Nowadays sequencing is done **altogether** in **one tube** in an **automated DNA sequencer**. The tube contains **all** the modified nucleotides, each with a **different coloured** fluorescent label, and a machine reads the sequence for you.

Practice Questions

Q1 What is a DNA probe?

Q2 Name the two techniques used together to determine the sequence of a gene.

Exam Question

Size of fragments | Total digest | Partial digest | Radioactive fragments of partial digest
7 kb, 6 kb, 4 kb, 3 kb, 2 kb

Q1 A piece of DNA 9 kb long is labelled at one end with a fluorescent nucleotide marker. The DNA is then digested using the restriction enzyme *Sal*I. The resulting DNA fragments are separated by electrophoresis to obtain the gel on the right.

a) How many times did the recognition sequence for *Sal*I appear in the original piece of DNA? [1 mark]

b) Use the gel to produce a restriction map of the DNA piece. [3 marks]

c) Explain why there are more DNA fragments in the partial digest lane than in the total digest lane. [2 marks]

Restriction mapping — I'm not very good with coordinates...

Okay, I won't deny it — there are a couple of difficult bits on this page, like restriction mapping. But just keep drawing the diagrams until the DNA pieces fall into place. All the techniques on this double page are perfect for exam questions, so make sure you learn them well. That way you'll be prepared for anything... in the exam that is.

DNA Probes in Medical Diagnosis

So a scientist could sequence a gene just for the fun of it — or they could do it to help diagnose a genetic disorder.

*Many **Human Diseases** are **Caused** by **Mutated Genes***

Some **mutated genes** can cause **diseases** such as **genetic disorders** and **cancer** (see page 175).
Other mutations can produce genes that are **useful** in **some situations** but **not** in others. For example:

Sickle-cell Anaemia

- Sickle-cell anaemia is a **recessive genetic disorder** caused by a **mutation** in the **haemoglobin gene**.
- The mutation causes an **altered haemoglobin protein** to be produced, which makes **red blood cells sickle-shaped** (**concave**).
- The **sickle** red blood cells **block capillaries** and **restrict blood flow**, causing **organ damage** and periods of **acute pain**.
- Some people are **carriers** of the disease — they have **one normal** and **one mutated copy** of the haemoglobin gene.
- Sickle-cell carriers are partially **protected** from **malaria**.
- This **advantageous** effect has caused an increase in the frequency of the sickle-cell **allele** (the mutated version of the gene) in areas where malaria is **common** (e.g. parts of Africa).
- However, this also **increases** the **likelihood** of people in those areas inheriting **two copies** of the sickle-cell allele, which means more people will **suffer** from the disease in these areas.

DNA Probes** Can be Used to **Screen** for **Mutated Genes

DNA probes (see page 190) can be used to look (**screen**) for clinically important genes (e.g. **mutated genes** that result in **genetic disorders**). There are two ways to do this:

1) The probe can be **labelled** and used to look for a **single gene** in a sample of DNA, as shown on page 190.
2) Or the probe can be used as part of a **DNA microarray**, which can screen **lots** of **genes** at the **same time**:

- A **DNA microarray** is a **glass slide** with **microscopic spots** of **different** DNA probes **attached** to it in **rows**.
- A sample of **labelled human DNA** is washed over the array.
- If the labelled human DNA **contains** any **DNA sequences** that **match** any of the **probes**, it will **stick** to the array.
- The array is **washed**, to remove any labelled DNA that **hasn't** stuck to it.
- The array is then **visualised** under **UV light** — any **labelled DNA attached** to a probe will **show up** (fluoresce).
- Any spot that fluoresces means that the person's DNA **contains** that specific **gene**. E.g. if the probe is for a mutated gene that causes a **genetic disorder**, this person has the gene and so **has** the disorder.

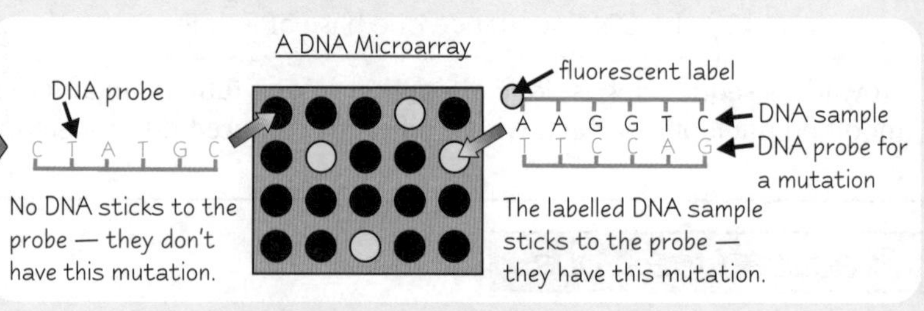

You also need to know how to **produce** a DNA probe — first the **gene** that you want to screen for is **sequenced** (see previous page). Then **PCR** (see p. 183) is used to produce **multiple copies** of **part** of the gene — these are the **probes**.

Scientific Methods** are **Continuously Updated** and **Automated

1) In the **past**, some of the gene technologies you've learnt about on the past few pages were **labour-intensive**, **expensive** and could only be done on a **small scale**.
2) Now these techniques are often **automated**, more **cost-effective** and can be done on a **large scale**.
3) For example, using a single **probe** to **screen** for a single **mutated gene** (see page 190) is **slow** and **small-scale**. Now we have **DNA microarrays** — they're **quick** and can screen for **thousands** of genes at once (see above).
4) Scientific methods like this are **constantly** being **updated** and **automated** to be **faster**, **cheaper**, and **more accurate** (because there's less human error). This means **medical diagnoses** become **faster** and **more accurate**.

DNA Probes in Medical Diagnosis

The **Results** of **Screening** can be used for **Genetic Counselling**...

1) **Genetic counselling** is **advising patients** and their **relatives** about the **risks** of **genetic disorders**.
2) It involves **advising** people about **screening** (e.g. looking for mutated genes if there's a **history** of **cancer**) and **explaining** the **results** of a screening. Screening can help to **identify** the **carrier** of a gene, the **type** of **mutated gene** they're carrying (indicating the type of genetic disorder or cancer) and the **most effective treatment**.
3) If the results of a screening are **positive** (an individual **has** the mutation) then genetic counselling is used to advise the patient on the **options** of **prevention** or **treatment** available. Here are two examples:

   EXAMPLE: A **woman** with a family history of **breast cancer** may have **genetic counselling** to help her **decide** whether or not to be **screened** for **known mutations** that can lead to breast cancer, e.g. a mutation in the BRCA1 **tumour suppressor gene** (see p. 175). If she is screened and the result is **positive**, genetic counsellors might explain that a woman with the mutated BRCA1 gene has an **85%** chance of developing **breast cancer** before the age of **70**. Counselling can also help the woman to **decide** if she wants to take **preventative** measures, e.g. a **mastectomy**, to prevent breast cancer developing.

   EXAMPLE: A couple who are **both carriers** of the **sickle-cell allele** (see previous page) may **like** to have **kids**. They may undergo genetic counselling to help them **understand** the **chances** of them having a child with sickle-cell anaemia (**one in four**). Genetic counselling also provides **unbiased advice** on the possibility of having **IVF** and **screening** their **embryos** for the alleles, so embryos **without the mutation** are **implanted** in the womb. It could also provide information on the **help** and **drugs** available if they have a child with sickle-cell anaemia.

...and **Deciding Treatment**

Cancers can be caused by **mutations** in **proto-oncogenes** (forming **oncogenes**) and mutations in **tumour suppressor genes** (see page 175). **Different mutations** cause **different cancers**, which **respond** to **treatment** in **different ways**. **Screening** using DNA probes for **specific** mutations can be used to help decide the **best** course of **treatment**. For example:

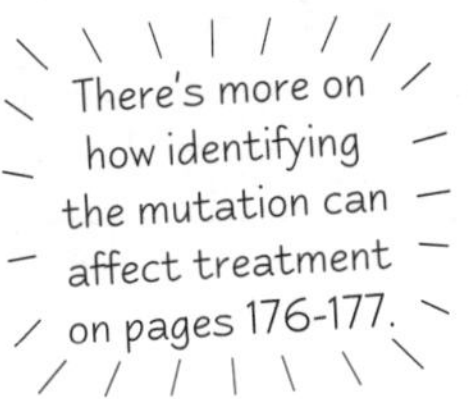
There's more on
how identifying
the mutation can
affect treatment
on pages 176-177.

Breast cancer can be caused by a **mutation** in the **HER2 proto-oncogene**. If a patient with breast cancer is screened and tests **positive** for the HER2 oncogene, they can be treated with the drug **Herceptin®**. This drug **binds** specifically to the **altered HER2 protein receptor** and **suppresses cell division**. Herceptin® is **only effective** against this type of breast cancer because it **only targets** the altered HER2 receptor. Studies have shown that **targeted treatment** like this, alongside less-specific treatment like chemotherapy, can **increase survival rate** and **decrease relapse rate** from breast cancer.

Practice Questions

Q1 Give an example of a mutation that is useful in one way but not in another.

Q2 Give an example of a scientific technique that has been automated.

Q3 What is genetic counselling?

Q4 Describe one situation where genetic counselling may be needed.

Exam Questions

Q1 a) Briefly describe how a DNA probe for a clinically important gene can be produced. [2 marks]

b) Describe how you could screen a person for this gene and many others at the same time. [4 marks]

Q2 Annette has colon cancer. A drug called Cetuximab is used to treat colon cancer caused by a mutation in the KRAS proto-oncogene. Annette is screened and tests negative for the KRAS oncogene.

a) Why is it unlikely that Annette will be treated with Cetuximab? [1 mark]

b) Suggest why Annette will undergo genetic counselling. [2 marks]

Information probes — they're called exams...

All of the techniques you've learnt earlier in this section (PCR, sequencing, DNA probes) come together nicely in this medical diagnosis stuff — it's good to know that what you've learnt has a point to it. It's also good to know that as techniques improve, better ways to diagnose some diseases are found.

Gene Therapy

Congratulations — you've made it to the last two pages of the section. I guess I'd better make them good 'uns then...

Gene Therapy Could be Used to Treat or Cure Genetic Disorders and Cancer

How it works:

1) Gene therapy involves **altering** the **defective genes** (mutated alleles) inside cells to treat **genetic disorders** and **cancer**.
2) How you do this depends on whether the disorder is caused by a mutated **dominant allele** or two mutated **recessive alleles** (see page 132):
   - If it's caused by two mutated **recessive** alleles you can **add** a working **dominant allele** to make up for them (you '**supplement**' the faulty ones).
   - If it's caused by a mutated **dominant** allele you can '**silence**' the **dominant allele** (e.g. by sticking a bit of DNA in the middle of the allele so it doesn't work any more).

Gene therapy isn't being used to treat people yet, but some gene therapy treatments are undergoing clinical trials.

A DNA-filled doughnut — surely the best way to deliver new alleles...

How you get the 'new' allele (DNA) inside the cell:

1) The allele is **inserted into cells** using **vectors** (see page 184).
2) Different **vectors** can be used, e.g. altered **viruses**, **plasmids** or **liposomes** (spheres made of lipid).

There are two types of gene therapy:

1) **Somatic therapy** — this involves **altering** the **alleles** in **body cells**, particularly the cells that are **most affected** by the disorder. For example, **cystic fibrosis** (CF) is a genetic disorder that's very **damaging** to the **respiratory system**, so somatic therapy for CF **targets** the epithelial cells lining the lungs. Somatic therapy doesn't affect the individual's **sex cells** (sperm or eggs) though, so any **offspring** could still **inherit** the disease.
2) **Germ line therapy** — this involves **altering** the **alleles** in the **sex cells**. This means that **every cell** of **any offspring** produced from these cells will be **affected** by the gene therapy and they **won't suffer from the disease**. Germ line therapy in humans is currently **illegal** though.

There are Advantages and Disadvantages to Gene Therapy

You need to be able to **evaluate** the **effectiveness** of **gene therapy** — this means being able to discuss the **advantages** and **disadvantages** of the technique, some of which are given in the table below:

ADVANTAGES	DISADVANTAGES
It could prolong the lives of people with genetic disorders and cancer.	The effects of the treatment may be short-lived (only in somatic therapy).
It could give people with genetic disorders and cancer a better quality of life.	The patient might have to undergo multiple treatments (only in somatic therapy).
Carriers of genetic disorders might be able to conceive a baby without that disorder or risk of cancer (only in germ line therapy).	It might be difficult to get the allele into specific body cells.
It could decrease the number of people that suffer from genetic disorders and cancer (only in germ line therapy).	The body could identify vectors as foreign bodies and start an immune response against them.
	An allele could be inserted into the wrong place in the DNA, possibly causing more problems, e.g. cancer.
	An inserted allele could get overexpressed, producing too much of the missing protein.
	Disorders caused by multiple genes (e.g. cancer) would be difficult to treat with this technique.

There are also many **ethical issues** associated with gene therapy. For example, some people are worried that the technology could be used in ways **other** than for **medical treatment**, such as for treating the **cosmetic effects** of **aging**. Other people worry that there's the potential to do **more harm** than good by using the technology (e.g. risk of overexpression of genes — see table).

Gene Therapy

You Might have to Interpret Some Data on the Effectiveness of Gene Therapy

In the **exam**, you could get a data **question** on the **effectiveness** of **gene therapy**. So, here's one I did earlier:

Background:

X-linked severe combined immunodeficiency disease (X-linked SCID) is an **inherited disorder** affecting the **immune system**. The disorder is caused by a **mutation** in the **IL2RG gene**, located on the **X chromosome**. The IL2RG gene codes for a **protein** that's **essential** for the **development** of some immune system cells, so the sufferer is **vulnerable** to **infectious diseases** and many **die** in **infancy**.

The study:

A study was designed to investigate the **effectiveness** of **gene therapy** in patients with **X-linked SCID**. **Ten patients** were treated with a **virus vector** carrying a **correct version** of the **IL2RG gene**. After gene transfer, the **patient's immune system** was **monitored** for **at least three years** and noted as **functional** (good) or not. Their **health** was also **monitored** for the same time. Bar charts of the results are shown on the right.

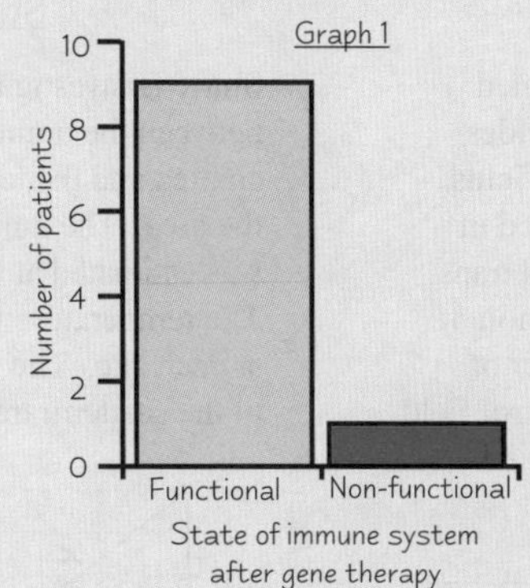

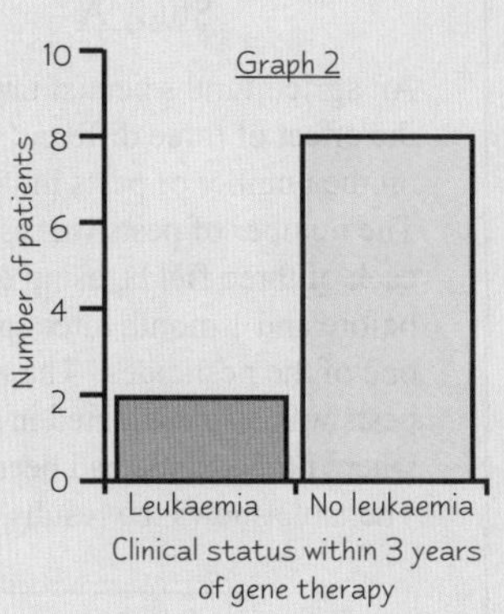

You could be asked to describe the data...

- The **first graph** shows that **nine** out of the **ten** patients had a **functional** immune system **after** gene therapy.
- The **second graph** shows that **two** out of the **ten** patients developed **leukaemia** within 3 years of the treatment.

...or draw conclusions...

- **Gene therapy** can be **used** to **correct** the symptoms of **X-linked SCID**, i.e. produce a **functioning immune system**. However, you **can't** say gene therapy can **cure** X-linked SCID as the study **doesn't** say **how long** the effects of the treatment lasted for.
- **Two out of the ten** patients developed **leukaemia** after the treatment, so there's a chance it's **linked** to the gene therapy (e.g. the vector could have **inserted** the gene into a **proto-oncogene** or **tumour suppressor**, see page 175). But you can only **suggest** a link as **other factors** may have been involved. For example, the patients could have been **more genetically predisposed** to develop cancer.

...or evaluate the methodology

- The **sample size** is **small** — only **ten** patients were treated. This makes the results **less reliable**.

Practice Questions

Q1 How could gene therapy be used to supplement mutated recessive alleles?

Q2 How are supplementary alleles added to human DNA?

Q3 What does germ line gene therapy involve?

Exam Question

Q1 A patient suffering from cystic fibrosis was offered gene therapy targeted at his lung epithelial cells to help treat the disease.

a) What does gene therapy involve? [1 mark]

b) What type of gene therapy was the patient offered? [1 mark]

c) Give three possible disadvantages of the treatment. [3 marks]

First counselling, now therapy — our genes are well looked after...

Now, you might think you need some therapy after this section, but don't worry, the buzzing in your head is normal — it's due to information overload. So go get yourself a cuppa and a biccie and have a break. Then go over some of the difficult bits in this section again. Believe me, the more times you go over it the more things will click into place.

How to Interpret Experiment and Study Data

If you're thinking this looks slightly familiar, then you're right... you had to be able to interpret the results of an experiment or study (and spot a badly designed one) at AS Level. But those pesky examiners have gone and decided that you need to know it for your A2 Level exams as well. So here I am with some more lovely examples to stoke the fires of your memory.

Here Are Some **Things** You Might be **Asked** to do...

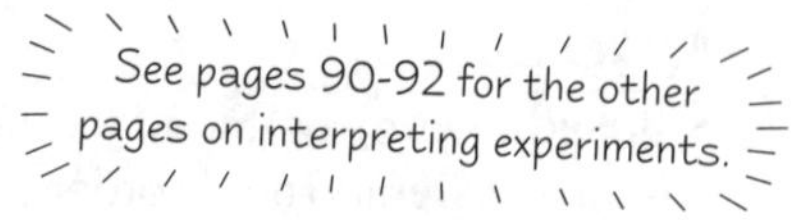

Here are three examples of the kind of data you could expect to get:

Study A

An agricultural scientist investigated the effect of three different pesticides on the number of pests in wheat fields. The number of pests was estimated in each of three fields, using ground traps, before and 1 month after application of one of the pesticides. The number of pests was also estimated in a control field where no pesticide had been applied. The table shows the results.

Pesticide	Number of pests	
	Before application	1 month after application
1	89	98
2	53	11
3	172	94
Control	70	77

Study B

Study B investigated the link between the number of bees in an area and the temperature of the area. The number of bees was estimated at ten 1-acre sites. The temperature was also recorded at each site. The results are shown in the scattergram below.

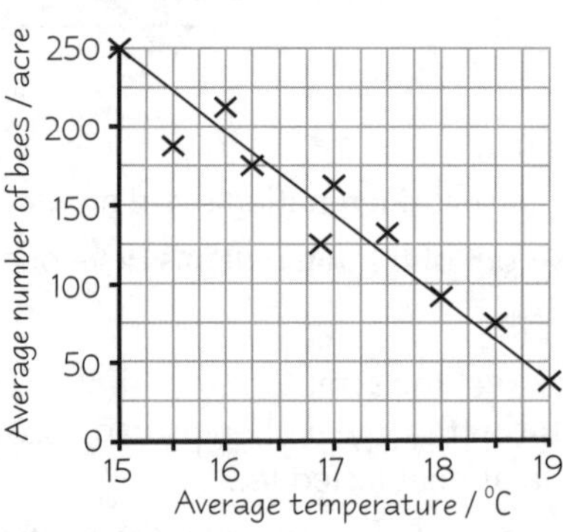

Experiment C

An experiment was conducted to investigate the effect of temperature on the rate of photosynthesis. The rate of photosynthesis in Canadian pondweed was measured at four different temperatures by measuring the volume of oxygen produced. All other variables were kept constant. The results are shown in the graph below.

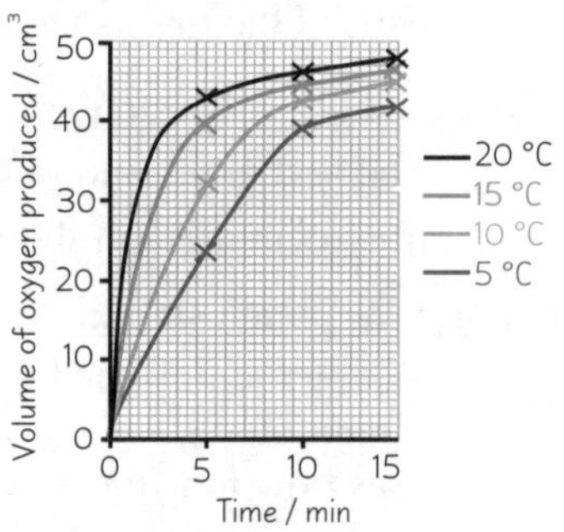

1) **Describe** and **Manipulate** the **Data**

First up again, **describing data**. There's a bit extra this time though, because you could also be asked to **manipulate** the data you're given (i.e. do some **calculations** on it). For the examples above:

Example — Study A

1) You could be asked to **calculate** the **percentage change** (**increase** or **decrease**) in the number of pests for each of the pesticides and the control. E.g. for pesticide 1: (98 – 89) ÷ 89 = 0.10 = **10% increase**.

2) You can then use these values to **describe** what the **data** shows — the **percentage increase** in pests in the field treated with **pesticide 1 was the same as for the control** (10% increase) (1 mark). **Pesticide 3 reduced** pest numbers by **45%**, but **pesticide 2** reduced the pest numbers the **most** (79% decrease) (1 mark).

Example — Study B

The data shows a **negative correlation** between the average number of bees and the temperature (1 mark).

Correlation describes the **relationship** between two variables — e.g. the one that's been changed and the one that's been measured. Data can show **three** types of correlation:

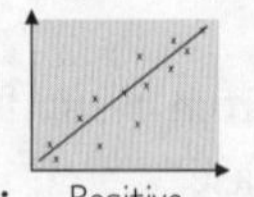

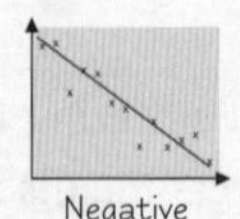

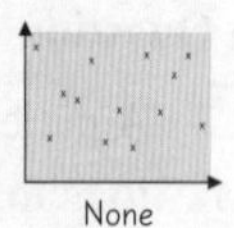

1) **Positive** — as one variable **increases** the other **increases**.
2) **Negative** — as one variable **increases** the other **decreases**.
3) **None** — there is **no relationship** between the two variables.

To tell if some data in a table **is correlated** — draw a **scatter diagram** of one variable against the other and **draw a line of best fit**.

Example — Experiment C

You could be asked to calculate the initial rate of photosynthesis at each temperature: The **gradient = the rate of photosynthesis**:

$$\text{Gradient} = \frac{\text{Change in Y}}{\text{Change in X}}$$

How to Interpret Experiment and Study Data

2) Draw or Check a Conclusion

1) Ideally, only **two** quantities would ever change in any experiment or study — everything else would be **constant**.
2) If you can keep everything else constant and the results show a correlation then you **can** conclude that the change in one variable **does cause** the change in the other.

Example — Experiment C

All other variables were **kept constant**. E.g. light intensity and CO_2 concentration **stayed the same** each time, so these **couldn't** have influenced the rate of reaction. So you **can say** that an increase in temperature up to 20 °C **causes** an increase in the rate of photosynthesis.

3) But usually all the variables **can't** be controlled, so other **factors** (that you **couldn't** keep constant) could be having an **effect**.
4) Because of this, scientists have to be very careful when **drawing conclusions**. Most results show a **link** (**correlation**) between the variables, but that **doesn't prove that a change in one causes the change in the other.**

Example — Study B

There's a **negative correlation** between the average number of bees and temperature. But you **can't** conclude that the increase in temperature **causes** the decrease in bees. **Other factors** may have been involved, e.g. there may be **less food** in some areas, there may be **more bee predators** in some areas, or **something else** you hadn't thought of could have caused the pattern...

5) The **data** should always **support** the conclusion. This may sound obvious but it's easy to **jump** to conclusions. Conclusions have to be **precise** — not make sweeping generalisations.

Example — Experiment C

A science magazine **concluded** from this data that the optimum temperature for photosynthesis is **20 °C**. The data **doesn't** support this. The rate **could** be greatest at 22 °C, or 18 °C, but you can't tell from the data because it doesn't go **higher** than 20 °C and **increases** of **5 °C** at a time were used. The rates of photosynthesis at in-between temperatures **weren't** measured.

3) Explain the Evidence

You could also be asked to **explain** the **evidence** (the data and results) — basically use your **knowledge** of the subject to explain **why** those results were obtained.

Example — Experiment C

Temperature increases the rate of photosynthesis because it **increases** the **activity** of **enzymes** involved in photosynthesis, so reactions are catalysed more quickly.

4) Comment on the Reliability of the Results

In order to draw a **valid conclusion**, the data you use has to be **reliable**. Remember, reliable means the results can be **consistently reproduced** in independent experiments. Here are some of the things that affect the reliability of data:

1) Size of the data set — For experiments, the **more repeats** you do, the **more reliable** the data. If you get the **same result** twice, it could be the correct answer. But if you get the same result **20 times**, it's much more reliable. The general rule for **studies** is the larger the **sample size**, the more **reliable** the **data** is.

E.g. Study B is quite **small** — they only studied ten 1-acre sites. The **trend** shown by the data may not appear if you studied **50 or 100 sites**, or studied them for a longer period of time.

2) The range of values in a data set — The **closer** all the values are to the **mean**, the **more reliable** the data set.

E.g. Study A is **repeated three more times** for pesticides 2 and 3. The percentage decrease each time is: 79%, 85%, 98% and 65% for **pesticide 2** (**mean = 82%**) and 45%, 45%, 54% and 43% for **pesticide 3** (**mean = 47%**). The data values are **closer to the mean** for **pesticide 3** than pesticide 2, so that data set is **more reliable**. The **spread of values about the mean** can be shown by calculating the **standard deviation** (SD).

The **smaller the SD** the **closer** the values to the **mean** and the **more reliable the data**. SDs can be shown on a graph using **error bars**. The ends of the bars show one SD **above** and one SD **below** the mean.

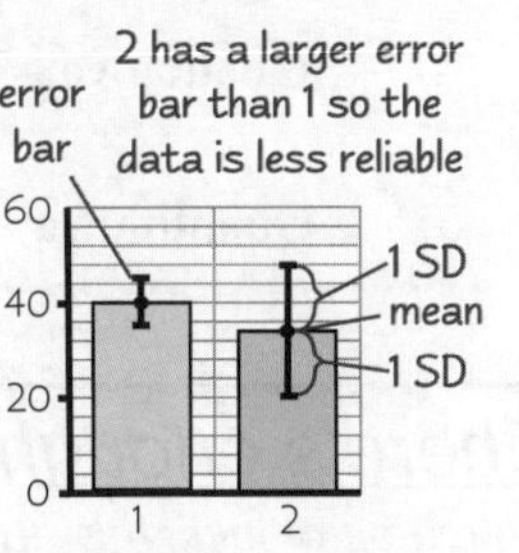

How to Interpret Experiment and Study Data

3) **Variables** — The **more variables** you **control**, the **more reliable** your data is. In an experiment you would control all the variables. In a study you try to control **as many as possible**.

E.g. ideally, all the sites in Study B would have a similar **type** of land, similar **weather**, have the same **plants** growing, etc. Then you could be more sure that the one factor being **investigated** (temperature) is having an **effect** on the thing being **measured** (number of bees).

The hat, trousers, shirt and tie variables had been well controlled in this study.

4) **Data collection** — think about all the **problems** with the **method** and see if **bias** has slipped in.

E.g. in Study A, the traps were placed on the **ground**, so pests like moths or aphids weren't included. This could have affected the results.

5) **Controls** — without controls, it's very difficult to **draw valid conclusions**. **Negative controls** are used to make sure that nothing you're doing in the experiment has an effect, **other than** what you're testing.

E.g. in Experiment C, the **negative control** would be all the equipment set up as normal but **without** the pondweed. If **no oxygen** was produced at any temperature it would show that the variation in the amount of oxygen produced when there was pondweed was due to the **effect** of temperature on the pondweed, and **not** the effect of temperature on **anything else** in the experiment.

6) **Repetition by other scientists** — for theories to become accepted as 'fact' other scientists need to **repeat** the work (see page 1). If **multiple studies** or **experiments** come to the same conclusion, then that conclusion is **more reliable**.

E.g. if a second group of scientists repeated Study B and got the same results, the results would be **more reliable**.

*There Are a Few **Technical Terms** You **Need to Understand***

Finally, **getting the basics right** is really important — so here's a bit about the most **commonly used** words when assessing and analysing experiments and studies:

1) **Variable** — A variable is a **quantity** that has the **potential to change**, e.g. weight. There are two types of variable commonly referred to in experiments:
   - **Independent variable** — the thing that's **changed** in an experiment.
   - **Dependent variable** — the thing that you **measure** in an experiment.

When drawing graphs, the dependent variable should go on the **y-axis** (the vertical axis) and the independent on the **x-axis** (the horizontal axis).

2) **Accurate** — Accurate results are those that are **really close** to the **true** answer. The true answer is **without error**, so if you can reduce error as much as possible you'll get a more accurate result. The most **accurate methods** are those that produce as **error-free** results as possible.

3) **Precise results** — These are results taken using **sensitive instruments** that measure in **small increments**, e.g. pH measured with a meter (pH 7.692) will be **more precise** than pH measured with paper (pH 8).

It's possible for results to be precise **but not** accurate, e.g. a balance that weighs to 1/1000 th of a gram will give precise results, but if it's not **calibrated** properly the results won't be accurate.

4) **Qualitative** — A **qualitative** test tells you **what's** present, e.g. an acid or an alkali.

5) **Quantitative** — A **quantitative** test tells you **how much** is present, e.g. an acid that's pH 2.46.

There's enough evidence here to conclude that data interpretation is boring...

With a bit of luck, this stuff has come flooding back and now you're a data interpreting pro. With even more luck, you'll have reached the reliable conclusion that interpreting data is great fun. Now see if your friends have reached the same conclusion.

AS-Level Answers

Unit 1: Section 1 — Disease and Immunity

Page 5 — Disease

1 *Maximum of 3 marks available.*
*A pathogen may rupture the host cells **[1 mark]**. It may break down and use nutrients in the host cells, so that the host cells starve **[1 mark]**. Or a pathogen may replicate inside host cells and burst them as it leaves **[1 mark]**.*

2 *Maximum of 2 marks available.*
*Axes correct way round with time spent sunbathing on x-axis and incidence of skin cancer on y-axis **[1 mark]**.*
*Positive correlation between two variables **[1 mark]**.*
E.g.

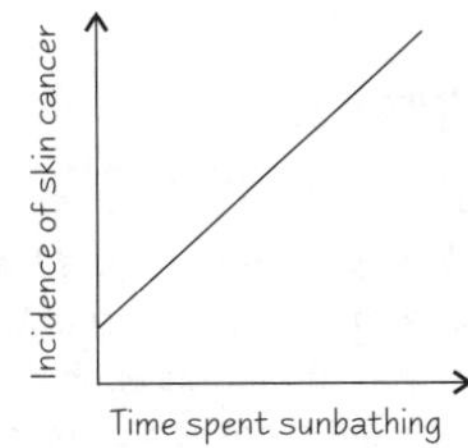

Page 7 — The Immune System

1 *Maximum of 3 marks available.*
*Antibodies coat pathogens, making it easier for phagocytes to engulf them **[1 mark]** and preventing them from entering host cells **[1 mark]**. They also bind to toxins to neutralise them **[1 mark]**.*
There are three marks available for this question, so you need to think of three different functions.

2 *Maximum of 6 marks available.*
*A secondary immune response is faster **[1 mark]** and produces a quicker, stronger response **[1 mark]** than the primary response. This is because memory cells are produced during the primary response **[1 mark]** which remember the foreign antigen **[1 mark]**. During the second infection, memory cell B-cells can quickly divide to form plasma cells, which secrete the correct antibody to the antigen **[1 mark]**. Memory T-cells quickly divide into the right type of T-cells to kill the cell carrying the antigen **[1 mark]**.*
You'll only get the full marks for this question if you <u>explain</u> (as well as describe) why the secondary response differs.

Page 9 — Vaccines and Antibodies in Medicine

1 *Maximum of 4 marks available.*
*The flu virus is able to change its surface antigens/shows antigenic variation **[1 mark]**. This means that when you're infected for a second time with a different strain, the memory cells produced from the first infection will not recognise the new/different antigens **[1 mark]**. The immune system has to carry out a primary response against these new antigens **[1 mark]**. This takes time and means you become ill **[1 mark]**.*

2 *Maximum of 4 marks available.*
*Monoclonal antibodies are made against antigens specific to cancer cells **[1 mark]**. An anti-cancer drug is attached to the antibodies **[1 mark]**. The antibodies bind to tumour markers on cancer cells because their binding sites have a complementary shape **[1 mark]**. This delivers the anti-cancer drug to the cells **[1 mark]**.*

Page 11 — Interpreting Vaccine and Antibody Data

1 a) *Maximum of 2 marks available.*
*Because people were immunised against Hib **[1 mark]** and also had the protection of herd immunity **[1 mark]**.*
b) *Maximum of 1 mark available.*
*The number of cases of Hib increased **[1 mark]**.*

Unit 1: Section 2 — The Digestive System

Page 13 — The Digestive System

1 a) *Maximum of 6 marks available.*
*The pancreas releases pancreatic juice into the small intestine/ duodenum **[1 mark]**. Pancreatic juice contains amylase, trypsin, chymotrypsin and lipase **[1 mark]**. Amylase breaks down starch into maltose **[1 mark]**. Chymotrypsin and trypsin break down proteins into peptides **[1 mark]**. Lipase breaks down lipids into fatty acids and glycerol **[1 mark]**. Pancreatic juice also neutralises acid from the stomach **[1 mark]**.*
b) *Maximum of 1 mark available.*
*E.g. salivary glands **[1 mark]***

Page 15 — Proteins

1 *Maximum of 5 marks available.*
*Two amino acids join together in a condensation reaction **[1 mark]**. A peptide bond **[1 mark]** forms between the carboxyl group **[1 mark]** of one amino acid and the amino group **[1 mark]** of the other amino acid. A molecule of water is released **[1 mark]**.*
If you find it difficult to explain a process, such as a dipeptide forming, learn the diagrams too because they may help you to explain the process.

2 *Maximum of 9 marks available.*
*Proteins are made from amino acids **[1 mark]**. The amino acids are joined together in a long (polypeptide) chain **[1 mark]**. The sequence of amino acids is the protein's primary structure **[1 mark]**. The amino acid chain/polypeptide coils or folds in a certain way **[1 mark]**. The way it's coiled or folded is the protein's secondary structure **[1 mark]**. The coiled or folded chain is itself coiled or folded into a specific shape **[1 mark]**. This is the protein's tertiary structure **[1 mark]**. Different polypeptide chains can be joined together in the protein molecule **[1 mark]**. The way these chains are joined together is the quaternary structure of the protein **[1 mark]**.*
The question specifically states that you don't need to describe the chemical nature of the bonds in a protein. So, even if you name them, don't go into chemical details of how they're formed — no credit will be given.

Page 17 — Carbohydrates

1 *Maximum of 4 marks available*
*Lactose intolerance is caused by a lack of the enzyme lactase **[1 mark]**. Sufferers don't have enough lactase to break down lactose, a sugar found in milk/milk products **[1 mark]**. Undigested lactose is fermented by bacteria **[1 mark]**. This can lead to intestinal complaints such as stomach cramps, flatulence and diarrhoea **[1 mark]**.*

AS-Level Answers

2 *Maximum of 6 marks available*
*Add dilute hydrochloric acid to the solution and boil **[1 mark]**.*
*Neutralise with sodium hydrogencarbonate **[1 mark]**.*
*Add blue Benedict's reagent to the solution and heat **[1 mark]**.*
*If a brick red precipitate forms this indicates that either non-reducing, or reducing sugars are present **[1 mark]**.*
*Carry out the reducing sugar test **[1 mark]** (heating with Benedict's reagent **[1 mark]**) to determine which of the two sugars are present.*
The question asks to describe a test for a non-reducing sugar. Remember, there are a couple more steps involved in the test for a non-reducing sugar than in the test for a reducing sugar.

Page 19 — Enzyme Action

1 *Maximum of 7 marks available.*
*In the 'lock and key' model the enzyme and the substrate have to fit together at the active site of the enzyme **[1 mark]**.*
*This creates an enzyme-substrate complex **[1 mark]**.*
*The active site then causes changes in the substrate **[1 mark]**.*
This mark could also be gained by explaining the change (e.g. bringing molecules closer together, or putting a strain on bonds).
*The change results in the substrate being broken down/joined together **[1 mark]**.*
*The 'induced fit' model has the same basic mechanism as the 'lock and key' model **[1 mark]**.*
*The difference is that the substrate is thought to cause a change in the enzyme's active site shape **[1 mark]**, which enables a better fit **[1 mark]**.*

2 *Maximum of 2 marks available.*
*A change in the amino acid sequence of an enzyme may alter its tertiary structure **[1 mark]**. This changes the shape of the active site so that the substrate can't bind to it **[1 mark]**.*

Page 21 — Factors Affecting Enzyme Activity

1 *Maximum of 8 marks available, from any of the 10 points below.*
*If the solution is too cold, the enzyme will work very slowly **[1 mark]**. This is because, at low temperatures, the molecules have little kinetic energy, so move slowly, making collisions between enzyme and substrate molecules less likely **[1 mark]**.*
*Also, fewer of the collisions will have enough energy to result in a reaction **[1 mark]**.*
The marks above could also be obtained by giving the reverse argument — a higher temperature is best to use because the molecules will move fast enough to give a reasonable chance of collisions and those collisions will have more energy, so more will result in a reaction.
*If the temperature gets too high, the reaction will stop **[1 mark]**.*
*This is because the enzyme is denatured **[1 mark]** — the active site changes shape and will no longer fit the substrate **[1 mark]**.*
*Denaturation is caused by increased vibration breaking bonds in the enzyme **[1 mark]**. Enzymes have an optimum pH **[1 mark]**. pH values too far from the optimum cause denaturation **[1 mark]**.*
Explanation of denaturation here will get a mark only if it hasn't been explained earlier.
*Denaturation by pH is caused by disruption of ionic and hydrogen bonds, which alters the enzyme's tertiary structure **[1 mark]**.*

2 a) *Maximum of 3 marks available.*
*Competitive inhibitor molecules have a similar shape to the substrate molecules **[1 mark]**. They compete with the substrate molecules to bind to the active site of an enzyme **[1 mark]**. When an inhibitor molecule is bound to the active site it stops the substrate molecule from binding **[1 mark]**.*

b) *Maximum of 2 marks available.*
*Non-competitive inhibitor molecules bind to enzymes away from their active site **[1 mark]**. This causes the active site to change shape so the substrate molecule can no longer fit **[1 mark]**.*

Unit 1: Section 3 — Cell Structure and Membranes

Page 23 — Animal Cell Structure

1 *Maximum of 4 marks available.*
*ribosomes **[1 mark]**, rough endoplasmic reticulum **[1 mark]**, Golgi apparatus **[1 mark]**, plasma membrane **[1 mark]***
This question really tests how well you know what each organelle does. The rough endoplasmic reticulum transports proteins that have been made in the ribosomes to the Golgi apparatus. At the Golgi apparatus the proteins are packaged and sent to the plasma membrane to be secreted or inserted in the membrane itself.

2 *Maximum of 2 marks available.*
*Ciliated epithelial cells have lots of mitochondria **[1 mark]** because they need lots of energy **[1 mark]**.*

Page 25 — Analysis of Cell Components

1 *Maximum of 6 marks available.*
*TEMs use electromagnets to focus a beam of electrons, which is transmitted through the specimen **[1 mark]**. Denser parts of the specimen absorb more electrons and appear darker **[1 mark]**. SEMs scan a beam of electrons across the specimen **[1 mark]**. This knocks off electrons from the specimen, which are gathered in a cathode ray tube, to form an image **[1 mark]**. TEMs can only be used on thin specimens **[1 mark]**. SEMs produce lower resolution images than TEMs **[1 mark]**.*

2 *Maximum of 8 marks available.*
*First, the cell sample is homogenised **[1 mark]** to break up the plasma membranes and release the organelles into solution **[1 mark]**. The cell solution is then filtered **[1 mark]** to remove any large cell debris or tissue debris **[1 mark]**. Next the solution is ultracentrifuged **[1 mark]** to separate out the different types of organelles **[1 mark]**. The organelles are separated according to mass, with the heaviest being separated first **[1 mark]**. Centrifugation is repeated at higher and higher speeds to separate out the lighter and lighter organelles **[1 mark]**.*
Make sure you remember to explain each step otherwise you won't be able to get full marks.

Page 27 — Plasma Membranes

1 *Maximum of 2 marks available.*
*The membrane is described as fluid because the phospholipids are constantly moving **[1 mark]**. It is described as a mosaic because the proteins are scattered throughout the membrane like tiles in a mosaic **[1 mark]**.*

AS-Level Answers

2 a) *Maximum of 3 marks available.*
Two fatty acid molecules ***[1 mark]*** *and a phosphate group* ***[1 mark]*** *attached to one glycerol molecule* ***[1 mark]****.*
Don't get phospholipids mixed up with triglycerides — a triglyceride has three fatty acids attached to one glycerol molecule.

b) *Maximum of 2 marks available.*
Saturated fatty acids don't have any double bonds between their carbon atoms ***[1 mark]****. Unsaturated fatty acids have one or more double bonds between their carbon atoms* ***[1 mark]****.*

Page 29 — Exchange Across Plasma Membranes

1 a) *Maximum of 3 marks available.*
The water potential of the sucrose solution was higher than the water potential of the potato ***[1 mark]****. Water moves by osmosis from a solution of higher water potential to a solution of lower water potential* ***[1 mark]****. So water moved into the potato, increasing its mass* ***[1 mark]****.*

b) *Maximum of 1 mark available.*
The water potential of the potato and the water potential of the solution was the same ***[1 mark]****.*

c) *Maximum of 4 marks available.*
– 0.4 g ***[1 mark]****. The potato has a higher water potential than the solution* ***[1 mark]*** *so net movement of water is out of the potato* ***[1 mark]****. The difference in water potential between the solution and the potato is the same as with the 1% solution, so the mass difference should be about the same* ***[1 mark]****.*

Page 31 — Exchange Across Plasma Membranes

1 *Maximum of 10 marks available.*
Some glucose moves across the intestinal epithelial cells and into the blood by diffusion ***[1 mark]****. This is because initially there is a higher concentration of glucose in the lumen of the small intestine than in the blood* ***[1 mark]****. The rest of the glucose moves into the blood by co-transport with sodium ions* ***[1 mark]****. There is a higher concentration of sodium ions in the lumen of the small intestine than inside the intestinal epithelial cell* ***[1 mark]****. This is because sodium ions are actively transported out of the cell into the blood by a sodium-potassium pump* ***[1 mark]****. So sodium ions diffuse from the small intestine lumen into the cell* ***[1 mark]*** *through sodium-glucose co-transporter proteins* ***[1 mark]****. These co-transporter proteins carry glucose into the cell along with the sodium* ***[1 mark]****. The concentration of glucose inside the cell increases* ***[1 mark]*** *and glucose diffuses out of the cell, into the blood* ***[1 mark]****.*
To answer this question it may help if you remember the diagram and then work through it step by step in your head, writing down each point. Also you need to remember the two methods of glucose transport (diffusion and co-transport) to get full marks.

2 *Maximum of 5 marks available.*
The hydrophobic tails ***[1 mark]*** *of the phospholipid bilayer prevent water-soluble molecules from diffusing through the membrane* ***[1 mark]****. Protein channels* ***[1 mark]*** *and carrier proteins* ***[1 mark]*** *control which of these water-soluble substances can enter and leave the cell* ***[1 mark]****.*

Page 33 — Cholera

1 *Maximum of 5 marks available.*
The bacterium produces a toxin ***[1 mark]*** *that causes chloride channels in the lining of the small intestine to open* ***[1 mark]****. Chloride ions diffuse into the small intestine* ***[1 mark]****. The small intestine now has a lower water potential than the blood* ***[1 mark]****, so water moves from the blood into the small intestine, causing diarrhoea and dehydration* ***[1 mark]****.*

2 *Maximum of 2 marks available.*
Against: e.g. children can't make their own decision to be part of the trial ***[1 mark]****. For: e.g. scientists believe treatment for a disease that mainly affects children must be tested on children* ***[1 mark]****.*

Unit 1: Section 4 — The Respiratory System

Page 35 — Lung Function

1 *Maximum of 6 marks available.*
There's a thin exchange surface ***[1 mark]*** *as the alveolar epithelium is only one cell thick* ***[1 mark]****. This means there's a short diffusion pathway, which increases the rate of diffusion* ***[1 mark]****. The number of alveoli provide a large surface area for gas exchange, which also increases the rate of diffusion* ***[1 mark]****. There's a steep concentration gradient between the alveoli and the capillaries surrounding them, which increases the rate of diffusion* ***[1 mark]****. This is maintained by the flow of blood and ventilation* ***[1 mark]****.*

Page 37 — How Lung Disease Affects Function

1 *Maximum of 7 marks available.*
Fibrosis is the formation of scar tissue in the lungs ***[1 mark]****. Scar tissue is less elastic than normal lung tissue, so lungs are less able to expand* ***[1 mark]*** *and tidal volume is reduced* ***[1 mark]****. It's also more difficult for gases to diffuse across the thicker scar tissue* ***[1 mark]*** *so less oxygen reaches the bloodstream* ***[1 mark]*** *and the rate of respiration in the cells is slower* ***[1 mark]****. Less energy is released, which results in the person feeling tired and weak* ***[1 mark]****.*

Page 39 — Interpreting Lung Disease Data

1 a) *Maximum of 3 marks available.*
The daily death rate increased rapidly after 4th December ***[1 mark]*** *peaking around the 7th then decreasing afterwards* ***[1 mark]****. Both pollutants followed the same pattern* ***[1 mark]****.*
You could also get the marks by saying it the other way round — the pollutants rose and peaked around the 7th then decreased, with the death rates following the same pattern.

b) *Maximum of 1 mark available.*
There is a link/correlation between the increase in sulfur dioxide and smoke concentration and the increase in death rate ***[1 mark]****.*
Don't go saying that the increase in sulfur dioxide and smoke caused the increase in death rate — there could have been another reason for the trend, e.g. there could have been other pollutants responsible for the deaths.

AS-Level Answers

Unit 1: Section 5 — The Circulatory System

Page 42 — The Heart

1 *Maximum of 6 marks available.*
*The valves only open one way **[1 mark]**. Whether they open or close depends on the relative pressure of the heart chambers **[1 mark]**. If the pressure is greater behind a valve (i.e. there's lots of blood in the chamber behind it) **[1 mark]**, it's forced open, to let the blood travel in the right direction **[1 mark]**. Once the blood has gone through the valve, the pressure is greater in front of the valve **[1 mark]**, which forces it shut, preventing blood from flowing back into the chamber **[1 mark]**.*
Here you need to explain how valves function in relation to blood flow, rather than just in relation to relative pressures.

2 a) *Maximum of 1 mark available.*
*0.2 - 0.4 seconds **[1 mark]**.*
The AV valves are shut when the pressure is higher in the ventricles than in the atria.

b) *Maximum of 1 mark available.*
*0.3 - 0.4 seconds **[1 mark]**.*
When the ventricles relax the volume of the chamber increases and the pressure falls. The pressure in the left ventricle was 16.5 kPa at 0.3 seconds and it decreased to 7.0 kPa at 0.4 seconds, so it must have started to relax somewhere between these two times.

Page 45 — Cardiovascular Disease

1 *Maximum of 4 marks available*
*Atheromas can lead to the formation of a blood clot/thrombosis **[1 mark]**. A blood clot could block blood flow to the heart muscle **[1 mark]**, causing a lack of oxygen, which damages the heart muscle **[1 mark]** and can lead to a heart attack **[1 mark]**.*
Be specific with the wording of your answers. Examiners won't award marks for unscientific phrases such as "putting strain on the heart".

2 *Maximum of 3 marks available.*
*Atheroma plaques damage and weaken arteries **[1 mark]** and can lead to increased blood pressure **[1 mark]**. When blood at high pressure travels through a weakened artery, the pressure can push the inner layers of the artery through the outer layer to form an aneurysm **[1 mark]**.*
This question is not asking about the consequences of an aneurysm, so no extra marks will be given if you write about it.

Unit 2: Section 1 — Variation

Page 47 — Causes of Variation

1 a) *Maximum of 3 marks available.*
*For species A, as the temperature increases the development time decreases **[1 mark]**. For species B the development time also decreases as the temperature increases **[1 mark]**. The development time of species B is less affected by temperature than species A **[1 mark]**.*

b) *Maximum of 4 marks available.*
*The variation between the species is mainly due to their different genes **[1 mark]**. Variation within a species is caused by both genetic and environmental factors **[1 mark]**. Individuals have different forms of the same genes (alleles), which causes genetic differences **[1 mark]**. Individuals may have the same genes, but environmental factors affect how they're expressed in their appearance (phenotype) **[1 mark]**.*

Page 49 — Investigating Variation

1 a) *Maximum of 1 mark available.*
*To provide a control against which the women who smoked could be compared **[1 mark]**.*

b) *Maximum of 4 marks available.*
*Environmental factors (smoking) affect birth mass **[1 mark]**. Women who smoked showed a mean reduction in the birth mass of their babies of 377 g **[1 mark]**. Genetic factors also affect birth mass of babies born to women who smoke **[1 mark]**. The reduction in birth mass was as much as 1285 g among women who smoked and had certain genotypes **[1 mark]**.*

c) *Maximum of 2 marks available, from any of the 5 points below. E.g. pre-pregnancy mass of the mothers **[1 mark]**, age **[1 mark]**, fitness levels **[1 mark]**, ethnic origin **[1 mark]**, if they'd had a previous pregnancy **[1 mark]**.*
Think of all the variables that need to be considered to isolate smoking as the only environmental factor that is influencing the variation.

Unit 2: Section 2 — Genetics

Page 51 — DNA

1 *Maximum of 4 marks available, from any of the 5 points below. Nucleotides are joined between the phosphate group of one nucleotide and the sugar of the next **[1 mark]** forming the sugar-phosphate backbone **[1 mark]**. The two polynucleotide strands join through hydrogen bonds **[1 mark]** between the base pairs **[1 mark]**. The final mark is given for at least one accurate diagram showing at least one of the above points **[1 mark]**.*
As the question asks for a diagram make sure you do at least one, e.g.:

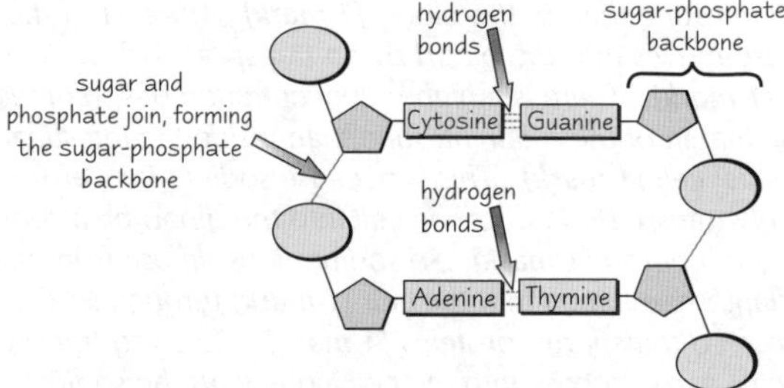

2 *Maximum of 5 marks available.*
*In eukaryotes, DNA is linear **[1 mark]** and wound around proteins (histones) **[1 mark]**.*
*In prokaryotes, DNA molecules are shorter **[1 mark]**, circular **[1 mark]** and not associated with proteins **[1 mark]**.*

Page 53 — Genes

1 a) *Maximum of 2 marks available.*
*A gene is a section of DNA **[1 mark]** that codes for a protein (polypeptide) **[1 mark]**.*

b) *Maximum of 1 mark available.*
*Tryptophan, proline, proline, glutamic acid **[1 mark]**.*

2 *Maximum of 4 marks available.*
*The DNA sequence codes for the sequences of amino acids in proteins **[1 mark]**. Enzymes are proteins, so DNA codes for all enzymes **[1 mark]**. Enzymes control metabolic pathways **[1 mark]**. Metabolic pathways help to determine nature and development **[1 mark]**.*

AS-Level Answers

Page 55 — Meiosis and Genetic Variation

1 *Maximum of 2 marks available, from any of the 3 points below.* *Normal body cells have two copies of each chromosome* ***[1 mark]****. Gametes have to have half the number of chromosomes so that when fertilisation takes place, the resulting embryo will have the correct diploid number* ***[1 mark]****. If the gametes had a diploid number, the resulting offspring would have twice the number of chromosomes that it should have* ***[1 mark]****.*

2 *Maximum of 6 marks available, from any of the 7 points below.* *The DNA unravels and replicates* ***[1 mark]****. The DNA condenses to form double-armed chromosomes* ***[1 mark]****. The chromosomes arrange themselves into homologous pairs* ***[1 mark]****. The pairs separate* ***[1 mark]****. The pairs of sister chromatids then separate* ***[1 mark]****. Four haploid, genetically different cells are produced* ***[1 mark]****. The final mark is given for at least one accurate diagram showing at least one of the above points* ***[1 mark]****.*
As the question asks for a diagram make sure you do at least one, e.g.:

Homologous pair
Cytoplasm
Nucleus
Cell
MEIOSIS I
MEIOSIS II

3 a) *Maximum of 4 marks available.* *During meiosis I, homologous pairs of chromosomes come together* ***[1 mark]****. The chromatids twist around each other and bits swap over* ***[1 mark]****. The chromatids now contain different combinations of alleles* ***[1 mark]****. This means each of the four daughter cells will contain chromatids with different combinations of alleles* ***[1 mark]****.*

b) *Maximum of 2 marks available.* *Independent segregation means the chromosome pairs can split up in any way* ***[1 mark]****. So, the daughter cells produced can contain any combination of maternal and paternal chromosomes with different alleles* ***[1 mark]****.*

Page 57 — Genetic Diversity

1 *Maximum of 3 marks available.* *An event that causes a big reduction in a population, e.g. many members of a population die* ***[1 mark]****. A small number of members survive and reproduce* ***[1 mark]****. Because there are fewer members, there are fewer alleles in the new population, so the genetic diversity is reduced* ***[1 mark]****.*

2 *Maximum of 3 marks available.* *Selective breeding involves humans selecting which organisms to breed until they produce one with the desired characteristics* ***[1 mark]****. Only organisms with similar traits and therefore similar alleles are bred together* ***[1 mark]****. So, the number of alleles in the population is reduced, resulting in reduced genetic diversity* ***[1 mark]****.*

Unit 2: Section 3 — Variation in Biochemistry and Cell Structure

Page 59 — Variation in Haemoglobin

1 *Maximum of 6 marks available.*

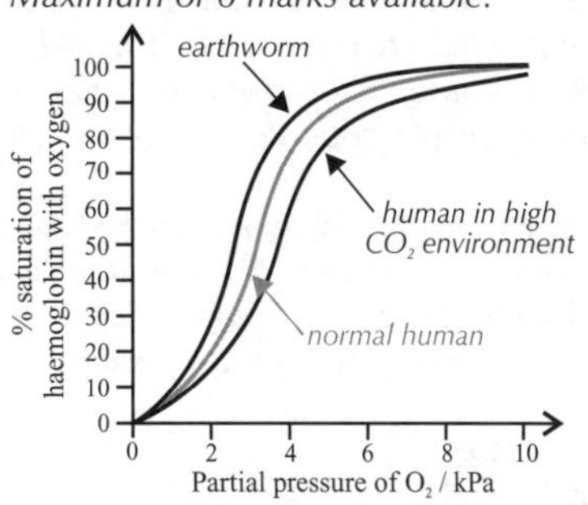

A curve to the left of the human one for the earthworm ***[1 mark]****. Dissociation curves to the left indicate a higher affinity for oxygen at lower partial pressures* ***[1 mark]****.* *This enables the earthworms' haemoglobin to be saturated at the lower partial pressures underground* ***[1 mark]****.* *A normal human dissociation curve that has shifted down for the human in a high carbon dioxide environment* ***[1 mark]****.* *This is the Bohr effect* ***[1 mark]****. High concentrations of carbon dioxide increase the rate of oxygen unloading and the saturation of blood with oxygen is lower for a given pO_2* ***[1 mark]****.*

2 *It is composed of more than one polypeptide chain* ***[1 mark]****.* The reason that haemoglobin has a quaternary structure is because it has more than one polypeptide chain. The fact that it's made up of four polypeptides isn't important.

Page 61 — Variation in Carbohydrates and Cell Structure

1 *Maximum of 6 marks available, from any of the 8 points below.* *Starch is made of two polysaccharides of alpha-glucose* ***[1 mark]****. Amylose is a long unbranched chain* ***[1 mark]*** *which forms a coiled shape* ***[1 mark]****. This coiled shape is very compact, making it good for storage* ***[1 mark]****. Amylopectin is a long, branched chain* ***[1 mark]****. Its side branches make it good for storage as the enzymes that break it down can reach the glycosidic bonds easily* ***[1 mark]****. Starch is insoluble in water* ***[1 mark]****. This means it can be stored in large quantities without bloating the cells by osmosis* ***[1 mark]****.*

Unit 2: Section 4 — The Cell Cycle and Differentiation

Page 63 — The Cell Cycle and DNA Replication

1 a) *Maximum of 1 mark available.*

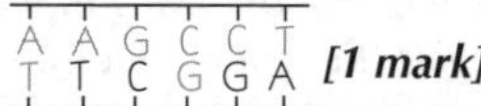

[1 mark]

b) *Maximum of 2 marks available. 1 mark for each new correct molecule with correct labels.*

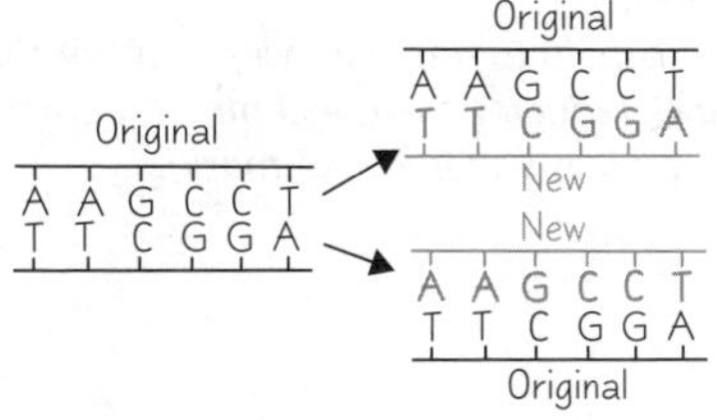

AS-Level Answers

Page 65 — Cell Division — Mitosis

1 a) *Maximum of 6 marks available.*
*A = Metaphase **[1 mark]**, because the chromosomes are lined up at the middle of the cell **[1 mark]**.*
*B = Telophase **[1 mark]**, because there are now two nuclei and the cytoplasm is dividing to form two new cells **[1 mark]**.*
*C = Anaphase **[1 mark]**, because the centromeres have divided and the chromatids are moving to opposite ends of the cell **[1 mark]**.*
If you've learned the diagrams of what happens at each stage of mitosis, this should be a breeze. That's why it'd be a total disaster if you lost three marks for forgetting to give reasons for your answers. Always read the question properly and do exactly what it tells you to do.

b) *Maximum of 3 marks available:*
*X = Chromatid **[1 mark]**.*
*Y = Centromere **[1 mark]**.*
*Z = Spindle fibre **[1 mark]**.*

Page 67 — Cell Differentiation and Organisation

1 *Maximum of 4 marks available.*
*The cell contains many microvilli/folds **[1 mark]** which increase the surface area for absorption **[1 mark]**. The cells form a layer just one cell thick **[1 mark]**, forming a short pathway for the nutrients to cross **[1 mark]**.*

2 *Maximum of 2 marks available.*
*It's best described as an organ **[1 mark]** as it is made of many tissues working together to perform a particular function **[1 mark]**.*

Unit 2: Section 5 — Exchange and Transport Systems

Page 69 — Size and Surface Area

1 *Maximum of 3 marks available.*
*Large mammals have a high demand for oxygen and glucose, which cannot be met by diffusion alone **[1 mark]**. This is because they have a small surface area:volume ratio **[1 mark]** and there is a large number of cells deep inside the body **[1 mark]**.*

Page 71 — Gas Exchange

1 *Maximum of 6 marks available.*
*Gaseous exchange surfaces have a large surface area **[1 mark]** — e.g. mesophyll cells in a plant (or any other suitable example) **[1 mark]**. They are thin, which provides a short diffusion pathway **[1 mark]** — e.g. the walls of tracheoles in insects (or any other suitable example not already mentioned) **[1 mark]**. There is a steep diffusion gradient, which is constantly maintained **[1 mark]** — e.g. blood flowing in the opposite direction to water in fish gills (the counter-current system) (or any other suitable example not already mentioned) **[1 mark]**.*

2 *Maximum of 2 marks available.*
*Sunken stomata and hairs help to trap any water vapour that does evaporate **[1 mark]**, reducing the concentration gradient from leaf to air, which reduces water loss **[1 mark]**.*

Page 73 — The Circulatory System

1 *Maximum of 6 marks available.*
*They have thick, muscular walls **[1 mark]** to cope with the high pressure produced by the heartbeat **[1 mark]**. They have elastic tissue in the walls **[1 mark]** so they can expand to cope with the high pressure produced by the heartbeat **[1 mark]**. The inner lining (endothelium) is folded **[1 mark]** so that the artery can expand when the heartbeat causes a surge of blood **[1 mark]**.*

2 *Maximum of 4 marks available.*
*At the start of the capillary bed, the pressure in the capillaries is greater than the pressure in the tissue fluid outside the capillaries **[1 mark]**. This means fluid from the blood is forced out of the capillaries **[1 mark]**. Fluid loss causes the water potential in the blood capillaries to become lower than that of the tissue fluid **[1 mark]**. So fluid moves back into the capillaries at the vein end of the capillary bed by osmosis **[1 mark]**.*

Page 75 — Water Transport in Plants

1 *Maximum of 4 marks available.*
*In the apoplast pathway **[1 mark]**, water passes through cell walls **[1 mark]**. In the symplast pathway **[1 mark]**, water passes from cell to cell through the plasmodesmata that connect the cytoplasm of adjacent cells **[1 mark]**.*

2 *Maximum of 4 marks available.*
*Loss of water from the leaves, due to transpiration, pulls more water into the leaves from the xylem **[1 mark]**. There are cohesive forces between water molecules **[1 mark]**. These cause water to be pulled up the xylem **[1 mark]**. Removing leaves means no transpiration occurs, so no water is pulled up the xylem **[1 mark]**.*
It's pretty obvious (because there are 4 marks to get) that it's not enough just to say removing the leaves stops transpiration. You also need to explain why transpiration is so important in moving water through the xylem. It's always worth checking how many marks a question is worth — this gives you a clue about how much detail you need to include.

Unit 2: Section 6 — Classification

Page 77 — Principles of Classification

1 a) *Maximum of 2 marks available.*
*Green monkey **[1 mark]** because it's the closest to humans on the tree **[1 mark]**.*

b) *Maximum of 1 mark available.*
*Pig **[1 mark]**.*

2 *Maximum of 2 marks available.*
*A group of similar organisms **[1 mark]** able to reproduce to give fertile offspring **[1 mark]**.*
Make sure you use the word fertile in your answer.

3 *Maximum of 5 marks available. 1 mark for each correct answer.*

Kingdom	Phylum	Class	Order	Family	Genus	Species
Animalia	Chordata	Mammalia	Primates	Hominidae	Homo	sapiens

AS-Level Answers

4 *Maximum of 3 marks available.*
E.g. you can't study the reproductive behaviour of extinct species ***[1 mark]****. Some species reproduce asexually* ***[1 mark]****. There are practical and ethical issues involved in studying some reproductive behaviour* ***[1 mark]****.*

Page 79 — Classifying Species

1 *Maximum of 5 marks available.*
DNA from two species is collected, separated into single strands and mixed together ***[1 mark]****. Where the DNA is similar, hydrogen bonds will form between the base pairs* ***[1 mark]****. The more similar the DNA the more hydrogen bonds will form* ***[1 mark]****. The strands are heated and the temperature at which they separate is recorded* ***[1 mark]****. A higher temperature will be needed to separate DNA strands from more similar species because more hydrogen bonds will have formed* ***[1 mark]****.*
You need to use the right terminology here to get the marks, e.g. hydrogen bonds (not just bonds) and base pairs (not just bases).

2 a) *Maximum of 1 mark available.*
Mouse and rat ***[1 mark]****.*
b) *Maximum of 1 mark available.*
Chicken ***[1 mark]****.*

Unit 2: Section 7 — Evolving Antibiotic Resistance

Page 81 — Antibiotic Action and Resistance

1 a) *Maximum of 3 marks available.*
A mutation occurs in the DNA of a bacterium ***[1 mark]****.*
If the mutation occurs in a gene it may alter the protein that gene codes for ***[1 mark]****, which may make the bacteria resistant to an antibiotic* ***[1 mark]****.*
b) *Maximum of 3 marks available.*
Resistance to antibiotics is spread between two bacteria by horizontal gene transmission ***[1 mark]****. The two bacteria join together by a process called conjugation* ***[1 mark]*** *and a copy of a plasmid carrying a gene for antibiotic resistance is transferred from one cell to the other* ***[1 mark]****.*
c) *Maximum of 4 marks available.*
Penicillin inhibits an enzyme involved in making the bacterial cell wall ***[1 mark]****. This prevents cell wall formation in growing bacteria and weakens the wall* ***[1 mark]****. Water moves into the cell by osmosis* ***[1 mark]****. The weakened cell wall can't withstand the increased pressure so bursts (lyses), killing the bacterium* ***[1 mark]****.*

Page 83 — Antibiotic Resistance

1 *Maximum of 6 marks available.*
As the use of the pesticide increased, the number of aphids fell ***[1 mark]*** *as they were being killed by the pesticide* ***[1 mark]****. Random mutations may have occurred in the aphid DNA, resulting in pesticide resistance* ***[1 mark]****. Any aphids resistant to the pesticide were more likely to survive and pass on their alleles* ***[1 mark]****. Over time, the number of aphids increased* ***[1 mark]*** *as those carrying pesticide-resistant alleles became more common* ***[1 mark]****.*

2 *Maximum of 3 marks available.*
Some bats in the population will carry a mutation for a longer tongue ***[1 mark]****. The bats with longer tongues will be able to feed from the flowers and so will be more likely to survive, reproduce and pass on their alleles* ***[1 mark]****. Over time, this feature will become common in the population* ***[1 mark]****.*

Page 85 — Evaluating Resistance Data

1 *Maximum of 2 marks available.*
Argument for: E.g. if a person has an infection that can be treated they should not be denied treatment. / If the person is not treated they may become very ill ***[1 mark]****.*
Argument against: E.g. a person suffering with dementia may forget to take their medication and so increase the risk of antibiotic resistant bacteria developing ***[1 mark]****.*

2 a) *Maximum of 2 marks available.*
E.g. the chest infection is mild ***[1 mark]****. Prescribing antibiotics for non-life threatening illnesses contributes to increased antibiotic resistance* ***[1 mark]****.*
b) *Maximum of 2 marks available.*
E.g. not prescribing antibiotics could reduce her son's quality of life ***[1 mark]****. He may take longer to get better without antibiotics, so she may have to take longer off work* ***[1 mark]****.*

3 *Maximum of 3 marks available, from any of the 4 points below.*
E.g. getting information from one area (East Anglia) would not show national trends ***[1 mark]****. It's a relatively small study (only 300 patients), which decreases its reliability* ***[1 mark]****. Patients don't always tell the truth on questionnaires, which reduces its reliability* ***[1 mark]****. Patients may not return the questionnaire* ***[1 mark]****.*

Unit 2: Section 8 — Species Diversity

Page 87 — Human Impacts on Diversity

1 a) *Maximum of 2 marks available.*
The number of different species ***[1 mark]*** *and the abundance of each species in a community* ***[1 mark]****.*
b) *Maximum of 4 marks available.*
Site 1 —
$51\ (51 - 1) = 2550$
$15\ (15 - 1) + 12\ (12 - 1) + 24\ (24 - 1) = 894$
Use of $N\ (N - 1) \div \Sigma n\ (n - 1)$ *to calculate diversity index of* $2550 \div 894 = 2.85$
[2 marks for correct answer, 1 mark for incorrect answer but correct working]*.*
Site 2 —
$132\ (132 - 1) = 17292$
$35\ (35 - 1) + 25\ (25 - 1) + 34\ (34 - 1) + 12\ (12 - 1) + 26\ (26 - 1) = 3694$
Use of $N\ (N - 1) \div \Sigma n\ (n - 1)$ *to calculate diversity index of* $17292 \div 3694 = 4.68$
[2 marks for correct answer, 1 mark for incorrect answer but correct working]*.*
It's always best if you put your working — even if the answer isn't quite right you could get a mark for correct working.
c) *Maximum of 2 marks available.*
The diversity of bumblebee species is greater at site 2 ***[1 mark]****. This suggests that field margins increase the diversity of bumblebee species* ***[1 mark]****.*

Page 89 — Interpreting Diversity Data

1 a) *Maximum of 2 marks available.*
Both woodland and farmland populations have declined since 1970 ***[1 mark]****. Farmland species have declined more than woodland species* ***[1 mark]****.*

A2-Level Answers

b) *Maximum of 3 marks available, from any of the 5 points below. (or other suitable answers).*
*Loss of habitat **[1 mark]**.*
*Fewer hedgerows/larger fields **[1 mark]**.*
*Deforestation/clearance of land **[1 mark]**.*
*Farming intensification/changes to farming practice **[1 mark]**.*
*Pesticides causing disruption in the food chain **[1 mark]**.*

c) *Maximum of 8 marks available, from any of the 14 points below. Maximum of 4 marks available for benefits. Maximum of 4 marks available for risks.*
*Agriculture benefits — more food can be produced/there is an increased yield **[1 mark]**. Food is cheaper to produce, so prices are lower **[1 mark]**. Local areas become more developed by attracting businesses **[1 mark]**.*
*Agriculture risks — natural beauty is lost **[1 mark]**. Diversity is reduced because of monoculture **[1 mark]**. Diversity is reduced because of land and hedgerow clearance **[1 mark]**. Diversity is reduced from use of pesticides/herbicides **[1 mark]**.*
*Deforestation benefits — wood is provided as well as access to other resources **[1 mark]**. More land is available for homes/agriculture **[1 mark]**. Local areas become more developed by attracting businesses **[1 mark]**.*
*Deforestation risks — less carbon dioxide is stored, which contributes to climate change **[1 mark]**. Potential medical/scientific discoveries are lost **[1 mark]**. Natural beauty is lost **[1 mark]**. Diversity is reduced/extinctions may occur **[1 mark]**.*
When the question asks you to discuss an issue, you need to make sure you talk about both sides — the benefits and the risks.

Unit 4: Section 1 — Populations

Page 95 — Populations and Ecosystems

1 a) *Maximum of 2 marks available.*
*They live on farmland, in open woodland, in hedgerows and urban areas **[1 mark]**, and roost in cracks in trees and buildings **[1 mark]**.*

b) *Maximum of 3 marks available.*
*Their wings are light and flexible, which allows them to catch fast and manoeuvrable insects. This increases their chances of catching enough food to survive **[1 mark]**. They use echolocation so they can catch insects that come out at night. This also increases their chances of catching enough food to survive **[1 mark]**. They make unique mating calls so they only attract a mate of the same species. This increases their chance of reproduction by making successful mating more likely **[1 mark]**.*
This question is only asking about the biotic conditions (the living features of the ecosystem), so you won't get any marks for talking about abiotic conditions (the non-living features of the ecosystem).

Page 97 — Investigating Populations

1 a) *Maximum of 1 mark available.*
*By taking random samples of the population **[1 mark]**.*

b) *Maximum of 4 marks available.*
*She could use quadrats to measure percentage cover of daffodils **[1 mark]**. Several would be placed on the ground at random locations within the field **[1 mark]**. The percentage of each quadrat that's covered by daffodils would be recorded **[1 mark]**. The percentage cover for the whole field could then be estimated by averaging the data collected in all of the quadrats **[1 mark]**.*
Make sure you don't write about quadrats in a transect — they're used to investigate the distribution of plant species, not the percentage cover in an area.

Page 99 — Investigating Populations

1 a) *Maximum of 5 marks available.*
*A group of snails would have been caught **[1 mark]**, marked in a way that wouldn't harm them, e.g. by painting a spot on their shell **[1 mark]**, then released back into the environment **[1 mark]**. After waiting a week a second sample would have been taken **[1 mark]** and the number marked in the second sample would have been counted **[1 mark]**.*

b) *Maximum of 2 marks available.*

$$\text{Total population size} = \frac{52 \times 38}{14} \quad \textbf{[1 mark]}$$

*Total population size = 141 **[1 mark]**.*
Award 2 marks for correct answer of 141 without any working.

Page 101 — Variation in Population Size

1 a) *Maximum of 7 marks available.*
*In the first three years, the population of prey increases from 5000 to 30 000. The population of predators increases slightly later (in the first five years), from 4000 to 11 000 **[1 mark]**. This is because there's more food available for the predators **[1 mark]**. The prey population then falls after year three to 3000 just before year 10 **[1 mark]**, because lots are being eaten by the large population of predators **[1 mark]**. Shortly after the prey population falls, the predator population also falls (back to 4000 by just after year 10) **[1 mark]**, because there's less food available **[1 mark]**. The same pattern is repeated in years 10-20 **[1 mark]**.*

b) *Maximum of 4 marks available.*
*The population of prey increased to around 40 000 by year 26 **[1 mark]**. This is because there were fewer predators, so fewer prey were eaten **[1 mark]**. The population then decreased after year 26 to 25 000 by year 30 **[1 mark]**. This could be because of intraspecific competition **[1 mark]**.*

Page 103 — Human Populations

1 *Maximum of 4 marks available.*
*At stage 1 of the DTM the population size is low and not increasing **[1 mark]**. At stage 5 the population is high but shrinking **[1 mark]**. At stage 1 the population structure is made up of a lot of young people and very few older people **[1 mark]**, but at stage 5 there are few young people and a lot of older people **[1 mark]**.*

Unit 4: Section 2 — Energy Supply

Page 105 — Photosynthesis, Respiration and ATP

1 *Maximum of 6 marks available, from any of the 8 points below.*
*In the cell, ATP is synthesised from ADP and inorganic phosphate/P_i **[1 mark]** using energy from an energy-releasing reaction, e.g. respiration **[1 mark]**. The energy is stored as chemical energy in the phosphate bond **[1 mark]**. ATP synthase catalyses this reaction **[1 mark]**. ATP then diffuses to the part of the cell that needs energy **[1 mark]**. Here, it's broken down back into ADP and inorganic phosphate/P_i **[1 mark]**, which is catalysed by ATPase **[1 mark]**. Chemical energy is released from the phosphate bond and used by the cell **[1 mark]**.*
Make sure you don't get the two enzymes confused — ATP **synth**ase **synth**esises ATP, and ATPase breaks it down.

A2-Level Answers

Page 109 — Photosynthesis

1 a) *Maximum of 1 mark available.*
*The thylakoid membranes **[1 mark]**.*
b) *Maximum of 1 mark available.*
*Photosystem II **[1 mark]**.*
c) *Maximum of 4 marks available.*
*Light energy splits water **[1 mark]**.*

H_2O ***[1 mark]*** $\rightarrow 2H^+ + \frac{1}{2}O_2$ ***[1 mark]**.*

*The electrons from the water replace the electrons lost from chlorophyll **[1 mark]**.*
The question asks you to explain the purpose of photolysis, so make sure you include why the water is split up — to replace the electrons lost from chlorophyll.
d) *Maximum of 1 mark available.*
*NADP **[1 mark]**.*

2 a) *Maximum of 6 marks available.*
*Ribulose bisphosphate/RuBP and carbon dioxide/CO_2 join together to form an unstable 6-carbon compound **[1 mark]**. This reaction is catalysed by the enzyme rubisco/ribulose bisphosphate carboxylase **[1 mark]**. The compound breaks down into two molecules of a 3-carbon compound called glycerate 3-phosphate/GP **[1 mark]**. Two molecules of glycerate 3-phosphate are then converted into two molecules of triose phosphate/TP **[1 mark]**. The energy for this reaction comes from ATP **[1 mark]** and the H^+ ions come from reduced NADP **[1 mark]**.*
b) *Maximum of 2 marks available.*
*Ribulose bisphosphate is regenerated from triose phosphate/TP molecules **[1 mark]**. ATP provides the energy to do this **[1 mark]**.*
This question is only worth two marks so only the main facts are needed, without the detail of the number of molecules.
c) *Maximum of 3 marks available.*
*No glycerate 3-phosphate/GP would be produced **[1 mark]**, so no triose phosphate/TP would be produced **[1 mark]**. This means there would be no glucose produced **[1 mark]**.*

Page 111 — Limiting Factors in Photosynthesis

1 a) *Maximum of 4 marks available.*
*By burning propane to increase air CO_2 concentration **[1 mark]**. By adding heaters to increase temperature **[1 mark]**. By adding coolers to decrease temperature **[1 mark]**. By adding lamps to provide light at night **[1 mark]**.*
b) *Maximum of 2 marks available.*
*Potatoes **[1 mark]** because the yield showed the smallest percentage increase of 25% (850 – 680 = 170, 170 ÷ 680 × 100 = 25%) **[1 mark]**.*

Page 113 — Respiration

1 *Maximum of 6 marks available, from any of the 7 points below.*
*First, the 6-carbon glucose molecule is phosphorylated **[1 mark]** by adding two phosphates from two molecules of ATP **[1 mark]**. This creates two molecules of triose phosphate **[1 mark]** and two molecules of ADP **[1 mark]**. Triose phosphate is oxidised (by removing hydrogen) to give two molecules of 3-carbon pyruvate **[1 mark]**. The hydrogen is accepted by two molecules of NAD, producing two molecules of reduced NAD **[1 mark]**. During oxidation four molecules of ATP are produced **[1 mark]**.*
When describing glycolysis make sure you get the number of molecules correct — one glucose molecule produces two molecules of triose phosphate. You could draw a diagram in the exam to show the reactions.

2 a) *Maximum of 2 marks available.*
*Pyruvate + reduced NAD **[1 mark]** → lactate + NAD **[1 mark]***
b) *Maximum of 2 marks available.*
*Lactate is produced to regenerate NAD **[1 mark]** so glycolysis can continue and ATP can be produced under anaerobic conditions to provide energy for biological processes **[1 mark]**.*

Page 115 — Respiration

1 a) *Maximum of 2 marks available.*
*The transfer of electrons down the electron transport chain stops **[1 mark]**. So there's no energy released to phosphorylate ADP/produce ATP **[1 mark]**.*
b) *Maximum of 2 marks available.*
*The Krebs cycle stops **[1 mark]** because there's no oxidised NAD/FAD coming from the electron transport chain **[1 mark]**.*
Remember that when the electron transport chain is inhibited, the reactions that depend on the products of the chain are also affected.

Unit 4: Section 3 — Energy Flow and Nutrient Cycles

Page 117 — Energy Transfer and Productivity

1 a) *Maximum of 2 marks available.*
Gross productivity = net productivity + respiratory loss
*1245 + 4165 = 5410 **[1 mark]***
*gross productivity = 5410 $kJm^{-2}yr^{-1}$ **[1 mark]***
Award 2 marks for correct answer of 5410 $kJm^{-2}yr^{-1}$ without any working.
b) *Maximum of 3 marks available.*
*Because not all of the energy is taken in by the Arctic hares **[1 mark]**. Some parts of the grass aren't eaten, so the energy isn't taken in **[1 mark]** and some parts of the grass are indigestible, so they'll pass through the hares and come out as waste **[1 mark]**.*
c) *Maximum of 2 marks available.*
*(11 ÷ 137) × 100 = 8 **[1 mark]***
*energy transfer efficiency = 8% **[1 mark]***
Award 2 marks for correct answer of 8% without any working.

Page 119 — Farming Practices and Productivity

1 *Maximum of 5 marks available.*
*Organic farmers might use biological agents **[1 mark]**. Biological agents reduce the numbers of pests, so crops lose less energy and biomass, increasing productivity **[1 mark]**. They include natural predators that eat the pest species to reduce their numbers **[1 mark]**. They could also use parasites that live in or lay their eggs on pest insects, killing the pests or reducing their ability to function **[1 mark]**. They could also introduce pathogenic bacteria or viruses that kill pests **[1 mark]**.*

A2-Level Answers

Page 121 — The Carbon Cycle and Global Warming

1 *Maximum of 6 marks available.*
*Carbon from CO_2 in the air and water becomes carbon compounds in plants when they photosynthesise **[1 mark]**. Carbon is then passed onto primary consumers when they eat the plants, and secondary and tertiary consumers when they eat the other consumers **[1 mark]**. When organisms die, the carbon in the dead organisms is digested by microorganisms called decomposers **[1 mark]**. Carbon is returned to the atmosphere as CO_2 because all living organisms carry out respiration, which produces CO_2 **[1 mark]**. When dead organic matter ends up in a place where there aren't any decomposers the carbon can be turned into fossil fuels over millions of years **[1 mark]**. Carbon in fossil fuels is released back into the atmosphere when they are burnt **[1 mark]**.*

2 a) *Maximum of 6 marks available.*
*Atmospheric CO_2 concentration has increased rapidly since the mid-19th century due to increased burning of fossil fuels, e.g. for industry or in cars **[1 mark]**, and the increased destruction of natural CO_2 sinks, e.g. by deforestation **[1 mark]**. Atmospheric methane concentration has also been increasing rapidly since the mid-19th century due to things like increased extraction of fossil fuels **[1 mark]**, more decaying waste in landfill sites **[1 mark]** and more cattle giving off methane as a waste gas **[1 mark]**. Methane can also be released when natural stores, such as frozen ground, thaw **[1 mark]**.*

b) *Maximum of 4 marks available.*
*Global warming is the term for the increase in average global temperature over the last century **[1 mark]**. Global warming is caused by enhancement of the greenhouse effect. The greenhouse effect is when atmospheric greenhouse gases absorb outgoing energy, so less is lost to space **[1 mark]**. The greenhouse effect helps to keep the planet warm, but too much greenhouse gas causes the planet to warm up **[1 mark]**. Increasing atmospheric concentrations of the greenhouse gases CO_2 and methane are enhancing the greenhouse effect, and so causing global warming **[1 mark]**.*

Page 123 — Effects of Global Warming

1 a) *Maximum of 4 marks available.*
*CO_2 concentration shows a general trend of increasing **[1 mark]** from around 338 ppm in 1980 to around 368 ppm in 2000 **[1 mark]**. The corn yield fluctuates but shows a general trend of increasing **[1 mark]** from around 105 bushels per acre in 1980 to around 135 bushels per acre in 2000 **[1 mark]**.*

b) *Maximum of 1 mark available.*
*There's a positive correlation between CO_2 concentration and corn yield / as CO_2 concentration increases so does corn yield **[1 mark]**.*
Even though there's a correlation you can't conclude that increasing CO_2 concentration is causing increases in corn yield, because there could be other factors involved, e.g. changing rainfall pattern.

c) *Maximum of 2 marks available.*
*CO_2 concentration is a limiting factor for photosynthesis **[1 mark]**, so increasing CO_2 concentration could mean crops grow faster, increasing crop yields **[1 mark]**.*

Page 125 — The Nitrogen Cycle and Eutrophication

1 a) *Maximum of 2 marks available.*
*A — ammonification **[1 mark]**, C — denitrification **[1 mark]***

b) *Maximum of 3 marks available.*
*Process B is nitrogen fixation **[1 mark]**. Nitrogen fixation is where nitrogen gas in the atmosphere is turned into ammonia **[1 mark]** by bacteria **[1 mark]**.*

Unit 4: Section 4 — Succession and Conservation

Page 127 — Succession

1 a) *Maximum of 6 marks available.*
*This is an example of secondary succession, because there is already a soil layer present in the field **[1 mark]**. The first species to grow will be the pioneer species, which in this case will be larger plants **[1 mark]**. These will then be replaced with shrubs and smaller trees **[1 mark]**. At each stage, different plants and animals that are better adapted for the improved conditions will move in, out-compete the species already there, and become the dominant species **[1 mark]**. As succession goes on, the ecosystem becomes more complex, so species diversity (the number and abundance of different species) increases **[1 mark]**. Eventually large trees will grow, forming the climax community, which is the final seral stage **[1 mark]**.*

b) *Maximum of 2 marks available.*
*Ploughing destroys any plants that were growing **[1 mark]**, so larger plants may start to grow, but they won't have long enough to establish themselves before the field is ploughed again **[1 mark]**.*

Page 129 — Conservation

1 a) *Maximum of 6 marks available.*
*Conserving rainforests is important for humans as they may provide lots of things that humans need such as clothes, food or drugs **[1 mark]**. If they're cut down, the source of these things will be lost and they won't be available in the future **[1 mark]**. Some people think rainforests should be conserved because it's the right thing to do — they think forests have a right to exist, and people don't have a right to cut them all down **[1 mark]**. The rainforests bring joy to lots of people who visit them. If they're cut down future generations won't be able to enjoy them **[1 mark]**. Conservation of rainforests will mean less trees are burnt. This helps to prevent climate change because burning trees releases CO_2 into the atmosphere, which contributes to global warming **[1 mark]**. Conserving rainforests will help to prevent the disruption of food chains. A decrease in one species could mean the loss of many more species in the food chain as there's less food, so more resources are lost **[1 mark]**.*

b) *Maximum of 2 marks available, from any of the 3 points below.*
*E.g. seedbanks **[1 mark]**, captive breeding programmes **[1 mark]**, relocation **[1 mark]**.*

Page 131 — Conservation Evidence and Data

1 a) *Maximum of 4 marks available.*
*The cod stock size increased from around 150 000 tonnes in 1963 to around 250 000 tonnes in 1971 **[1 mark]**. The stock size then decreased (apart from a couple of smaller increases) to around 30 000 tonnes in 2006 **[1 mark]**. The fishing mortality rate fluctuated but showed a trend of increasing **[1 mark]** from around 0.5 in 1963 to just below 0.8 in 2006 **[1 mark]**.*

A2-Level Answers

b) *Maximum of 2 marks available.*
There's a link between fishing mortality rate and the cod stock size ***[1 mark]****. As the fishing mortality rate increases, the cod stock size decreases/there's a negative correlation between fishing mortality rate and cod stock size* ***[1 mark]****.*

c) *Maximum of 1 mark available.*
1978/1979 ***[1 mark]***

d) *Maximum of 1 mark available.*
It could be used by governments to make decisions about cod fishing quotas (the amount of cod allowed to be removed from the sea by fishermen each year), to try to keep the cod stock above 150 000 tonnes ***[1 mark]****.*

Unit 4: Section 5 — Inheritance, Selection and Speciation

Page 133 — Inheritance

1 a) *Maximum of 3 marks available.*
Parents' genotypes identified as RR and rr ***[1 mark]****. Correct genetic diagram drawn with gametes' alleles identified as R, R and r, r* ***[1 mark]*** *and gametes crossed to show Rr as the only possible genotype* ***[1 mark]****.*
The question specifically asks you to draw a genetic diagram so make sure that you include one in your answer, e.g.

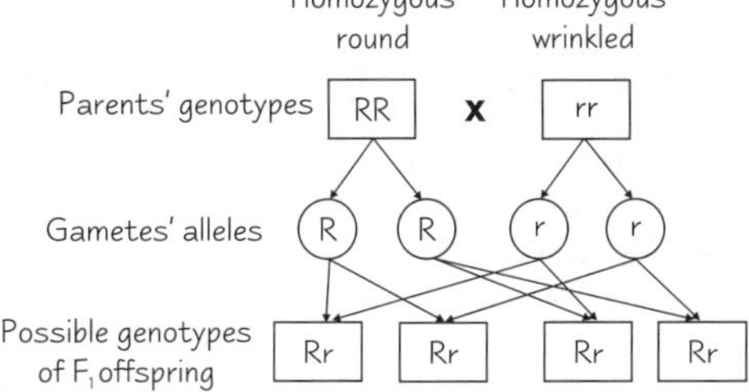

b) *Maximum of 3 marks available.*
Gametes' alleles (produced by F_1 generation) identified as R, r and R, r ***[1 mark]****. Gametes crossed to show genotypes RR, Rr and rr in a 1:2:1 ratio* ***[1 mark]****. RR and Rr genotypes identified as giving a round phenotype and rr as wrinkled phenotype, giving a 3:1 ratio of round : wrinkled seeds* ***[1 mark]****. Award three marks for a correct ratio of 3:1 for round : wrinkled seeds.*
The question doesn't ask for a genetic diagram but it can help you work out the answer, e.g.

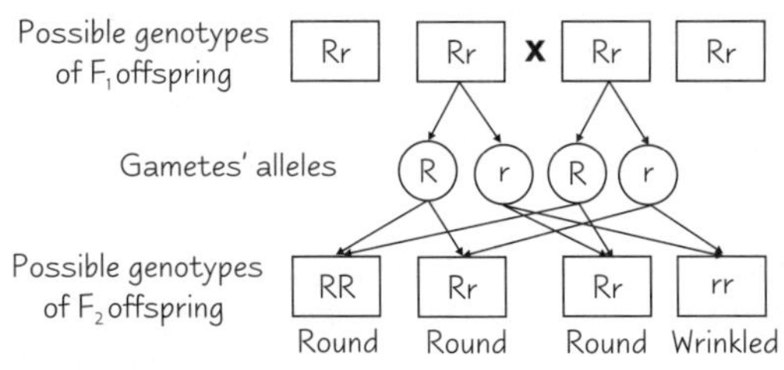

Page 135 — Inheritance

1 *Maximum of 3 marks available.*
Men only have one copy of the X chromosome (XY) but women have two (XX) ***[1 mark]****. Haemophilia A is caused by a recessive allele so females would need two copies of the allele for them to have haemophilia A* ***[1 mark]****. As males only have one X chromosome they only need one recessive allele to have haemophilia A, which makes them more likely to have haemophilia A than females* ***[1 mark]****.*

2 *Maximum of 4 marks available.*
Genotypes of parents identified as I^AI^O and I^BI^B ***[1 mark]****. Correct genetic diagram drawn with gametes' alleles identified as I^A, I^O and I^B, I^B* ***[1 mark]*** *and gametes crossed to show genotypes I^AI^B and I^BI^O in a 1:1 ratio* ***[1 mark]****. The probability of the couple having a child with blood group B is 0.5 (or 50%)* ***[1 mark]****.*
The question specifically asks you to draw a genetic diagram so make sure that you include one in your answer, e.g.

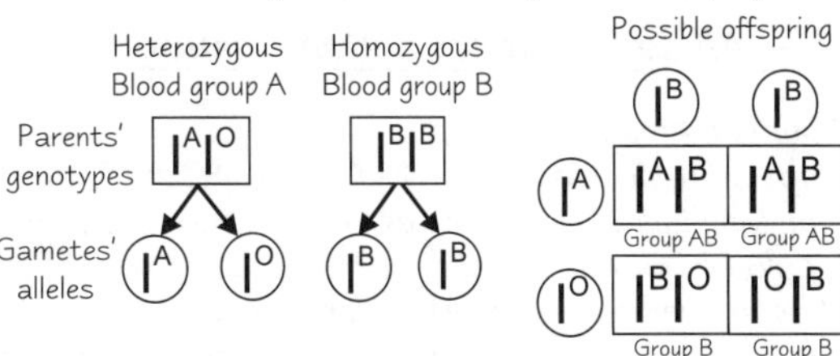

3 a) *Maximum of 1 mark available.*
Individual 2 could have genotype AA or Aa ***[1 mark]****.*
Individual 2 is an unaffected male so he must have at least one A allele (AA or Aa). But you can't say for sure if he has AA or Aa.

b) *Maximum of 2 marks available*
Individual 6's genotype is Aa ***[1 mark]****. The offspring of individuals 5 and 6 has ADA deficiency (aa), so both parents must be carriers of the recessive allele (Aa)* ***[1 mark]****. Or, individual 3 has ADA deficiency (aa), so must have passed a recessive allele onto individual 6. Individual 6 is unaffected so must have a dominant allele as well (Aa)* ***[1 mark]****.*

c) *Maximum of 4 marks available*
Parents' genotypes identified as both Aa ***[1 mark]****. Correct genetic diagram drawn with gametes' alleles identified as A, a and A, a* ***[1 mark]*** *and gametes crossed to show genotypes AA, Aa and aa in a 1:2:1 ratio* ***[1 mark]****. aa genotype identified as causing ADA deficiency, giving a probability of 0.25 (25%)* ***[1 mark]****.*
The question asks you to show your working so you could draw a genetic diagram, e.g.

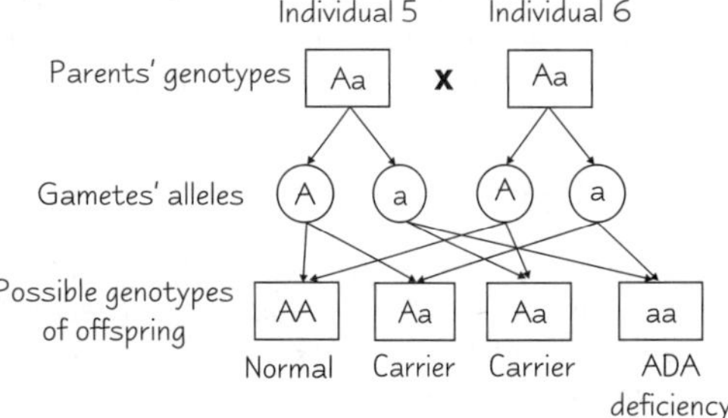

Page 137 — The Hardy-Weinberg Principle

1 a) *Maximum of 3 marks available.*
Frequency of genotype CC = p^2 = 0.14 ***[1 mark]*** *so the frequency of the dominant allele C = p = $\sqrt{0.14}$ = 0.37* ***[1 mark]****. The frequency of the recessive allele c = q = 1 – p = 1 – 0.37 = 0.63* ***[1 mark]****. Award three marks for a correct answer of 0.63.*

b) *Maximum of 1 mark available.*
Frequency of homozygous recessive genotype cc = q^2 = 0.63^2 = 0.40 ***[1 mark]****.*

A2-Level Answers

c) *Maximum of 2 marks available.*
*Those that don't have a cleft chin are homozygous recessive cc = 40% **[1 mark]**, so the percentage that do have a cleft chin, Cc or CC, is 100% – 40% = 60% **[1 mark]**. Award two marks for a correct answer of 60%.*
There are other ways of calculating this answer, e.g. working out the value of 2pq and adding it to p^2. It doesn't matter which way you do it as long as you get the right answer.

Page 139 — Allele Frequency and Speciation

1 a) *Maximum of 4 marks available.*
*As temperature decreases from 22 °C to 16 °C the frequency of h, the long hair allele, increases from 0.11 to 0.23 **[1 mark]**. This could be because the allele for long hair is more beneficial at colder temperatures **[1 mark]**. Hamsters with the h allele will have a greater chance of surviving, reproducing and passing on their genes, including the beneficial h allele **[1 mark]**. So a greater proportion of the next generation will inherit the beneficial allele and the frequency of the h allele will increase **[1 mark]**.*
b) *Maximum of 1 mark available.*
*Directional selection **[1 mark]**.*

Unit 5: Section 1 — Responding to the Environment

Page 141 — Nervous and Hormonal Communication

1 a) *Maximum of 1 mark available.*
*Nervous communication is more suitable because electrical impulses travel faster than hormones, so the (protective) response is quicker **[1 mark]**.*
b) *Maximum of 5 marks available.*
*Receptors detect the stimulus **[1 mark]**, e.g. light receptors (photoreceptors) in the animal's eyes detect the bright light **[1 mark]**. The receptors send impulses along neurones (via the CNS) to the effectors **[1 mark]**. The effectors bring about a response **[1 mark]**, e.g. the circular iris muscles contract to constrict the pupils and protect the eyes **[1 mark]**.*

Page 143 — Receptors

1 *Maximum of 6 marks available, from any of the 7 points below.*
*A tap on the arm is a mechanical stimulus that's detected by pressure receptors/mechanoreceptors called Pacinian corpuscles **[1 mark]**. The stimulus deforms the layers of connective tissue (lamellae) **[1 mark]**, which press on the sensory nerve ending **[1 mark]**. This causes deformation of stretch-mediated sodium ion channels in the neurone cell membrane **[1 mark]**. Sodium ion channels open and sodium ions diffuse into the cell **[1 mark]**. This creates a generator potential **[1 mark]**. If the generator potential reaches the threshold it triggers a nerve impulse/action potential **[1 mark]**.*

2 *Maximum of 5 marks available.*
*The human eye has high sensitivity because many rods join one neurone **[1 mark]**, so many weak generator potentials combine to reach the threshold and trigger an action potential **[1 mark]**. The human eye has high acuity because cones are close together and one cone joins one neurone **[1 mark]**. When light from two points hits two cones, action potentials from each cone go to the brain **[1 mark]**. So you can distinguish two points that are close together as two separate points **[1 mark]**.*

Page 146 — Nervous System — Neurones

1 a) *Maximum of 1 mark available.*
*Stimulus **[1 mark]**.*
b) *Maximum of 3 marks available.*
*A stimulus causes sodium ion channels in the neurone cell membrane to open **[1 mark]**. Sodium ions diffuse into the cell **[1 mark]**, so the membrane becomes depolarised **[1 mark]**.*
c) *Maximum of 2 marks available.*
*The membrane was in the refractory period **[1 mark]**, so the sodium ion channels were recovering and couldn't be opened **[1 mark]**.*

2 *Maximum of 5 marks available.*
*Transmission of action potentials will be slower in neurones with damaged myelin sheaths **[1 mark]**. This is because myelin is an electrical insulator **[1 mark]**, so increases the speed of action potential conduction **[1 mark]**. The action potentials 'jump' between the nodes of Ranvier/between the myelin sheaths **[1 mark]**, where sodium ion channels are concentrated **[1 mark]**.*
Don't panic if a question mentions something you haven't learnt about. You might not know anything about multiple sclerosis but that's fine, because you're not supposed to. All you need to know to get full marks here is how myelination affects the speed of action potential conduction.

Page 149 — Nervous System — Synaptic Transmission

1 *Maximum of 6 marks available, from any of the 8 points below.*
*The action potential arriving at the presynaptic membrane stimulates voltage-gated calcium ion channels to open **[1 mark]**, so calcium ions diffuse into the neurone **[1 mark]**. This causes synaptic vesicles, containing acetylcholine, to fuse with the presynaptic membrane **[1 mark]**. The vesicles release acetylcholine into the synaptic cleft **[1 mark]**. The acetylcholine diffuses across the synaptic cleft **[1 mark]** and binds to cholinergic receptors on the postsynaptic membrane **[1 mark]**. This causes sodium ion channels in the postsynaptic membrane to open **[1 mark]** and the influx of sodium ions triggers a new action potential to be generated at the postsynaptic membrane **[1 mark]**.*

2 *Maximum of 4 marks available.*
*They might have weaker muscular responses than normal **[1 mark]**. If receptors are destroyed at neuromuscular junctions then there will be fewer receptors for acetylcholine/ACh to bind to **[1 mark]**, so fewer sodium ion channels will open **[1 mark]**, meaning fewer muscle cells can be stimulated **[1 mark]**.*

3 *Maximum of 3 marks available.*
*Galantamine would stop acetylcholinesterase/AChE breaking down acetylcholine/ACh, so there would be more ACh in the synaptic cleft **[1 mark]** and it would be there for longer **[1 mark]**. This means more nicotinic cholinergic receptors would be stimulated **[1 mark]**.*

Page 151 — Effectors — Muscle Contraction

1 *Maximum of 3 marks available.*
*Muscles are made up of bundles of muscle fibres **[1 mark]**. Muscle fibres contain long organelles called myofibrils **[1 mark]**. Myofibrils contain bundles of myofilaments **[1 mark]**.*

A2-Level Answers

2 a) *Maximum of 3 marks available.*
*A = sarcomere **[1 mark]**.*
*B = Z-line **[1 mark]**.*
*C = H-zone **[1 mark]**.*
b) *Maximum of 2 marks available.*
*The A-bands stay the same length during contraction **[1 mark]**. The I-bands get shorter **[1 mark]**.*
c) *Maximum of 3 marks available.*
*Drawing number 3 **[1 mark]** because the M-line connects the middle of the myosin filaments **[1 mark]**. The cross-section would only show myosin filaments, which are the thick filaments **[1 mark]**.*
The answer isn't drawing number 1 because all the dots in the cross-section are smaller, so the filaments shown are thin actin filaments — which aren't found at the M-line.

Page 153 — Effectors — Muscle Contraction

1 *Maximum of 3 marks available.*
*Muscles need ATP to relax because ATP provides the energy to break the actin-myosin cross bridges **[1 mark]**. If the cross bridges can't be broken, the myosin heads will remain attached to the actin filaments **[1 mark]**, so the actin filaments can't slide back to their relaxed position **[1 mark]**.*

2 *Maximum of 3 marks available.*
*The muscles won't contract **[1 mark]** because calcium ions won't be released into the sarcoplasm, so troponin won't be removed from its binding site **[1 mark]**. This means no actin-myosin cross bridges can be formed **[1 mark]**.*

Page 155 — Responses in Animals

1 a) *Maximum of 5 marks available.*
*High blood pressure is detected by pressure receptors in the aorta called baroreceptors **[1 mark]**. Impulses are sent along sensory neurones to the medulla **[1 mark]**. Impulses are then sent to the SAN along a parasympathetic neurone **[1 mark]**. The parasympathetic neurone secretes acetylcholine, which binds to receptors on the sinoatrial node/SAN **[1 mark]**. This slows the heart rate (reducing blood pressure) **[1 mark]**.*
b) *Maximum of 2 marks available.*
*No impulses sent from the medulla would reach the SAN **[1 mark]**, so the heart rate wouldn't increase or decrease/ control of the heart rate would be lost **[1 mark]**.*

Page 157 — Survival and Responses in Plants

1 a) *Maximum of 2 marks available.*
*The data shows that the plants provided with auxins grew more than those not given auxins **[1 mark]**. This is because auxins stimulate plant growth (by cell elongation) **[1 mark]**.*
b) *Maximum of 5 marks available.*
*Auxin is redistributed to the shaded side of the plant **[1 mark]**. In shoots, auxin stimulates cell elongation **[1 mark]** so the shoots bend to grow towards the light **[1 mark]**. In roots, high concentrations of auxin inhibit cell elongation **[1 mark]** so the roots bend to grow away from the light **[1 mark]**.*
c) *Maximum of 2 marks available.*
*The students should repeat the experiment to see if their results are reliable **[1 mark]**. They should also make sure all other variable conditions are the same for each group **[1 mark]**.*
d) *Maximum of 1 mark available.*
*The results suggest that auxins stimulate plant growth, so auxins could be used to increase tomato yield **[1 mark]**.*

Unit 5: Section 2 — Homeostasis

Page 159 — Homeostasis Basics

1 a) *Maximum of 2 marks available.*
*Statement A **[1 mark]** because body temperature continues to increase from the normal level and isn't returned **[1 mark]**.*
b) *Maximum of 2 marks available.*
*It makes metabolic reactions less efficient **[1 mark]** because the enzymes that control metabolic reactions may denature **[1 mark]**.*

2 *Maximum of 2 marks available.*
*Multiple negative feedback mechanisms give more control over changes in the internal environment than just having one feedback mechanism **[1 mark]**. This is because you can actively increase or decrease a level so it returns to normal **[1 mark]**.*

Page 161 — Control of Body Temperature

1 *Maximum of 4 marks available, from the 8 points below.*
1 mark for each method, up to a maximum of 2 marks.
1 mark for each explanation, up to a maximum of 2 marks.
*Vasoconstriction of blood vessels **[1 mark]** reduces heat loss because less blood flows through the capillaries in the surface layers of the dermis **[1 mark]**. Erector pili muscles contract to make hairs stand on end **[1 mark]**, trapping an insulating layer of air to prevent heat loss **[1 mark]**. Muscles contract in spasms to make the body shiver **[1 mark]**, so more heat is produced from increased respiration **[1 mark]**. Adrenaline and thyroxine are released **[1 mark]**, which increase metabolism so more heat is produced **[1 mark]**.*

2 *Maximum of 2 marks available.*
*Thermoreceptors in the skin detect a higher external temperature than normal **[1 mark]**. The thermoreceptors send impulses along sensory neurones to the hypothalamus **[1 mark]**.*

3 *Maximum of 4 marks available.*
*Snakes are ectotherms **[1 mark]**. They can't control their body temperature internally and depend on the temperature of their external environment **[1 mark]**. In cold climates, snakes will be less active **[1 mark]**, which makes it harder to catch prey, avoid predators, find a mate, etc. **[1 mark]**.*
You need to use a bit of common sense to answer this question — you know that the activity level of an ectotherm depends on the temperature of the surroundings, so in a cold environment it won't be very active. And if it can't be very active it'll have trouble surviving.

Page 163 — Control of Blood Glucose Concentration

1 *Maximum of 5 marks available, from the 7 points below.*
*High blood glucose concentration is detected by cells in the pancreas **[1 mark]**. Beta/β cells secrete insulin into the blood **[1 mark]**, which binds to receptors on the cell membranes of liver and muscle cells **[1 mark]**. This increases the permeability of the cell membranes to glucose, so the cells take up more glucose **[1 mark]**. Insulin also activates glycogenesis **[1 mark]** and increases the rate that cells respire glucose **[1 mark]**. This lowers the concentration of glucose in the blood **[1 mark]**.*
You need to get the spelling of words like glycogenesis right in the exam or you'll miss out on marks.

A2-Level Answers

2 *Maximum of 3 marks available.*
*They have Type II diabetes **[1 mark]**. They produce insulin, but the insulin receptors on their cell membranes don't work properly, so the cells don't take up enough glucose **[1 mark]**. This means their blood glucose concentration remains higher than normal **[1 mark]**.*

Page 165 — Control of the Menstrual Cycle

1 a) *Maximum of 5 marks available.*
*Follicle-stimulating hormone/FSH stimulates a follicle to develop **[1 mark]**. FSH also stimulates the ovary to release oestrogen **[1 mark]**, and oestrogen is also released by the follicle **[1 mark]**. Oestrogen inhibits the release of FSH from the anterior pituitary **[1 mark]**, so no more follicles are stimulated to develop **[1 mark]**.*

b) *Maximum of 3 marks available.*
*High oestrogen concentration stimulates the anterior pituitary to release luteinising hormone/LH **[1 mark]**. LH stimulates the ovary to release more oestrogen **[1 mark]**, which further stimulates the anterior pituitary in a positive feedback loop **[1 mark]**.*

2 *Maximum of 3 marks available.*
*Oestrogen and progesterone inhibit follicle-stimulating hormone/FSH release from the anterior pituitary **[1 mark]**. Because there's no FSH, the follicle isn't stimulated to develop **[1 mark]**, so there is no ovulation **[1 mark]**.*

Unit 5: Section 3 — Genetics

Page 167 — DNA and RNA

1 a) *Maximum of 2 marks available.*
*The sugar in DNA is a deoxyribose sugar whilst in RNA it is a ribose sugar **[1 mark]**. The bases in DNA are adenine, thymine, guanine and cytosine. In RNA, thymine is replaced by uracil **[1 mark]**.*
You've only been asked to describe the differences in composition so you don't need to write about the shape of the molecules.

b) i) *Maximum of 1 mark available.*
*tRNA **[1 mark]***
ii) *Maximum of 1 mark available.*
*DNA **[1 mark]***
iii) *Maximum of 1 mark available.*
*tRNA **[1 mark]***

Page 169 — Protein Synthesis

1 *Maximum of 2 marks available.*
*The drug binds to DNA, preventing RNA polymerase from binding, so transcription can't take place and no mRNA can be made **[1 mark]**. This means there's no mRNA for translation and so protein synthesis is inhibited **[1 mark]**.*

2 a) *Maximum of 2 marks available.*
*10 × 3 = 30 nucleotides long **[1 mark]**. Each amino acid is coded for by three nucleotides (a codon), so the mRNA length in nucleotides is the number of amino acids multiplied by three **[1 mark]**.*

b) *Maximum of 3 marks available.*
*Greater **[1 mark]**. DNA contains introns and exons **[1 mark]** but the introns are removed to form mRNA by splicing, so the DNA sequence will have more nucleotides than the mRNA sequence **[1 mark]**.*

Page 171 — The Genetic Code and Nucleic Acids

1 a) *Maximum of 2 marks available.*
*The mRNA sequence is 18 nucleotides long and the protein produced from it is 6 amino acids long **[1 mark]**. 18 ÷ 6 = 3, suggesting three nucleotides code for a single amino acid **[1 mark]**.*

b) *Maximum of 2 marks available.*
*The protein produced was leucine-cysteine-glycine. This would only be produced if the code is non-overlapping, e.g. UUGUGUGGG = UUG-UGU-GGG = leucine-cysteine-glycine **[1 mark]**.*
If the code was overlapping the codons would be UUG-UGU-GUG-UGU, which would give the protein leucine-cysteine-valine-cysteine.
*Also the protein produced is only 6 amino acids long, which is correct if the code is non-overlapping — the protein would be longer if the code overlapped **[1 mark]**.*

2 a) *Maximum of 2 marks available. Award 2 marks if all four amino acids are correct and in the correct order. Award 1 mark if three amino acids are correct and in the correct order.*
GUG = valine
UGU = cysteine
CGC= arginine
GCA = alanine
Correct sequence = valine, cysteine, arginine, alanine.

b) *Maximum of 2 marks available. Award 2 marks if all four codons are correct and in the correct order. Award 1 mark if three codons are correct and in the correct order.*
arginine = CGC
alanine = GCA
leucine = UUG
phenylalanine = UUU
Correct sequence = CGC GCA UUG UUU.

c) *Maximum of 3 marks available.*
valine = GUG
arginine = CGC
alanine = GCA
mRNA sequence = GUG CGC GCA.
*DNA sequence = CAC **[1 mark]** GCG **[1 mark]** CGT **[1 mark]**.*

Page 173 — Regulation of Transcription and Translation

1 a) *Maximum of 4 marks available.*
*The results of tubes 1 and 2 suggest that oestrogen affects the expression of the gene for the Chi protein **[1 mark]** because mRNA and active protein production only occur in the presence of oestrogen **[1 mark]**. When oestrogen is present it binds to the oestrogen receptor (transcription factor), forming an oestrogen-oestrogen receptor complex **[1 mark]**. This complex works as an activator, helping RNA polymerase to bind to the DNA, activating transcription and resulting in protein production in the presence of oestrogen **[1 mark]**.*

b) *Maximum of 3 marks available.*
*The mutant could have a faulty oestrogen receptor **[1 mark]**. Oestrogen might not bind to the receptor, or the oestrogen-oestrogen receptor complex might not work as an activator **[1 mark]**. This would mean even in the presence of oestrogen transcription wouldn't be activated, so no mRNA or protein would be produced **[1 mark]**.*
This is a pretty tricky question — drawing a diagram of how oestrogen controls transcription would help you figure out the answer.

A2-Level Answers

c) Maximum of 3 marks available.
E.g. the results would be no full length mRNA and no protein produced **[1 mark]**. The siRNA and associated proteins would attach to the mRNA of the Chi protein and cut it up into smaller portions, resulting in no full length mRNA **[1 mark]**. No mRNA would be available for translation, so no protein would be produced **[1 mark]**.

Page 175 — Mutations, Genetic Disorders and Cancer

1 a) Maximum of 1 mark available.
AGGTATGAGGCC

b) Maximum of 5 marks available.
The original gene codes for the amino acid sequence serine-tyrosine-glutamine-alanine **[1 mark]**. The mutated gene codes for the amino acid sequence arginine-tyrosine-glutamic acid-alanine **[1 mark]**. Even though there are three mutations there are only two changes to the amino acid sequence **[1 mark]**. This is because of the degenerate nature of the DNA code, which means more than one codon can code for the same amino acid **[1 mark]**. So the substitution mutation on the last triplet doesn't alter the amino acid (GCT and GCC both code for alanine) **[1 mark]**.

c) Maximum of 2 marks available.
Acquired **[1 mark]**, because they weren't present before exposure to a mutagenic agent / they weren't present in the gametes **[1 mark]**.

d) Maximum of 3 marks available.
The mutations could result in the gene becoming an oncogene **[1 mark]**. When they are functioning normally, proto-oncogenes stimulate cell division by producing proteins that make cells divide **[1 mark]**. However, when proto-oncogenes mutate to form oncogenes the gene can become overactive. This stimulates the cells to divide uncontrollably (the rate of division increases), resulting in a tumour (cancer) **[1 mark]**.

Page 177 — Diagnosing and Treating Cancer and Genetic Disorders

1 Maximum of 25 marks available.
HINTS:
- Start off by describing what hereditary and acquired mutations are and the types of disorders they cause, e.g. genetic disorders and cancer.
- Then explain how knowing that some cancers are caused by acquired mutations affects prevention, e.g. if you know that mutagenic agents cause acquired mutations you can try to avoid them and so help prevent cancer.
- Next do the same for how knowing that some cancers are caused by acquired mutations affects diagnosis, e.g. high risk individuals can be screened more frequently to try to diagnose cancer earlier. Don't forget to include that early diagnosis increases their chances of recovery.
- Then repeat the previous two bits for hereditary cancer and genetic disorders.
- Use plenty of examples from pages 176-177 to back up your points.

Page 179 — Stem Cells

1 a) Maximum of 1 mark available.
Totipotent cells are stem cells that can mature into any cell type in an organism **[1 mark]**.

b) Maximum of 3 marks available.
The totipotent cells grew and divided at all pHs, but they grew the most at pH 4 (up to 30 g in mass) **[1 mark]**. The totipotent cells only specialised into shoot cells at pH 4 **[1 mark]**. This suggests that the pH helps to control the specialisation of cells for this type of plant **[1 mark]**.

Page 181 — Stem Cells in Medicine

1 Maximum of 4 marks available.
E.g. stem cell therapies are currently being used for some diseases affecting the blood and immune system **[1 mark]**. Bone marrow contains stem cells that can become specialised to form any type of blood cell **[1 mark]**. Bone marrow transplants can be used to replace faulty bone marrow in patients with leukaemia (a cancer of the blood or bone marrow) **[1 mark]**. The stem cells in the transplanted bone marrow divide and specialise to produce healthy blood cells **[1 mark]**.

2 Maximum of 2 marks available.
Obtaining embryonic stem cells involves the destruction of an embryo **[1 mark]**. Some people believe that embryos have a right to life and that it's wrong to destroy them **[1 mark]**.

Unit 5: Section 4 — Gene Technology

Page 183 — Making DNA Fragments

1 a) Maximum of 2 marks available.
There's a BamHI recognition sequence at either side of the DNA fragment, so you could use this restriction endonuclease to isolate the fragment **[1 mark]**. BamHI would be incubated with the bacterial DNA, so that it cuts the DNA at each of these recognition sequences **[1 mark]**.
You could include a diagram to help explain your answer.

b) Maximum of 5 marks available, from any of the 6 points below.
The bacterial DNA is mixed with free nucleotides, primers and DNA polymerase **[1 mark]**. The mixture is heated to 95 °C to break the hydrogen bonds **[1 mark]**. The mixture is then cooled to between 50 – 65 °C to allow the primers to bind/anneal to the DNA **[1 mark]**. The primers bind/anneal to the DNA because they have a sequence that's complementary to the sequence at the start of the DNA fragment (e.g. CAA) **[1 mark]**. The mixture is then heated to 72 °C and DNA polymerase lines up free nucleotides along each template strand, producing new strands of DNA **[1 mark]**. The cycle would be repeated over and over to produce lots of copies **[1 mark]**.
This question asks you to describe and explain, so you need to give the reasons why each stage is done to gain full marks.

Page 185 — Gene Cloning

1 a) Maximum of 2 marks available.
Colony A is visible/fluoresces under UV light, but Colony B isn't visible/doesn't fluoresce **[1 mark]**. So only Colony A contains the fluorescent marker gene, which means it contains transformed cells **[1 mark]**.

A2-Level Answers

b) *Maximum of 3 marks available.*
*The plasmid vector DNA would have been cut open with the same restriction endonuclease that was used to isolate the DNA fragment containing the target gene **[1 mark]**. The plasmid DNA and gene (DNA fragment) would have been mixed together with DNA ligase **[1 mark]**. DNA ligase joins the sticky ends of the DNA fragment to the sticky ends of the plasmid DNA **[1 mark]**.*

Page 187 — Genetic Engineering

1 a) *Maximum of 3 marks available.*
*The drought-resistant gene could be inserted into a plasmid **[1 mark]**. The plasmid is then inserted into a bacterium **[1 mark]**, which is used as a vector to get the gene into the plant cells **[1 mark]**.*

b) *Maximum of 2 marks available.*
*The transformed wheat plants could be grown in drought-prone regions **[1 mark]**, where they would reduce the risk of famine and malnutrition **[1 mark]**.*

c) *Maximum of 1 mark available.*
*They could be concerned that the large agricultural company will have control over the recombinant DNA technology used to make the drought-resistant plants, which could force smaller companies out of business **[1 mark]**.*

Page 189 — Genetic Fingerprinting

1 a) *Maximum of 5 marks available, from any of the 8 points below.*
*A sample of DNA is obtained, e.g. from a person's blood/saliva/skin etc. **[1 mark]**. PCR is used to amplify multiple areas containing different sequence repeats **[1 mark]**, using primers that anneal to either side of a repeat so the whole repeat is amplified **[1 mark]**. A fluorescent tag is added to all the DNA fragments **[1 mark]**. The DNA mixture undergoes electrophoresis **[1 mark]** — the DNA mixture is placed into a well in a slab of gel and an electric current is applied **[1 mark]**. The DNA fragments move towards the positive electrode at the other end of the gel and separate out **[1 mark]**. The separated bands produce the genetic fingerprint, which is viewed under UV light **[1 mark]**.*

b) *Maximum of 2 marks available.*
*Genetic fingerprint 1 is most likely to be from the child's father **[1 mark]** because five out of six of the bands on his genetic fingerprint match that of the child's, compared to only one on fingerprint 2 **[1 mark]**.*

c) *Maximum of 1 mark available, from any of the 4 points below.*
*E.g. they can be used to link a person to a crime scene (forensic science) **[1 mark]**. To prevent inbreeding between animals or plants **[1 mark]**. To diagnose cancer or genetic disorders **[1 mark]**. To investigate the genetic variability of a population **[1 mark]**.*

Page 191 — Locating and Sequencing Genes

1 a) *Maximum of 1 mark available.*
*Two **[1 mark]**.*
Three fragments were produced in the total digest, so it must have cut one piece of DNA in two places.

b) *Maximum of 3 marks available. 1 mark for 1 fragment in the correct place, 2 marks for 2/3 fragments in the correct place and 1 mark for labelling the Sal1 restriction sites.*

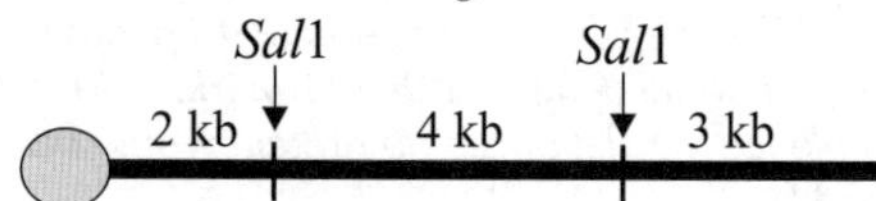

You can tell that the first cut must be after 2 kb because it's the smallest radioactive fragment in the partial digest. You can tell that the next fragment is 4 kb and not 3 kb because the other radioactive fragment in the partial digest is 6 kb long (2 kb + 4 kb). If the middle fragment was 3 kb then the radioactive fragment in the partial digest would be 5 kb long (2 + 3).

c) *Maximum of 2 marks available.*
*Because Sal1 has not been left long enough to cut at all of its recognition sequences **[1 mark]**, so there are other lengths of DNA present, i.e. 2 + 4 = 6 kb, 4 + 3 = 7 kb **[1 mark]**.*

Page 193 — DNA Probes in Medical Diagnosis

1 a) *Maximum of 2 marks available.*
*The gene that you want to screen for is sequenced **[1 mark]**. Multiple copies of parts of the gene are made by PCR to be used as DNA probes **[1 mark]**.*

b) *Maximum of 4 marks available.*
*Microscopic spots of different DNA probes are attached in series to a glass slide, producing a microarray **[1 mark]**. A sample of the person's labelled DNA is washed over the array and if any of the DNA matches any of the probes, it will stick to the array **[1 mark]**. The array is washed and visualised, under UV light/X-ray film **[1 mark]**. Any spot that shows up means that the person's DNA contains that specific gene **[1 mark]**.*
This question asks you to describe how many genes can be screened for at once (which is a microarray), but you could be asked how you can use DNA probes to look for a single gene too (see page 190).

2 a) *Maximum of 1 mark available.*
*Because she tested negative for the mutated gene (KRAS oncogene) that the drug specifically targets **[1 mark]**.*

b) *Maximum of 2 marks available.*
*So the results of her screening can be explained to her **[1 mark]** and so her treatment options can also be explained **[1 mark]**.*

Page 195 — Gene Therapy

1 a) *Maximum of 1 mark available.*
*Gene therapy involves altering/supplementing defective genes (mutated alleles) inside cells to treat genetic disorders and cancer **[1 mark]**.*

b) *Maximum of 1 mark available.*
*Somatic gene therapy **[1 mark]**.*

c) *Maximum of 3 marks available, from any 6 of the points below.*
*E.g. the effect of the treatment may be short-lived **[1 mark]**. The patient might have to undergo multiple treatments **[1 mark]**. It might be difficult to get the allele into specific body cells **[1 mark]**. The body may start an immune response against the vector **[1 mark]**. The allele may be inserted into the wrong place in the DNA, which could cause more problems **[1 mark]**. The allele may become overexpressed **[1 mark]**.*

Acknowledgements

Page 5 — Data used to construct the graph of BMI and cancer reproduced with kind permission from Gillian K Reeves, Kristin Pirie, Valerie Beral et al. Copyright © 2007, BMJ Publishing Group Ltd.

Page 10 — MMR graph adapted from H. Honda, Y. Shimizu, M. Rutter. No effect of MMR withdrawal on the incidence of autism: a total population study. Journal of Child Psychology and Psychiatry 2005; 46(6):572-579.

Page 10 — Data used to construct Herceptin® graph from M.J. Piccart-Gebhart, et al. Trastuzumab after Adjuvant Chemotherapy in HER2-positive Breast Cancer. NEJM 2005; 353: 1659-72.

Page 11 — Data used to construct the Hib graph reproduced with kind permission from the Health Protection Agency.

Page 38 — With thanks to Cancer Research UK for permission to reproduce the graphs.
Cancer Research UK, http://info.cancerresearchuk.org/cancerstats/types/lung/mortality/, January 2008.
Cancer Research UK, http://info.cancerresearchuk.org/cancerstats/types/lung/smoking/, January 2008.

Page 39 — Data used to construct asthma and sulfur dioxide graphs, source: National Statistics website: www.statistics.gov.uk Crown copyright material is reproduced with the permission of the Controller Office of Public Sector Information (OPSI).

Page 39 — Exam question graph: The Relationship Between Smoke And Sulphur Dioxide Pollution And Deaths During The Great London Smog, December 1952, Source: Wilkins, 1954.

Page 45 — Data used to construct the graph from R. Doll, R. Peto, J. Boreham, I Sutherland.
Mortality in relation to smoking: 50 years' observations on male British doctors. BMJ 2004; 328:1519.

Page 84 — Data used to construct the graph from National Statistics online. Reproduced under the terms of the Click-Use licence.

Page 88 — Graph of skylark population from BTO/JNCC Breeding Birds of the Wider Countryside.

Page 89 — Graph of rainforest diversity from Schulze et al. Biodiversity Indicator Groups of Tropical Land-Use Systems: Comparing Plants, Birds and Insects. Ecological Applications 2004; 14(5) Ecological Society of America.

Page 89 — Exam question graph from Defra. Reproduced under the terms of the Click-Use licence.

Page 120 — Data used to construct the graph of daily CO_2 concentration reproduced with kind permission from a study at Griffin Forrest, Perthshire performed by the University of Edinburgh and supported by the Natural Environment Research Council.

Page 120 and 123 — Data used to construct the graphs of yearly CO_2 concentration and average yearly CO_2 concentration reproduced with kind permission from Atmospheric CO_2 at Mauna Loa Observatory, Scripps Institution of Oceanography, NOAA Earth System Research Laboratory.

Page 121 — Data used to construct the graph of temperature change over the last 1000 years reproduced with kind permission from Climate Change 2001: The Scientific Basis, Contribution of Working Group I to the Third Assessment Report of the Intergovernmental Panel on Climate Change, SPM Figure 1. Cambridge University Press.

Page 121 — Data used to construct the graph of methane concentration © CSIRO Marine and Atmospheric Research, reproduced with permission from www.csiro.au.

Page 121 — Data used to construct the graph of CO_2 concentration reproduced with kind permission from U.S. Global Change Research Program, http://www.usgcrp.gov/usgcrp/nacc/background/scenarios/images/co2hm.gif

Page 122 — Data used to construct the graph of wheat yield from Global scale climate-crop yield relationships and the impacts of recent warming. D. B. Lobell and C. B. Field. Environmental Research Letters 2 (2007) 014002 (7pp). IOP Publishing.

Page 123 — Data used to construct the graph of average global temperature adapted from Crown Copyright data supplied by the Met Office.

Page 123 — Data used to construct the graph of global sea temperature reproduced with kind permission from NASA Goddard Institute for Space Studies.

Page 123 — Diagram showing the distribution of subtropical plankton reproduced with kind permission from Plankton distribution changes, due to climate changes – North Sea. (February 2008). In UNEP/GRID-Arendal Maps and Graphics Library. http://maps.grida.no/go/graphic/plankton-distribution-changes-due-to-climate-changes-north-sea.

Page 123 — Data used to construct the graph of corn yield provided by the U.S. Department of Agriculture – National Agricultural Statistics Service.

Page 131 — Data used to construct the graph showing the stock of spawning cod in the North Sea and the rate of mortality caused by fishing since 1960 from the International Council for the Exploration of the Sea.

Page 195 — Data used to construct the graphs from S. Hacein–Bey–Abina et al. SCIENCE 302: 415–419 (2003).

Every effort has been made to locate copyright holders and obtain permission to reproduce sources. For those sources where it has been difficult to trace the originator of the work, we would be grateful for information. If any copyright holder would like us to make an amendment to the acknowledgements, please notify us and we will gladly up date the book at the next reprint. Thank you.

Index

A

B

C

Index

Index

Index

Index